Social Geographies

Pearson Education

We work with leading authors to develop the strongest educational materials in geography, bringing cutting-edge thinking and best learning practice to a global market.

Under a range of well-known imprints, including Prentice Hall, we craft high-quality print and electronic publications which help readers to understand and apply their content, whether studying or at work.

To find out more about the complete range of our publishing please visit us on the World Wide Web at: www.pearsoneduc.com

Social Geographies

Space and society

Gill Valentine

Prentice
Hall

An imprint of **Pearson Education**

Harlow, England · London · New York · Reading, Massachusetts · San Francisco · Toronto · Don Mills, Ontario · Sydney
Tokyo · Singapore · Hong Kong · Seoul · Taipei · Cape Town · Madrid · Mexico City · Amsterdam · Munich · Paris · Milan

Pearson Education Ltd
Edinburgh Gate
Harlow
Essex CM20 2JE
England

and Associated Companies around the World

Visit us on the World Wide Web at:
www.pearsoneduc.com

First edition 2001

© Pearson Education Limited 2001

ISBN 0582-35777-2

British Library Cataloguing-in-Publication Data
A catalogue record for this book can be obtained from the British Library

Library of Congress Cataloging-in-Publication Data
Available from the publisher

10 9 8 7 6 5 4 3 2 1
05 04 03 02 01

Typeset in 10/12.5pt Sabon by 35
Produced by Pearson Education Asia Pte Ltd
Printed in Singapore

For Ray Short and Sue Reading who first inspired my interest in Geography

Contents

List of plates

List of figures

Acknowledgements

Whenever I'm struggling to write I always fantasize about this moment: the moment when there is nothing more to say other than 'Thanks' to all those who have helped and contributed along the way. This book has had an unnaturally long gestation period, in that the contract to write it was first signed as far back as 1996, and now that I have finally reached the end it is harder than I imagined to pull together acknowledgements which span four years.

First I owe a huge debt to Tristan Palmer (who commissioned this title for John Wiley) for his tremendous support, not only for this project but more generally too. I'm grateful also to Linda McDowell and the others who reviewed the original proposal. When John Wiley scrapped its Geography list, this title was rescued by Matthew Smith and Longman. I want to thank him for being such a fun and easy-going editor to work with and for the many lunches I conned him out of to discuss my rather sluggish 'progress'. I am also very grateful to Julie Knight, Senior Editor at Pearson Education for all her help in chasing permissions and producing the book.

The fact that I did not finish this manuscript until 2000 is in part due to my tendency to take on more work than I can handle. Special thanks are due to David Bell, Jon Binnie, Sarah Holloway, Tracey Skelton and Ruth Butler, who have put up with my attempts to juggle writing this book with my commitments to work with them on other projects. Their patience is much appreciated.

This book is developed from a course I have taught at the University of Sheffield since 1995. Thanks to all the students who have taken this module for their enthusiasm and suggestions along the way – especially Laura Peacock, Hal Parker and Kerry O'Connor for nagging me over my deadlines more than I nagged them over theirs.

A number of people helped me along the way by supplying information, photographs or other material. I am especially grateful to Andy Horrocks for finding me up-to-date statistics, Jacqui Moore for references to sacred spaces, Hester Parr for sending me the papers on institutions, and Chris Philo for forwarding the page proofs of *Animal Spaces, Beastly Places*. Becky Kennison (Plates 6.3, 6.4, 7.3 and 9.1), Andy and Trude Hodgson (Plate 8.4), Andrew Milne (Plate 8.3), April Veness (Plates 3.2 and 3.3); Jonathon Brookes (Plate 9.2), Giles Wiggs (Plate 8.5), Sarah O'Hara (Figure 5.2) and Ruth Butler (Figure 2.2) all generously researched and supplied me with their photographs, cartoons or other illustrative materials. Thanks also to Graham Allsopp for redrawing Figures 3.1, 4.1, 7.3, 9.1, 9.2 and 9.3. Thanks to Becky Ellis for allowing me to use Plate 3.1, a self-directed photograph by an interviewee from her PhD titled Constructions of Home: Subjectivity, Identity and Performance in the Home Space (Dept of Geography, Sheffield University).

The following also gave their permission to reproduce material: Figure 2.2 © Angela Martin from *Able Lives*, edited by Jenny Morris, 1989, The Women's Press, 34 Great Sutton Street, London, EC1V 0LQ; Figure 3.1 Pluto Press; Figure 3.2 by permission of The Advertising Archives Ltd.; Figure 4.1 redrawn from Park, R., Burgess, E. and McKenzie, R. (1967) *The City* (4th edn) by permission of The University of Chicago Press; Plate 5.1 © Jacky Fleming from *Be A Bloody Train Driver* (Penguin 1991); Plate 5.2 © Jacky Fleming; Figures 5.1 and A1 reprinted with permission of Universal Press Syndicate © Watterson. All rights reserved; Plate 6.1 © Peter Erland; Figure 6.1 By permission of Bob Gorrell and Creators Syndicate, Inc.; Figures 7.1 and 7.2 redrawn from Pile, S., Brook, C., and Mooney, G. (eds) *Unruly Cities?* (1999) by permission of Routledge and the Open University Press; Box 7.1 R. Frankenberg from 'Growing up white: feminism, racism and the social geography of childhood' in *Feminist Review*, 45, 51–4 by permission of Taylor & Francis Ltd.; Figure 9.4 © Dave Chisholm. Some material on schools in Chapter 5 was first published in *Geoforum*, **31**, 257–67 and material on prisons also within Chapter 5 was first published in the *Journal of Material Culture*, 3, 131–52. Every attempt has been made to obtain permission to reproduce copyright mater-ial. If any proper acknow-ledgement has not been made, I would invite copyright holders to inform me of the oversight.

Finally, I would like to acknowledge my close friends for their support and friend-ship through both good and bad times.

Space and society

■ 1.1 About this book

Social geography is an inherently ambiguous and eclectic field of research and writing. It is perhaps best summed up as 'the study of social relations and the spatial structures that underpin those relations' (Johnston *et al.* 2000: 753). While social geography's intellectual roots can be traced back to the nineteenth-century French school of '*la géographie humaine*', it was in the 1960s and 1970s that it developed rapidly as a subdiscipline. It was in this period of social and political turbulence, when civil rights protests were at their height, that social geography came into its own. Informed by the development of Marxist and feminist approaches, social geography engaged with social inequalities and questions of social justice (Harvey 1973). Subsequently, in the late 1980s and 1990s this subdiscipline was influenced by the '**cultural turn**' in geography, leading to a shift in emphasis away from issues of structural inequality towards one of identity, meanings, representation and so on. It is therefore increasingly difficult to distinguish between social geography and cultural geography.

Although this book is entitled *Social Geographies*, it makes no claims to occupy a discrete intellectual space which can be identified or sealed off from other traditional subdisciplinary areas such as cultural geography or political geography. Rather, the plural social geographies which emerge here are a porous product – an expression of the many connections and interrelationships that exist between different fields of geographical inquiry. Indeed, they are perhaps more appropriately characterized by the subtitle: *space and society*.

Given the intellectual heritage of the 1960s and 1970s, social issues such as poverty, housing and crime, rather than spaces, have commonly provided the structuring framework for social geography textbooks, with social identities such as gender, race sexuality and disability each discussed within discrete chapters. Instead of replicating this format, this book adopts geographical scale as an organizing device to think about how social identities and relations are constituted in and through different spaces. Through this device it both addresses conventional social geography topics such as urban social segregation, class, fear of crime and so on, while also making more explicit the ways in which social relations and space are mutually constituted.

This chapter sets the scene for what follows by briefly outlining the ways in which geographers have conceptualized space and society, by explaining how 'scale' is employed in this text, and by offering some guidelines as to how this book might be read and used.

■ 1.2 Space and society

Space is a central organizing concept within geography. As the discipline developed in the nineteenth, twentieth and early twenty-first centuries the ways in which geographers have conceptualized space have become increasingly sophisticated.

In the late nineteenth and early twentieth centuries geography was concerned with the identification and description of the earth's surface. Space was conceived by explorers, cartographers, and geographers as something to be investigated, mapped and classified (a process enhanced by the development of instrumental, mathematical and graphical techniques). Indeed, this understanding of space underpinned the subjugation and exploitation of territories and populations through the process of colonialism.

After the Second World War a recognition of the deficiencies of regional geography and the need for more systematic approaches to research, combined with geographers' increasing engagement with quantitative methods, led to a shift in the focus of the discipline. The emphasis on the description of uniqueness was replaced by a concern with similarity. Specifically, **positivist** approaches to geography were concerned with uncovering universal spatial laws in order to understand the way the world worked. The focus was on spatial order and the use of quantitative methods to explain and predict human patterns of behaviour (Johnston 1991, Unwin 1992).

Within social geography attempts were made to understand the complexity of society and social relations by mapping and exploring spatial patterns. The emphasis was on scientific techniques such as social area analysis which used key variables (e.g. about occupation, ethnicity, etc) from the census to describe the characteristics of social areas; and factorial ecology which involved the uses of multivariate statistical techniques to produce areal classifications. These, and other such methods, were used to map patterns of urban social segregation, most notably of racial segregation. In other words, social differences were understood in terms of spatial separation. Some of this work and its legacy for the study of social geography is discussed in Chapter 4 and Chapter 7.

Within this explanatory framework space was conceptualized as an objective physical surface with specific fixed characteristics upon which social identities and categories were mapped out. Space was, in effect, understood as the container of social relations and events. Likewise, social identities and categories were also taken for granted as 'fixed' and mutually exclusive. The emphasis on understanding social relations in terms of the way they were mapped out in space also meant that those social relations (such as gender and sexuality) that were not easily studied within this spatial framework were overlooked (Smith 1999).

In the 1970s this positivist approach to human geography was subject to many-stranded critiques, of which **humanistic** geography and radical geography were two. Their focus on the theoretical and methodological limitations of positivism (see Cloke *et al.* 1991) also drew attention to the dualistic way in which society and space were conceptualized in this approach. Humanistic geography, for example, 'rejected the exclusivity and pretensions to objectivity of positivist science, and proposed the importance of subjective modes of knowing. Geographical space was not simply an objective structure but a social experience imbued with interwoven layers of social meaning . . . In humanistic geography "social space", not physical or objective space, was made the object of inquiry' (Smith 1990: 76). Radical approaches, most notably those inspired by **Marxism**, sought to understand space as the product of social forces, observing that different societies use and organize space in different ways; and to explain the processes through which social differences become spatial patterns of inequalities (Smith 1990). In turn, geography's subsequent engagement with **postmodernism** also produced a new sensitivity to 'the myriad variations that exist between the many "sorts" of human beings studied by human geographers – the variations between women and men, between social classes, between ethnic groups, between human groups defined on all manner of criteria – and to recognise (and in some ways represent) the very different inputs and experiences these diverse populations have into, and of, "socio-spatial" processes' (Cloke *et al.* 1991: 171).

Against this philosophical backdrop understandings of space and society have been reassessed. Social categories (such as class, gender, sexuality and race) are no longer taken for granted as given or fixed but rather are understood to be socially constructed. As such they can also be contested, resisted and (re)negotiated. These ideas are central to much of the work discussed throughout this book. For example, see Chapter 2, sections 2.2.2, 2.6.1 and 2.7, and Chapter 7, section 7.2.1. These sections show that 'gender', 'age', 'disability' and 'race' are social constructions rather than 'natural' or biologically given differences. Further, identities are now understood as always relational in that they are constructed in terms of their sameness to, and difference from 'others'. For example, Chapter 2 examines how 'man' and 'woman' are defined reciprocally, while Chapter 9 draws on the work of Edward Said (1978), in which he shows how 'the Orient' as a social and cultural construction has provided Europeans with a sense both of 'Otherness' and of 'Self', to demonstrate how national identities are often defined not on the basis of their own intrinsic properties but in terms of what they are not. In such writings the 'Other' is simultaneously understood as desirable, 'exotic' and fascinating while also provoking emotions of fear and dread. In this way, the concept of otherness has helped geographers to understand why and how particular bodies such as the disabled, the homeless, those with mental ill-health and so on, are socially and spatially excluded (see, for example, Chapter 2 and Chapter 6).

However, geographers have also come under criticism for their tendency to focus on, and often to appropriate, the experiences of 'others' rather than to examine privileged and powerful identities. Derek Gregory (1994: 205), for example, questions 'By what right and on whose authority does one claim to speak for those "others"? On whose terms is a space created in which "they" are called upon to speak? How are they (and we) interpellated?' Such questions about **positionality** have in turn stimulated new interest in **hegemonic** identities such as whiteness (see, for example, Chapter 2, Chapter 7 and Chapter 8), masculinity (Chapter 2) and heterosexuality (Chapter 3 and Chapter 7). At the same time geographers have also woken up to the fact that class, race, sexuality, and disability cannot be examined in isolation – rather, individuals and groups have multiple identities, occupying positions along many separate lines of difference at the same time; indeed, people's identities also develop in between the boundaries of social categories such as black and white (see, for example, Chapter 7, section 7.2.1). These 'old' fixed categories and identities can be, and are being, displaced by new forms of social identification – or hybrid identities – in which difference is no longer viewed in terms of social and cultural hierarchies.

Judith Butler's theory of gender as a performance – outlined in Chapter 2 – has played a particularly important part in shaping geographers' understandings of the production of identities. She rejects the notion that biology is a bedrock which underlies social categories such as gender. Rather, she theorizes gender (and implicitly other identities too) as performative, arguing that 'gender is the repeated stylisation of the body, a set of repeated acts within a highly rigid regulatory framework that congeal over time to produce the appearance of substance, of a natural sort of being' (Butler 1990: 33). This approach has helped geographers to think about social identities as per-manently contested and unstable categories and has inspired work that looks at how such performances and contests take place in lived space (see, for example, Chapter 5).

Just as social identities are no longer regarded as fixed categories but are understood as multiple, contested and fluid, so too space is no longer understood as having particular fixed characteristics. Nor is it regarded as being merely a backdrop for social relations, a pre-existing terrain which exists outside of, or frames everyday life. Rather, space is understood to play an active role in the constitution and reproduction of social identities; and social identities, meanings and relations are recognized as producing material and symbolic or metaphorical spaces. Thus Doreen Massey (1999: 283) explains that space 'is the product of the intricacies and the complexities, the interlockings and the non-interlockings, of relations from the unimaginably cosmic to the intimately tiny. And precisely because it is the product of relations,

relations which are active practices, material and embedded, practices which have to be carried out, space is always in a process of becoming. It is always being made.'

Each chapter of this book focuses on how a space, from the body to the nation, is invested with certain meanings, how these meanings shape the way these spaces are produced and used, and, in turn, how the use of these spaces can feed back into shaping the way in which people categorize others and identify themselves. In other words, space and society do not merely interact with or reflect each other but rather are mutually constituted. To give an example from Chapter 7, lesbian and gay sexualities are inherently spatial in that they depend on particular spaces for their construction. For example, spatial visibility (e.g. in terms of the establishment of so-called gay ghettos or various forms of street protest or Mardi Gras) has been important to the development of lesbian and gay rights. In turn, these performances of sexual dissidents' identities (re)produce these spaces as lesbian and gay spaces in which sexual identities can be, and are forged (Mitchell 2000). Likewise, in Chapter 8, section 8.3 explores the way in which rurality is constructed by both people living in the countryside and those outside it. Section 8.4 illustrates how the dominant image of the English countryside as a white landscape characterized by close-knit social relations and heterosexual family life obscures 'other' meanings and experiences of the rural, and how this, in turn, contributes to alienating social groups such as black people, travellers, and lesbians and gay men who feel uncomfortable or 'out of place' in these environments.

As this second example demonstrates, dominant sets of power relations often mask the complexities, multiplicities and ambiguities of the social activities, meanings and behaviours associated with the production of particular spaces. In some cases hegemonic **discourses** are literally inscribed in the landscape. Chapter 2, section 2.6.1 examines their role in creating socio-spatial environments which dis-able people with bodily impairments by marginalizing them economically, socially and politically. Likewise, Chapter 3 examines some of the ways housing designs are instrumental in shaping our everyday experiences.

Discourses can also be more invisibly imposed across space, influencing what assumptions, expectations and social behaviours are expected or deemed appropriate for particular spaces. By focusing on acts of transgression Tim Cresswell's (1996) work has exposed the way that these normative landscapes are often 'taken for granted', only becoming apparent when they are disrupted. Chapter 6 examines how normative or moral landscapes are produced in public space. Here, Butler's (1990) notion of performativity has also been used by geographers to understand how spaces, like identities, are produced through repetitive acts that take place within a regulatory framework. Chapter 6 shows how the performance of lesbian and gay sexual identities in public space disrupts and therefore exposes the way in which the street is commonly produced as 'naturally' or 'normally' a heterosexual space.

Space has also become a resource for those who are marginalized or excluded. In particular, the spaces of the margin (real material locations, but also symbolic spaces of oppression) have been reclaimed by black writers such as bell hooks as spaces from which to speak. She (hooks 1991: 149) describes marginality 'as a site one stays in, clings to even, because it nourishes one's capacity to resist. It offers the possibility of radical perspectives from which to see and create, to imagine altern-atives, new worlds.' Chapter 3 explores the way in which the home acts as a site of resistance for black people in the face of white hegemony.

In writing about space geographers have often drawn on a number of **dualisms** which are significant in Western thought such as mind/body, public/private, work/home, human/animal, white/black, etc. (see Chapter 2). These dualisms are invested with power in that they are not two sides of unrelated terms 'A' and 'B'. Rather '[w]ithin this structure, one term A has a positive status and an existence independ-ent of the other; the other term B is purely negatively defined, and has no contours of its own; its limiting boundaries are those which define the positive term' (Grosz 1989: xvi). This dualistic way of thinking has structured the way geographers have come to analyse and understand some spaces. For example, the study of the work-place has often been privileged over the space of the home, and clear boundaries have been assumed to be drawn between dualisms such as public and private space or urban and rural space. However, these dualistic ways of thinking about, and analysing, space are increasingly being challenged and resisted. Chapter 3 questions the boundaries which are often drawn between home and work, while Chapter 3 and Chapter 6 both expose the fact that 'private' and 'public' spaces are not fixed and stable categories but rather that the boundaries between them are blurred and fluid. Similar arguments are also made in Chapter 7 and Chapter 8 in relation to the boundaries which have been imagined between the urban and rural and between culture and nature.

Spatial metaphors such as 'inside' and 'outside', 'centre' and 'margins' are fre-quently employed not only by geographers but also by social scientists in thinking about social relations. Yet such positions do not represent marked or differentiated positions. Rather, Gillian Rose (1993: 140) argues that, paradoxically, we can simultaneously occupy 'spaces that would be mutually exclusive if charted on a two-dimensional map – centre and margin, inside and outside spaces'. She cites the work of Patricia Hill Collins (1990), who describes how black women employed as domestic workers in white households occupy such a contradictory position. On the one hand, they are familiar with and close to the children within the white family; on the other hand, she explains, they are also made to feel that they do not belong. They are present but also absent. For writers such as Rose (1993: 155) these para-doxical spaces 'threaten the polarities which structure the dominant geographical imagination'.

In challenging dichotomies, geographers are increasingly 'imagining a some-where else' (Johnston et al. 2000: 771). While Rose (1993) describes this as a 'para-doxical space', other writers have talked in terms of hybrid space (Bhabha 1994) or Thirdspace (Soja 1996). These different conceptualizations of space represent ways of thinking about the world which focus on 'the production of heterogeneous spaces of "radical openness"' (Johnston *et al.* 2000: 771). Susan Smith (1999: 21) argues that the concept of Thirdspace:

> turns our attention away from the givens of social categories and towards the strategic process of identification. It forces us to accept the complexity, ambiguity and multi-dimensionality of identity and captures the way that class, gender and 'race' cross-cut and intersect in different ways at different times and places.

Further, she goes on to argue that 'Thirdspace may provide an opportunity to move beyond our historic preoccupation with social divisions – with what holds people apart – and think about what is gained from a discourse of belonging.'

■ Summary

- The way in which geographers have conceptualized space has become increasingly sophisticated.

- In the past, it was conceptualized as an objective physical surface with specific fixed characteristics upon which social categories were mapped out. Likewise, social identities were taken for granted as 'fixed' and mutually exclusive.

- Understandings of space and society have been reassessed. Space is now understood to play an active role in the constitution and reproduction of social identities, and social identities and relations are recognized as producing material and symbolic or metaphorical spaces.

- Thus space and society do not merely interact with, or reflect, each other but rather are mutually constituted.

- It is now acknowledged that individuals and groups have multiple identities, occupying positions along many separate lines of difference at the same time; and that identities also develop in between the boundaries of social categories.

- Dualisms have frequently structured geographical analyses of space. In challenging dichotomies geographers are now opening up new ways of thinking about space: Thirdspace.

■ 1.3 Boundaries and connections

This book is organized around a loosely nested set of scales from the body to the nation. As Neil Smith has argued, scale is a useful way of organizing or thinking about geographical differentiation. He writes that scale is 'the criteria of difference not so much between places as between different *kinds* of places' (Smith 1993: 99). In this sense he suggests that each scale is effectively a platform for thinking about specific kinds of socio-spatial activities, for example the home as the site of 'family' relationships, the community as the site of neighbourhood activism and so on (although it is important to remember that social relations cut across scales too: see below).

Yet, this is not to suggest that there is anything 'natural' or inevitable about the geographical scales which provide the framework for this book. Each is a product of different moments and contexts. Nation states were rare before the seventeenth century in Europe, in the nineteenth and twentieth centuries they became a powerful scale of social organization but at the beginning of the twenty-first century there is increased speculation that globalization poses a threat to their legitimacy (see Chapter 9). In other words, 'scale is produced in and through social activity which, in turn, produces and is produced by geographical structures of social interaction' (Smith 1993: 97). Geographical scales are thus fluid and pliable.

The social context of, and ideological assumptions within, the discipline of geography also shape which particular practices of geography and geographical knowledges become legitimated at any one time – although this is not to deny that geographical knowledges are always diverse, negotiated and contested (Livingstone 1992, Driver 1995). While scales such as the nation and community have been the subject of much research and writing, others, such as the body and the home, have only recently made it onto the geographical agenda. Turner (1992) credits **feminism** with awakening a greater academic awareness of, and sensitivity to, how the body is socially constructed; and to the significance of different (gendered, sexed, racialized) bodies. Within geography, feminists have questioned the production of geographical knowledge, specifically the way that the discipline (and indeed the social sciences) has been founded on a Cartesian mind–body split, in which the mind which has been associated with rationality and masculinity, has been privileged over the irrational, emotional, feminine body (Longhurst 1997). This 'othering' is now being challenged by a new generation of feminist geographers who are producing embodied geographies which are contributing to the re-theorization of the discipline (see Chapter 2).

Smith (1993) argues that the production of geographical scale is a process that has great political significance. He writes '[i]t is geographical scale that defines the boundaries and bounds the identities around which control is exerted *and* contested' (Smith 1993: 101). Scale can thus act as a form of empowerment or containment. Smith argues his case by pointing to the way in which a political movement can develop a power base at the scale of the city and use this to challenge a national government. This was the case in the 1980s in Britain when Margaret Thatcher's national Conservative government came under attack from several metropolitan authorities which were controlled by the opposition Labour Party. In turn, Prime Minister

Thatcher contained this political threat by abolishing this tier of metropolitan government; in other words, she eliminated the scale at which her power was being contested (Johnston *et al.* 2000).

This is not to suggest, however, that each scale represents a rigidly bounded spatial sphere, nor that there is any fixed hierarchy or ordering of scales. Rather, Smith (1993) uses the term 'jumping scales' to describe the way power at one geographical scale can be expanded to another. Here he uses the example of the invention of a Homeless Vehicle. This is an elaborate development of the sort of shopping trolley that homeless people sometimes use to carry their possessions, which not only includes sections for carrying belongings but can also be slept in. Without a home and with nowhere to store their possessions, homeless people are normally relatively spatially immobile; they are therefore quite easily contained in prescribed locations by the police and city authorities and so can be erased from the public gaze. The mobility afforded by the Homeless Vehicle, however, provides people on the streets with more opportunities to go panhandling and to find new sleeping locations; it offers a means to escape violence on the streets or police harassment and it increases the visibility of the homeless within the city. In this sense it is empowering because it enables homeless people to carve out a space for themselves within the exclusionary landscape of the city and to challenge definitions of 'community'. In doing so, it allows evicted people to 'jump scales' 'to organise the production and reproduction of daily life and to resist oppression and exploitation at a higher scale, over a wider geographical field' (Smith 1993: 90). In this way, scale is a means not only of containment and exclusion but also expansion and inclusion (Smith 1993).

The eight scales outlined in this text, while offering some form of ordering for the discussion in this book, do not represent a coherent spatial structure. They are not discrete, independent, compartmentalized and opposing spaces. As section 1.2 explained, all spaces are formed out of numerous performances and practices. Scales such as the body or the home are not bounded, fixed or stable locations but are constituted in and through their relations and linkages with 'elsewhere', with spaces which stretch beyond them (Massey 1991). In other words, these locations are open, porous and provisional spaces. Massey (1991) uses the example of Kilburn, a district of London – describing the presence of Italian restaurants, saris on sale in shop windows, graffiti about the IRA, advertisements for Hindi concerts and so on – to show how every place is formed out of intersecting social relations which are stretched more widely. In this sense each of the spatial scales which make up this book represents the intersection of a whole range of connections, interrelations and movements, and of different people who have very different ways of participating in, understanding or belonging to them.

Yet, this is not to suggest that boundaries are completely irrelevant. The multiplicity of times and spaces (and their superimposition) which exist within geographical scales, and the relationships and interdependencies which exist between them, can be sources of tension and conflict. One response to such tensions is for groups to attempt to form boundaries which close off or homogenize certain spaces on the basis of particular social identities rather than to be open to difference: to disconnect rather than to connect. Each of the chapters throughout this book therefore reflects on the juxtapositions of socio-spatial relations, on homogeneity and difference, and on disputes over control and disorder.

The book begins by thinking about how social identities are constructed in and through the site of the body, reflecting on how **corporeal** differences can serve as a basis for socio-spatial forms of exclusion and oppression at other scales. It explores some of the ways in which the body is inflected by regimes of power at home, within the community and at the level of the nation state. The next chapter turns to the home, which, after the body, is one of the most important sites for the production of personal identity. Here, the focus is on the home not only as a physical location but also as a matrix of social relations, a place which has multiple meanings and which is experienced very differently by different social groups. Like the body, the home is a contested space which is shaped through connections to community relations, state power and so on.

The chapter on community understands it not only in terms of homes/neighbourhoods but also as a structure of meaning which provides a useful way of thinking of social relations at a national or even global scale. Following on from the discussion of community, institutions are conceptualized as spaces which seek to 'place' the body geographically and temporally (Parr 1999: 196) and to discipline it. Although the four institutions (school, the workplace, the prison and the asylum) discussed here represent spaces or power structures which are designed to achieve particular ends, this chapter demonstrates that they are not pre-given, stable structures but rather are dynamic, fluid and precarious achievements.

The street and the city, which are discussed next, are both sites of proximity where different social groups and worlds come together. As such, they are sites of political expression and celebration, but they are also sites of struggle and exclusion, in which the right to take up or command space is contested and resisted. Chapter 8 explores debates about whether rural society can be distinguished from urban society because of its strong sense of 'community' and whether rural life is closer to nature than city life. In this sense this chapter resonates strongly with Chapter 4 and Chapter 7. It concludes by recognizing that such universal conceptualizations are now outdated and by focusing on the multiple meanings ascribed to living in the countryside. The final chapter examines understandings of nations and national identities and the role of the nation state in institutionalizing social differences through its definitions of citizenship. It then goes on to explore some of the forms of nationalism which both facilitate and threaten contemporary nation states and to think about other forms of social identification, such as sexual citizenship and **diasporic** citizenship, which are claimed to be replacing national identity as a master narrative. The chapter concludes by focusing on processes of globalization and how individuals' lives around the world are being woven together in increasingly complex ways in a global economy, and by reflecting on the possibilities for global forms of citizenship.

■ **Summary**

- Scale is a useful way of organizing or thinking about geographical differentiation.

- There is nothing 'natural' or inevitable about the geographical scales which provide the framework for this book. Scales are fluid and pliable.

- The production of geographical scale is a process that has political significance in that it defines the boundaries, and bounds the identities, around which control is exerted and contested.

- The chapters of this book do not represent discrete, compartmentalized, stable or opposing spaces, nor a fixed hierarchy or ordering of scales.

- Each spatial scale is constituted in and through its relations and linkages with the spaces which stretch beyond it.

- The multiplicity of times and spaces which exist within each geographical scale, and the relationships and interdependencies which exist between them, are sources of tension and conflict.

■ **1.4 Using this book**

The contents of each chapter are highlighted in a series of bullet points on the chapter's title page and are also outlined in opening summary paragraphs. Bullet point summaries of key points are also included at the end of each of the sections within the chapters. Exercises and essay questions are provided at the end of the chapters (except this one) as one way of helping you to think about the issues each chapter has addressed. The guides to further reading (which include academic references and some fiction) are also intended to act as stimulus to encourage you to explore further some of the topics introduced.

Some of the issues covered in this book are complex and many of the concepts employed have complicated academic biographies in that they have been understood and conceptualized in different ways at different times. Boxed examples and quotations from my own and others' empirical work are therefore used to try to illustrate in everyday terms some of the academic ideas being presented. A definition of key words is also provided in a glossary at the end of this book. All the words included in it are highlighted in bold the first time they are used in the text. *The Dictionary of Human Geography* (Johnston *et al.* 2000) would also serve as a useful companion to guide you through some of the jargon you might encounter.

While each of the following eight chapters has a coherent spatial theme, the contents of all the chapters cross-cut and intersect with each other in complex ways because each geographical scale is constituted in and through its relations with the others (as section 1.3 above explains). Some of these connections are signposted for you in the text to draw your attention to them. Others are there for you to find. This book is not necessarily intended to be read in a linear way from cover to cover. Rather, the reader can dip in and out as different themes are pursued between the chapters. For example, if you are interested in homelessness you might begin with the discussion of the definition, nature and causes of homelessness in Chapter 3 and then move to Chapter 5, which examines in more detail policies of deinstitutional-ization which have contributed to putting people with mental ill-health on the streets. From here you might skip to Chapter 2 and Chapter 6 to think about how the bod-ies of those with mental ill-health and groups such as the homeless are 'othered' as dirty or dangerous and how they are excluded from certain public spaces. Or you might make the connection with Chapter 9, which discusses the sense of displace-ment and 'homelessness' experienced by some migrants who are forced to leave their homes and homelands against their will.

I hope you find this book an interesting read and a useful resource. If so, it might help to inspire you to think about choosing a social geography topic as the subject of your dissertation or other form of assessed work. The book therefore concludes with a guide to writing a project. This is intended to help you choose, research and write up a non-invigilated assessment. Good luck!

■ Further Reading

- Good introductions to (and criticisms of the limitations of) the different philosophical approaches to human geography are found in a number of texts such as: Cloke, P. *et al.* (1991) *Approaching Human Geography*, Paul Chapman, London, Johnston, R.J. (1991) *Geography and Geographers*, Edward Arnold, London, and Unwin, T. (1992) *The Place of Geography*, Longman, Harlow.

- Examples of the efforts of social geographers to measure and quantify urban social segregation include: Peach, C. (ed.) (1975) *Urban Social Segregation*, Longman, London, and Peach, C., Robinson, V. and Smith, S. (eds) (1981) *Ethnic Segregation in Cities*, Croom Helm, London.

- There are a wide range of books which deal with the social construction of categories such as 'race', disability, gender and sexuality. Some examples include: Jackson, P. and Penrose, J. (eds) (1993) *Constructions of Race, Place and Nation*, UCL Press, London, Butler, R. and Parr, H. (1999) (eds) *Mind and Body Spaces: Geographies of Illness, Impairment and Disability*, Routledge, London, Laurie, N., Dwyer, C., Holloway, S. and Smith, F. (1999) *Geographies of New Femininities*, Longman, Harlow, and Bell, D. and Valentine, G. (eds) (1995) *Mapping Desire: Geographies of Sexualities*, Routledge, London.

- Given the importance of space to geographers, it is difficult to pick out just a few key books. However, good starting points are: Massey, D. and Jess, P. (eds) (1995) *A Place in the World? Places, Cultures and Globalization*, Open University Press, Milton Keynes, and Massey *et al.* (1999) *Human Geography Today*, Polity Press, Cambridge. Both of these texts deal with issues of difference and identity, imaginative geographies and the nature of space and place. More specific studies of social exclusion and transgression are found in the work of: Sibley, D. (1995a) *Geographies of Exclusion: Society and Difference in the West*, Routledge, London, and Cresswell, T. (1996) *In Place/Out of Place: Geography, Ideology and Transgression*, University of Minnesota Press, Minneapolis, MN.

- Neil Smith has published a number of important articles about geographical scale. For example: Smith, N. (1992a) 'Geography, difference and the politics of scale', in Doherty, J., Graham, E. and Malek, M. (eds) *Postmodernism and the Social Sciences*, Macmillan, London, and Smith, N. (1993) 'Homeless/global: scaling places', in Bird, J., Curtis, B., Putnam, T., Robertson, G. and Tickner, L. (eds) *Mapping the Futures: Local Cultures, Global Change*, Routledge, London. See also Marston, S. (2000) 'The social construction of scale', in *Progress in Human Geography*, vol. 24, pp. 219–42.

- More advanced critiques of geographers' understandings of space and society – which you may find quite difficult to follow – include: Rose, G. (1993) *Feminism and Geography*, Polity, Cambridge, Soja, E. (1996) *Thirdspace*, Blackwell, Oxford, and Crang, M. and Thrift, N. (eds) (2000) *Thinking Space*, Routledge, London.

2 The body

■ 2.1 The body

Geographers have focused on the body as a space. Adrienne Rich (1986: 212) describes the body as 'the geography closest in'. It marks a *boundary* between self and other, both in a literal physiological sense but also in a social sense. It is a personal *space*. A sensuous organ, the site of pleasure and pain around which social definitions of wellbeing, illness, happiness and health are constructed, it is our means for connecting with, and experiencing, other spaces. It is the primary *location* where our personal identities are constituted and social knowledges and meanings inscribed. For example, social identities and differences are constructed around bodily differences such as gender, race, age, and ability (Smith 1993). These can form the basis for exclusion and oppression (Young 1990a). The body then is also a *site* of struggle and contestation. Access to our bodies, control over what can be done to them, how they move, and where they can or cannot go, are the source of regulation and dispute between household members, at work, within communities, at the level of the state, and even the globe (see Chapter 3, Chapter 4, Chapter 5, Chapter 6 and Chapter 9). The following sections of this chapter outline geographical work in relation to each of these different dimensions of the body.

The chapter begins by exploring debates about the nature of the body, focusing specifically on understandings about the relationship between the body and mind and whether the body is a 'natural' or social entity. Second, it examines the body as a space – a space made up of surface, senses and psyche. Third, it explores tensions between the body as an individual project over which we have control versus the body as a site regulated or inflected by other regimes of power. Fourth, it considers how bodies take up space by looking at how we physically occupy and connect with surrounding spatial fields. Fifth, it addresses how our bodies make a difference to our experience of place, examining how corporeal difference can become the basis of discrimination and oppression. Sixth, it focuses on the social construction of age. Finally, it examines how the development of new technologies and possibilities for medical intervention in the body are producing new uncertainties about what the body is and what future it has.

■ 2.2 What is the body?

From the ancient Greeks and the Romans, through Judaeo-Christian thought, from the Renaissance to the present, the body has fascinated and preoccupied philosophers. There is even no universal agreement about where the body begins and ends (Synott 1993). Anthony Synott (1993) asks, for example, whether the shadow is part of the body and what about nail clippings and faeces – are they merely the body in another place? Neither is there is any consensus about the meanings of the body. Different philosophers throughout the ages have defined the body 'as good or bad; tomb or temple; machine or garden; cloak or prison; sacred or secular; friend or enemy; cosmic or mystical; at one with mind and soul or separate; private or public; personal or the property of the state; clock or car; to varying degrees plastic, bionic, communal; selected from a catalogue, engineered; material or spiritual; a corpse or a self' (Synott 1993: 37). These different understandings of corporeality are not confined to particular moments in time, but rather have been (re)produced as competing, sometimes complementary and often contradictory paradigms.

Of all these philosophical debates about the meaning and nature of the body, two in particular are important to understanding how geographers have looked at, and thought about, the body: the relationship between the mind and the body; and whether the body is a 'natural' or social entity.

■ 2.2.1 Mind/body dualism

The seventeenth-century philosopher Descartes established a dualistic concept of mind and body. He argued that only the mind had the power of intelligence, spirituality, and therefore selfhood. The corporeal body was nothing but a machine (akin to a car or a clock) directed by the soul (Turner 1996). His philosophy is captured in his famous phrase: *Cogito ergo sum* – I think therefore I am. Although his view was contested by other philosophers both at the time and since, the Cartesian division and subordination of the body to the mind and the emphasis placed on dualistic thinking and scientific rationalization had a profound impact on Western thought. Indeed, the Cartesian view of the world is said to have laid the foundations for the development of modern science and, in particular, medicine by establishing the body as a site of **objective** intervention to be mapped, measured and experimented on (Turner 1996).

This distinction between the mind and the body has been gendered. Whereas the mind has been associated with positive terms such as rationality, consciousness, reason and masculinity, the body has been associated with negative terms such as emotionality, nature, irrationality and femininity. Although both men and women have bodies, in Western culture, white men transcend their **embodiment** (or at least have their bodily needs met by others) by regarding the body as merely the container of their consciousness (Longhurst 1997). In contrast, women have been understood as being more closely tied to, and ruled by, their bodies because of natural cycles of menstruation, pregnancy and childbirth. Whereas Man is assumed to be able to separate himself from his emotions, experiences and so on, Woman has been presumed to be 'a victim of the vagaries of her emotions, a creature who can't think straight as a consequence' (Kirby 1992: 12–13).

Feminist geographers such as Gillian Rose (1993) and Robyn Longhurst (1997) have argued that these dualisms are important because they have shaped geographers' understandings of society and space and the way geographical knowledge has been produced. Drawing on the work of feminists such as Michele Le Doeuff (1991), Rose (1993) suggests that theorists from Descartes onwards have defined rational knowledge as a form of knowledge which is masculinist. It 'assumes a knower who believes he can separate himself from his body, emotions, values and past experiences so that he and his thought are autonomous, context-free and objective' (Rose 1993: 7). As a result of this belief in the objectivity of masculinist rationality – that it is untainted by bodily identity and experience – Rose claims that it is assumed to be universal, the only form of knowledge available. In other words, she argues that white, bourgeois, heterosexual man tends to see other people who are not like himself only in relation to himself. She writes: 'He understands femininity, for example, only in terms of its difference from masculinity. He sees other identities only in terms of his own self-perception; he sees them as what I shall term his Other' (Rose 1993: 6).

Applying these arguments to geography, Rose (1993) shows how white, heterosexual men have tended to exclude or marginalize women as producers of geographical knowledge, and what are considered women's issues as topics to study. The mind/body dualism has therefore played a key role in determining what counts as legitimate knowledge in geography with the consequence that topics such as embodiment and sexuality were, until the mid–late 1990s, regarded as inappropriate topics to teach and research. They have been 'othered' within the discipline (Longhurst 1997).

Fortunately, these sorts of critiques have played an important part in stimulating geographers at the end of the twentieth, and at the beginning of the twenty-first, centuries to challenge the privileging of the mind over the body within the discipline. As a result, what Longhurst (1997: 494) terms 'dirty topics' are being put on the map and geographers are beginning to think about ways of writing (for example, using autobiographical material and personal testimonies) and methodological practices which recognize that all knowledges are embodied and situated (Rose 1997).

■ 2.2.2 The natural and the social

Historically, women have been stereotypically defined in terms of their biology. The notion that women were closer to nature and the animal world than men because they menstruate and give birth gained important currency in the sixteenth and seventeenth centuries. Women's periods were read as signs of women's inherent lack of control over their bodies. Women leaked, while men were self-contained (although see Grosz's 1994 discussion of seminal fluid). Their role in reproduction was also understood to mean that they were 'naturally' more nurturing and therefore more closely linked to Mother Earth than men. The other side of this association between women and nature was an assumption that, just as nature was wild and potentially uncontrollable (except by rational male science), so women were less able to control their emotions and passions than men (Merchant 1990). Indeed, women's unstable bodies were considered to be a threat to their minds (Jordanova 1989). In the late nineteenth century, when the suffragette movement with its campaigns for women's right to vote and to education began to gain momentum, opponents used scientific claims that women had naturally smaller brains than men and that education might damage their ovaries to justify excluding them from public life (Shilling 1993). In other words, women's bodies were used to justify what was regarded as a 'natural inequality' between the sexes.

These notions that women's bodies are both different and inferior to men's persisted into the twentieth century (see also Chapter 5). Chris Shilling (1993) notes that even in the 1960s, the argument that women's hormones meant that they were inherently intellectually and emotionally unstable was used to prevent women being allowed to train as pilots in Australia. He writes:

> There have been repeated attempts to limit women's civil, social and political rights by taking the male body, however defined, as 'complete' and the norm and by defining women as different and inferior as a result of their unstable bodies. Women were supposedly confined by their biological limitations to the private sphere, while only men were corporeally fit for participating in public life (Shilling 1993: 55).

Similar naturalistic views have been used to legitimize the subordination of black bodies. Like women, black people have also been defined through their bodies. The black body has been understood to be pre-social, to be driven by biology, in opposition to the civilized and rational white body (Shilling 1993). In particular, colonialization and slavery have played an important part in defining and developing understandings of black bodies as driven by insatiable sexual appetites, 'dangerous', uncivilized, uncontrollable, and a threat to whites (Mercer and Race 1988). For example, Shilling (1993: 57) argues that in the USA white slave owners developed myths about animal nature and black sexuality to justify the atrocities perpetuated against black people; 'defining the worth of black people through their bodies was also used to justify the treatment of blacks as commodities and the use of black women for slave breeding'. He notes, for example, that one in three of all the black men lynched in the USA between 1885 and 1900 were accused of rape.

These claims about the 'natural' differences between men and women, white and black, are what are known as **essentialist** arguments. They assume that sexual and racial differences are determined by biology, that bodies are 'natural' or pre-discursive entities – in other words, that bodies have particular stable, fixed properties or 'essences' (Fuss 1990). Essentialist explanations have been challenged by social construction- ists. They argue that there is no 'natural' body, rather, the body is always 'cultur- ally mapped; it never exists in a pure or uncoded state' (Fuss 1990: 6), so that what essentialists 'naturalize' or portray as 'essence' is actually socially constructed dif- ference. These differences are produced through material and social practices, dis- courses and systems of representation rather than biology. Social constructionists demonstrate this by pointing to the fact that what is understood by 'man' and 'woman', 'black' and 'white' (for discussions of whiteness see Chapter 7 and Chapter 9) varies historically and in different cultural contexts.

Feminists have made a distinction between *sex* – the biological difference between men and women – and *gender* (masculinity and femininity) – the social meanings which are ascribed to men and women (see Women and Geography Study Group 1997, Laurie *et al*. 1999). The term 'gender' incorporates a recognition of the way differences between the sexes are socially constructed in a hierarchical way (see Figure 2.1). Likewise, other writers have argued that meanings ascribed to black and white bod- ies are also socially produced in ways which exaggerate and hierarchize the differ- ences between them. Focusing on the wider associations of blackness in white societies, David Sibley (1995a) points out how black is used to describe dirt, dis- ease, death and decay. These associations not only emphasize the threatening qual- ity of blackness but also carry implicit suggestions of contamination. This fear of infection, that racialized minority groups carry disease and threaten white society, was evident in 'moral panics' (see also Chapter 6) which occurred in the UK in 1905 following an outbreak of smallpox amongst Pakistani immigrants (Shilling 1993). More recently, the same anxieties about the black body have been reproduced in racist claims that AIDS originated from Africa.

Another example is found in the work of Vron Ware (1992: 3) who opens her book *Beyond the Pale* with the following 'story' or urban myth about a white English woman holidaying in New York, USA:

> Nervous about travelling as a single woman and alarmed at the prospect of being in a city renowned for violent crime, she booked into an expensive hotel where she thought she would be safe. One day she stepped into an empty elevator to go up to her room, and was startled when a tall black man accompanied by a large ferocious-looking dog came in and stood besides her just as the lift doors were closing. Since he was wearing

Figure 2.1 Gender incorporates a recognition of the way differences between the sexes are socially constructed in a hierarchical way

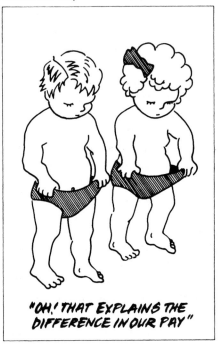

Source: LIBERTY

shades she could not be sure whether he was looking at her, but she nearly leapt out of his skin when she heard his voice: 'Lady, lie down'. Terrified, she moved to obey him, praying that someone would call the elevator and rescue her in time. But instead of touching her the man stepped back in confusion. 'I was talking to my dog', he explained, almost as embarrassed as she was' . . . later it transpires the man in the lift was Lionel Ritchie.

This 'story' shows how the black body continues to be constructed as an object of dread by white women (see also Chapter 6). Ware (1992) unpacks why this is so, explaining how this anecdote can only be understood within the context of slavery and colonialism. She argues that colonial ideologies about black masculinity (as dangerous, criminal and uncontrollable) and white femininity (pure, vulnerable, etc) continue to underpin social relations between black men and white women in contemporary Western societies.

In the same way, if you go back and reread the examples above about the suffragettes, Australian women pilots, and black masculinity, you should be able to see how these understandings of the female and the black male body are also a product of discourses, representations and material practices rather than being the product of 'natural' essences.

It is important to note, however, that although essentialist explanations for sexual and racial differences have often been used to justify sexism and racism, radical feminists and the black power movement have also celebrated women's closeness to nature and black corporeality respectively (Shilling 1993). For example, in the 1970s radical feminists countered the way essentialist arguments about women's closeness to nature were used to subordinate women by employing similar essentialist arguments but reversing their meaning. They celebrated the power of women's biology, their connectedness to Mother Earth, arguing that women's reproductive role provided them with privileged knowledge and power (Griffin 1978, O'Brien 1989). Their goal was not to achieve equality with men by challenging and changing the social meanings of masculinity and femininity; rather they sought to achieve complete autonomy from men and the man-made world by creating women-only communities (see Chapter 8).

Other groups have employed what is termed 'strategic essentialism'. They have mobilized a belief in a shared identity and experience in order to achieve a particular political aim. This tactic involves recognizing, but suspending, differences between those involved in order to form a strategic alliance. For example, lesbian and gay activists have sometimes used the argument that their sexuality is a product of a gay gene rather than being a social choice (even though they believe this to be untrue) in order to establish lesbians and gay men as a legitimate minority group that deserve the same protection and civil rights as ethnic minorities (Epstein 1987). The danger of adopting such an approach, however, is that it imposes an assumption of homogeneity on lesbians and gay men which obscures other differences (such as gender, class, age, ethnicity) between them; and it also makes lesbians and gay men a more visible, and therefore easier, target for opponents (Epstein 1987).

The debate between essentialism and constructionism has gradually been overtaken by a recognition that the distinction between biology and social meanings – between sex and gender – does not hold up. Julia Cream (1995) identifies three bodies which disrupt traditionally accepted notions of sex and gender: the *transsexual*, the *intersex baby* and the *XXY female*.

- *Transsexuals* believe their body does not correspond with their gender identity. They believe themselves to be either women trapped in men's bodies or men trapped in women's bodies. The development of medical technologies has enabled transsexuals to change their sexed bodies to fit their gender identities (see Ekins and King 1999).

- *Intersexed babies* have genitals which are neither clearly male nor female. Usually, doctors assign these children a sex soon after birth according to the best genital fit. In this way, the anomalous body is made to conform in order to maintain the fiction of a binary distinction between male and female.

- *XXY females* are women who have XXY chromosomes rather than XX chromosomes which women are assumed to carry, or XY chromosomes which men are assumed to have. This chromosome pattern, which disrupts the assumed male/female binary, first came to light when chromosome or what was termed 'sex testing' was introduced at the Olympics to prevent men competing in women's events. It led to suggestions that perhaps women with XXY chromosomes were 'really' men.

Cream's (1995) three examples neatly demonstrate that we do not all necessarily fit into one of only two bodies: male or female. There are no coherent 'natural' categories: man/woman. Yet, despite all the biological ambiguity evident in Cream's examples, our bodies are still allowed to be one of two sexes only. Ambiguous bodies such as transsexuals, intersex babies and XXY females are contained and medicalized in order to conform to our culture's demand for only two categories of sex (male/female) and to confirm or maintain the association between gender and sex (Cream 1995). For example, intersex babies are usually operated on soon after birth so that they can be identified as either male or female. 'What this means is that our understanding of gender (man and woman) is not determined by sex (male and female) but that our understandings of sex itself are dictated by an understanding that man and woman should inhabit distinct and separate bodies. So, sex does not make gender; gender makes sex' (Women and Geography Study Group 1997: 195). In other words, the dichotomous distinction (outlined earlier) which some academics have made between sex (biological or essentialist) and gender (social) is collapsed; both are exposed as social. The body can never be understood as a pure, neutral or pre-social form onto which social meanings are projected. It is always a social and discursive object (Grosz 1994).

Understandings of gender have been fundamentally challenged and reworked by the philosopher Judith Butler (1990, 1993a). She rejects the notion that biology is a bedrock which underlies the categories of gender and sex. Rather, she theorizes gender (and implicitly other identities too) as performative, arguing that 'gender is the repeated stylisation of the body, a set of repeated acts within a highly rigid regulatory framework that congeal over time to produce the appearance of substance, of a natural sort of being' (Butler 1990: 33). In other words, gender is an *effect* of dominant discourses and matrices of power. There is no 'real' or original identity behind any gender performance. Butler (1993a) suggests that social and political change within the performance of identity lies in the possible displacement of dominant discourses. In a reading of drag balls she argues that the parodic repetition and mimicry of heterosexual identities at these events disrupts dominant sex and gender identities because the performers' supposed 'natural' identities (as male) do not correspond with the signs produced within the performance (e.g. feminine body language and dress). '[B]y disrupting the assumed correspondence between a 'real' interior and its surface markers (clothes, walk, hair, etc), drag balls make explicit the way in which all gender and sexual identifications are ritually performed in daily life' (Nelson 1999: 339). In other words, they expose the fact that all identities are fragile and unstable fictions.

Butler's (1990) writing has become important within social and cultural geo-graphy. The notion of performativity has been used to frame geographical studies, and to talk, not only about bodily identities, but also about space (e.g. Bell *et al.* 1994, McDowell and Court 1994, Sharp 1996, Kirby 1996, Lewis and Pile 1996, Valentine 1996a, Rose 1997 and Delph-Januirek 1999). Instead of thinking about space and place as pre-existing sites in which performances occur, some of these studies argue that bodily performances themselves constitute or (re)produce space and place. However, geographers have also been criticized for overlooking the problematic aspects of Butler's work (particularly in relation to her assumptions about **subjectivity**, agency and change) when employing her theorization of performativity (Walker 1995, Nelson 1999).

■ Summary

- The Cartesian division and subordination of the body to the mind has shaped geographers' understandings of society and space.

- Historically, the notion that women's bodies are different from, and inferior to, men's has been used to justify their limited participation in public space.

- Similar 'naturalistic' views have also been used to legitimate the subordination of black bodies.

- Arguments that assume sexual and racial differences are a product of biology or natural essences are termed essentialist.

- Essentialist explanations have been contested by social constructionists, who argue that bodies are the product of discourses, representations and material practices rather than biology.

- A distinction has been made between *sex* (a natural category based on biological difference) and *gender* (the social meanings ascribed to sex).

- Geographers, however, are now moving away from thinking about sex and gender in this way. Instead the focus is on embodied performance.

■ 2.3 The body as a space

Adrienne Rich describes the body as 'the geography closest in'. The body is not just in space, it *is a space*. There are three ways in which we can think of the body as a space. It is a surface which is marked and transformed by our culture. It is a sen-suous being, the material basis for our connection with, and experience of, the world. It is what bounds the space of the psychic.

■ 2.3.1 Surface

The body is a surface of inscription – a surface on which we inscribe our identities, and a surface upon which cultural values, morality and social laws are written, marked, scarred or transformed by various institutional regimes. This process Elizabeth Grosz (1993: 12) has termed 'social tattooing'. She writes:

> In our own culture, inscriptions occur both violently and in more subtle forms. In the first case, violence is demonstrable in social institutions, keeping the body confined, constrained, supervised, and regimented, marked by implements such as handcuffs . . . the straitjacket, the regimen of drug habituation and rehabilitation, chronologically regulated time and labor divisions, cellular and solitary confinement, the deprivation of mobility, the bruising of bodies in police interrogations, etc. Less openly violent but no less coercive are the inscriptions of cultural and personal values, norms, and commitments according to the morphology and categorization of the body into socially significant groups – male and female, black and white, and so on. The body is involuntarily marked, but it is also incised through 'voluntary' procedures, life-styles, habits, and behaviours. Makeup, stilettos, bras, hair sprays, clothing, underclothing mark women's bodies, whether black or white, in ways in which hair styles, professional training, personal grooming, gait, posture, body building, and sports may mark men's. There is nothing natural or ahistorical about these modes of corporeal inscription. Through them, bodies are made amenable to the prevailing exigencies of power. They make the flesh into a particular type of body – pagan, primitive, medieval, capitalist, Italian, American, Australian. What is sometimes loosely called body language is a not inappropriate description of the ways in which culturally specific grids of power, regulation, and force condition and provide techniques for the formation of particular bodies (Grosz 1994: 141–2).

The work of three social theorists – Norbert Elias, Pierre Bourdieu and Michel Foucault – has been particularly important in helping social scientists to think about the body as a surface which is inscribed and regulated.

■ 2.3.1.1 Elias: the civilized body

From the Middle Ages to the present there have been fundamental changes in what are considered appropriate forms of bodily expression. In medieval times life was short, food was scarce, emotions were freely expressed, people even took pleasure from watching the torture and mutilation of others. There were few social prohibitions about appropriate bodily behaviour – it was commonplace for people to eat, belch, fart, shit and spit in public. The emphasis was on satiating the self, not on moderation or self-restraint. Then, in 1530, Erasmus produced a short treatise on manners, *De Civilitate Morum Puerilium*, which set out new codes of bodily propriety which required that the body should be controlled and hidden and that all signs of bodily functions should take place in 'private' rather than 'public' space.

According to the German sociologist Norbert Elias (1978/1982), this marked a gradual but fundamental shift in public behaviour. Body management and control of the emotions increased as taboos developed around bodily functions such as spitting and defecating, culminating in the emergence of concepts such as self-restraint, embarrassment and shame. Elias documents the role of Renaissance court society in promoting 'civility'. He argues that an individual's survival in court depended not on their physical force or strength but on their impression management and etiquette. An individual's social identity or status could be read from their bodily deportment and manners. Those who regarded themselves as socially superior attempted to distinguish themselves from others through their superior bodily control, while those who wanted to be climb the social ladder had to imitate their superiors. These moral 'codes' eventually trickled down the social strata and became practised to different degrees by all citizens, becoming part of accepted everyday behaviour. Thus parents have to 'civilize' children by teaching them how to manage their bodies, to use the toilet, not to spit, and so on. Through such processes children's bodies are turned into adult bodies (see also Chapter 5). As Elias (1982: 38) notes, contemporary children now have 'in the space of a few years to attain the advanced level of shame and revulsion that [adults have] developed over many centuries'.

This gradual civilizing of the body has produced five significant cultural shifts: first, in the nature of fear. While individuals in unregulated medieval societies feared attack, in modern societies social fears of shame and embarrassment are, for most people, more pressing everyday concerns. The second shift is in the construction of the social in opposition to biology/nature: by controlling our bodies we suppress our 'nature', transcend our animality and emphasize our humanness. The third lies in the importance of rational thought and the control of emotions. The fourth is in social differentiation, namely that increased emphasis on self-control has encouraged greater reflexivity about our bodies, individualization, and a desire to distance ourselves from others. The fifth shift is in the social distance which has developed between adults and children (Shilling 1993).

■ 2.3.1.2 Bourdieu: the body as the bearer of symbolic value

In a famous book titled *Distinction: a Social Critique of the Judgement of Taste*, the French sociologist Pierre Bourdieu (1994: 190) argues that 'the body is the most indisputable materialization of class taste'. He suggests that class becomes imprinted upon our bodies in three ways (Shilling 1993). First, through our *social location* in that our material circumstances shape the way we can look after our bodies, dress, etc. Second, through what he termed **habitus**. This is the class-oriented, unintentional pre-disposed ways we have of behaving which often betray our class origins. Bourdieu argues that every aspect of our embodiment, from the way we hold our cutlery to the way we walk, articulates and reproduces our social location in this way. Third, through *taste*. This refers to 'the processes whereby individuals appropriate as voluntary choices and preferences, lifestyles which are actually rooted in material constraints' (Shilling 1993: 129). Thus Shilling (1993: 129) argues, 'The development of taste, which can be seen as a conscious manifestation of habitus, is embodied and deeply affects people's orientations to their bodies.' For example, working-class people usually have limited material resources to spend on food so they tend to buy cheaper foods. These are commonly foods that are high in fat. In turn, their preference for and consumption of these foods affect their bodily shape, health and ultimately mortality.

Bourdieu's (1984) work clearly shows how different classes produce different bodily forms. For example, the working classes have little free time and so the body is for them a means to an end. Working-class sports, such as boxing, football and weight-lifting, reflect this in that they are about excitement – a temporary release from the tensions of everyday life – and involve developing strength and skills which are oriented towards manual work (Shilling 1993). In contrast, the middle classes have more leisure time and resources to invest in the body as a project. Their emphasis is often on slimness and looking good because appearance or presentation rather than strength is important in many middle-class occupations (see Chapter 5). There is also significant differentiation within the middle classes, with different groups being oriented differently towards their bodies. While the upwardly mobile middle class tend towards activities such as fitness training, the elite bourgeoisie tend to engage in leisure pursuits such as polo, riding, yachting or golf, which combine sport with socializing. These activities develop physical, social and **cultural capital**.

The distinct bodily forms produced through the different relationships the working classes and middle classes have with their bodies are important because they are valued differently and, so Bourdieu (1984) argues, contribute to the development of social inequalities. The physical capital of the working classes has less economic value than that of the middle classes. Likewise, the working classes find it more difficult to convert their cultural capital into other resources because their bodily comportment (in terms of manners, ways of speaking, etc) is often judged as negative at employment interviews. Shilling (1993) suggests that, in the case of social capital, the aggressive management of bodies which enables young working-class men to gain respect amongst their peers is not valued in other contexts. In contrast, the middle classes, especially the elite, have more opportunities to turn physical capital into other resources. For example, elite sports which have strict codes of etiquette enable individuals to acquire or demonstrate that they have the sort of bodily competence which is often crucial in employment and educational selection processes. These sports can also enable individuals to develop social networks which may allow them to make important professional contacts and to meet appropriate partners (Shilling 1993).

There are two further important points to note about Bourdieu's analysis of the body as the bearer of symbolic value. Firstly, social differences in bodily forms are often mistakenly assumed to be 'natural' differences. Secondly, the value attached to different bodily forms is not fixed, but rather there are struggles and conflicts between and within social groups to define and control which bodily forms are valued.

Geographers have drawn on Bourdieu's work specifically in relation to studies of consumption (e.g. Bell and Valentine 1997), and of gentrification and middle-class formation (e.g. Savage *et al.* 1993, Ley 1996) (see Chapter 7).

■   2.3.1.3 Foucault: the disciplining gaze

Michel Foucault is a French philosopher whose work has been particularly influential in shaping understandings of how the body is produced by, and exists in, discourse and how it has historically been disciplined and subdued (Driver 1985). He coined the term 'biopower' (power over life) to describe a diverse range of techniques through which bodies are subjugated and populations controlled (Foucault 1978).

In his work on sexuality Foucault (1978) argued that sex is not a 'natural' or biological urge (see also section 2.2.2) but rather is an historical construct. He traced the history of the way sexualities have been produced through discourses, identifying how heterosexual relations have been produced as the 'norm' and other forms of sexuality classified as 'deviant'. In this way, Foucault showed how these discourses have played an important role not only in controlling individual bodies but also in the management of whole populations.

Foucault also had a particular interest in the regulation of populations and societies through the disciplining power of the surveillant gaze. His book *Discipline and Punish* (1977) opens with a graphic description of an eighteenth-century prisoner being tortured to death. Foucault uses this example to explain how in traditional societies the body was the target of penal repression. Punishment for social wrongdoing was literally acted out on the body of individual prisoners through these public acts of ritual torture. Foucault then goes on to trace the development of penology and to show how, following penal reform in the eighteenth century, more subtle tactics and new technologies of power were used to punish offenders, which relied on using institutional space as a way of controlling prisoners. Foucault focuses on the example of a prison designed by Jeremy Bentham. Known as Bentham's **Panopticon**, this prison had a circular design to ensure that all the cells were under potential constant surveillance from a central watchtower, although the occupants of the cells could not tell if, or when, they were actually being watched. (There is some dispute, however, as to whether such a prison was ever built.)

Foucault (1977: 177) argued that the fact of being able always to be seen meant that individuals would exercise self-surveillance and self-discipline. He writes: '[T]here is no need for arms, physical violence, material constraints. Just a gaze. An inspecting gaze, a gaze which each individual under its weight will end by interiorising to the point that he is his own overseer, each individual thus exercising surveillance over and against himself' (Foucault 1977: 155). Through this control and regulation of movements, time and everyday activities, Foucault argues, the body becomes invested with relations of power and domination, resulting in obedient 'docile bodies' (Foucault 1977: 177). He also pointed out, though, that where there is power there is always resistance. These ideas have been applied by social scientists beyond the realm of institutional settings such as prisons, to think about the ways 'normal' social life and activities takes place within a system of 'imperfect Panopticism' (Hannah 1997).

Foucault's work has therefore become very important in the social sciences. First, it demonstrates that the body is constituted within discourse and that different discursive regimes produce different bodies. Second, it demonstrates that the transition from traditional to modern societies was marked by a shift in the *target of discourse* away from controlling the body through brute force towards the use of the mind as the surface of the inscription of power. Third, it reveals a shift in the *scope of discourse*, away from controlling individual bodies towards controlling the population as a whole. Fourth, it emphasizes the individual body as the effect of an endless circulation of power and knowledge. In doing so it makes an important connection between everyday practices and the operation of power at a wider scale (Shilling 1993).

Geographers in particular have used Foucault's ideas to think about the way bodies are disciplined in a range of spatial contexts, for example: the asylum (Philo 1989), the workhouse (Driver 1993), the workplace (McDowell 1995), the prison (see Valentine 1998), and the state (Robinson 1990) (see Chapter 5). They have also drawn upon the concept of 'imperfect Panopticism' (Hannah 1997) in discussing new technologies (such as CCTV) and the emergence of a surveillance society (Hannah 1997, Fyfe and Bannister 1998, Graham 1998) (see Chapter 6).

■   2.3.2 The sensuous body

Section 2.3.1 focused on how the body is produced through discourse. However, Bryan Turner (1984: 245) argues that such work ignores the material body. He writes that the 'immediacy of personal sensuous experience of embodiment which is involved in the notion of *my* body receives scant attention. My authority, possession and occupation of a personalised body through sensuous experience are minimised in favour of an emphasis on the regulatory controls exercised from outside.'

Among those who have focused on the lived body is the phenomenologist Merleau-Ponty (1962). For Merleau-Ponty (1962) subjectivity is located in the body rather than the mind. He argues that the body is the locus of experience, it exists prior to all reflection and knowledge. The body, and the way it acts and orientates itself to its surroundings, is the basis of everything else (Young 1990b). His emphasis is therefore on perception. The subject exists for him insofar as the body can approach and connect with its surroundings and thus bring them into being (Young 1990b). As Langer (1989: 41) writes, 'my awareness of my body is inseparable from the world of my perception. The things which I perceive, I perceive always in reference to my body, and this is so only because I have an immediate awareness of my body itself as it exists *"towards them"*.' In other words, for Merleau-Ponty, the mind 'is always embodied, always based on corporeal and sensory relations' (Grosz 1994: 86).

For Merleau-Ponty (1962) the body is the subject that constitutes space because without the body there would be no space, but it is also the object of spatial relations. He grants the body intentionality, arguing that bodily movements can be directed by the body's intelligent connections with the world which surrounds it.

The geographer David Seamon (1979) has drawn on Merleau-Ponty's work to understand the everyday nature of environmental experience. Using focus groups conducted with students, he explored the way in which people move through and occupy everyday spaces. Like Merleau-Ponty, Seamon does not conceive of the body as a passive object, but rather as capable of its own thought and action. To illustrate this he draws on extracts from his focus group discussions in which participants describe how they would do things 'automatically', such as walking home without thinking about the route they were taking or reaching for a clean towel under the sink before they have remembered there is not one there (Box 2.1). Seamon (1979: 56) argues that underlying and guiding these everyday movements is a bodily intentionality which he terms 'body subject'. He describes this as 'the inherent capacity of the body to direct behaviours of the person intelligently, and thus function as a special kind of subject which expresses itself in a preconscious way usually described by such words as "automatic", "habitual", "involuntary" and "mechanical"' (Seamon 1979: 41).

Developing the theme of habitual movement, Seamon (1979: 55) goes on to argue that individuals build up 'time–space routines'. This is the series of behaviours or rituals of everyday life we habitually repeat as part of daily or weekly schedules – for example, getting up at 7.30 a.m., having a bath, dressing, having breakfast, leaving the home at a set time, buying a paper from the nearby store, catching the same bus to work, and so on. Where many individuals' time–space routines come together they fuse to create 'place ballets' (Seamon 1979: 56). These can occur in all sorts of environments from the street, to the university, a marketplace or a station. For Seamon (1979) the power of place ballets to generate a sense of place has important implications for planning and design.

While Seamon's work focuses on the way bodies move through and occupy space, other geographers have been concerned with the role of the senses in forming a dialogue between the body and its surroundings (Rodaway 1994). According to Rodaway (1994: 31), there are four dimensions to understanding the sensuous geography of the body. First, the body's geometry (awareness of front, back, up, down, etc) and its senses give us an *orientation* in the world. Second, the senses provide a *measure* of the world in that they enable us to appreciate distance, scale, etc. Third, the *locomotion* of the body enables us to explore and evaluate our surroundings with all our senses. Fourth, the body is a *coherent system* integrating and co-ordinating the information and experiences gathered by each of the senses.

Box 2.1: Bodily intentionality

'When I was living at home and going to school, I couldn't drive to the university directly – I had to go around one way or the other. I once remembered becoming vividly aware of the fact that I always went there by one route and back the other – I'd practically always do it. And the funny thing was that I didn't have to tell myself to go there one way and back the other. Something in me would do it automatically; I didn't have much choice in the matter.'

'I know where the string switches to the lights in my apartment are now. In the kitchen, even in the dark, I walk in, take a few steps, my hand reaches for the string, pulls and the light is on. The hand knows exactly what to do. It happens fast and effortlessly – I don't have to think about it at all.'

'Sometimes for an early class I'll get to the class and wonder how I got there – you do it so mechanically. You don't remember walking there. You get up and go without thinking – you know exactly where you have to go and you get there but you don't think about getting there while you're on the way.'

'I operated an ice-cream truck this past summer. On busy days I'd work as fast as I could . . . As I worked, I'd get into a rhythm of getting ice cream and giving change. My actions would flow, and I'd feel good. I had about twenty kinds of ice cream in my truck. Someone would order, and automatically I would reach for the right container, make what the customer wanted, and take his [sic] money. Most of the time I didn't have to think about what I was doing. It all became routine.'

Cited in Seamon 1979: 163–4, 165, 170.

While everyday perception is, for most people, multi-sensual, involving some combination of touch, taste, hearing, smell and sight (though see geographical work on sensory impairment, notably the experiences of the visually impaired and blind: Butler 1994, Butler and Bowlby 1997), geographers have tended to privilege the visual over the other senses. The emphasis has often been on how bodies are dressed and on how they appear as they move through and occupy space. The 'surveillant gaze' (see section 2.3.1.3 above) has been an important means through which bodies are interpreted and regulated. Recently, however, some geographers (e.g. Thrift 1996) have begun to stress the importance of valorizing all the senses and forms of social knowledge, not just the visual and the act of looking. Our understandings of our bodies and our attempts to manage them are based not just on visual information but also other sensual information. Space and time are perceived through all the different senses of the body. Our bodies might look active, young and dynamic, yet we might feel heavy, slow, immobile, have an inhibited sense of our own spatiality – or vice-versa (Valentine 1999a).

The body may also be considered as a site of pleasure. Lynda Johnston's (1998) work on women body builders captures some of the corporeal pleasures of pumping iron (see also section 2.4.1.1) while David Crouch (1999) describes some of the physical and sensual pleasures of caravanning. Yet, this is not to forget that the body is also a site of pain. Liz Crow (1996) warns that by focusing on the social construction of disability and challenging stereotypes of the disabled as dependent or vulnerable, researchers are in danger of ignoring the very real bodily experiences of pain that an illness or impairment can cause. The complex and contradictory bodily realities of a chronic illness are evident in Pam Moss and Isabel Dyck's (1999) study of women with *myalgic encephalomyelitis* (ME). Here, they describe how the women's experiences of fatigue, pain and cognitive dysfunction may be followed by days of being symptom-free, showing how these contradictory bodily experiences can position the women as simultaneously 'ill' and 'healthy'. This theme is also pursued in Moss's (1999) autobiographical account of having ME, while elsewhere Dyck (1999) considers how women with multiple sclerosis (MS) negotiate changes in their bodies' corporeality.

■ 2.3.3 The psyche

The previous sections have distinguished between surface presentations – how we might appear – and sensuous experiences – how we might feel. This section pursues the distinction between 'surface' and 'depth' or 'inside' and 'outside' further by focusing on the psyche – how we really feel (Kirby 1996: 14).

Geographers have used psychoanalytical theory and its concern with the unconscious to try to understand the relationship between the individual and the external world. Analysts believe that it is the unconscious (mental processes of which we are unaware) which generates the thoughts and feelings which motivate or inhibit our actions. Consequently, it is the unconscious, rather than consciousness, which they regard as the key to understanding individual and group behaviour (Pile 1996). In summarizing the complex and highly contested 'notion' of the unconscious, Steve Pile (1996: 7) explains that 'For most analysts . . . the unconscious is made up of residues of infantile experiences and the representatives of the person's (particular sexual) drives. Although there is considerable disagreement about how children develop increasingly intricate and dynamic psychological structures, the experiences of early childhood are generally accepted to be critical.' He goes on to explain that we cope with early upsetting or painful experiences by repressing them from our consciousness. This causes a split within the mind into the conscious and the unconscious. Although the unconscious does not dictate what goes on in the mind, it does continually struggle to express itself, for example through dreams.

There are many different schools of psychoanalytic thought, each of which is highly contested. Geographers have drawn on a range of approaches, including the work of Freud, Lacan, Winnicott, Kristeva and Klein, in order think about the relationships between subjectivity, society and space (see for example: Aitken and Herman 1997, Blum and Nast 1996, Bondi 1997, Pile 1996, Rose 1996). David Sibley's work (1995a, 1995b) is a particularly good example of how psychoanalytic theory can be used to understand how we locate ourselves and others in the world.

In his book *Geographies of Exclusion* David Sibley (1995a) examines why some groups, including gypsies, children, lesbian and gay men and ethnic minorities, are both socially and spatially marginalized. He argues that each of these groups is feared or loathed and goes on to shows how this attitude to cultural difference is played out in space, giving examples of how each of these groups has been subject to spatial regulation and segregation. In trying to understand why these groups are feared and how socio-spatial boundaries are constructed along axes of difference such as race and sexuality Sibley (1995a, 1995b) draws on the work of object relations psychoanalysts such as Melanie Klein and Julia Kristeva. He uses their work to understand how people make the distinction, and mark boundaries between self and other.

Pile (1996) explains – in his own words, very crudely – that in object relations theory the child's sense of itself is developed through its relation to objects (which include the mother, toys, etc). Specifically, the child develops a sense of itself as a single bounded identity as a result of its gradual recognition that it has a separate body from that of its mother. The mother is both a 'good' object because she gives the child food, love, and so on, and a 'bad' object because she is not always available or does not always respond to the child's desires. This tension between these experiences of desire and insecurity produces a sense of ambivalence within the child towards the mother (Pile 1996). Sibley (1995b: 125) writes: 'Aversion and desire, repulsion and attraction, play against each other in defining the border which gives the self identity and, importantly, these opposed feelings are transferred to others during childhood.'

Sibley then uses Kristeva's concept of abjection to explain this displacement of these contradictory feelings onto those regarded as different. For Kristeva, the subject feels a sense of repulsion at its own bodily residues (excrement, decay, etc). To maintain the purity of the self, the boundaries of the body must be constantly defended against the impure. Sibley (1995a: 8) argues that the 'urge to make separations between clean and dirty, ordered and disordered, "us" and "them", that is, to expel the abject, is encouraged in western cultures, creating feelings of anxiety because such separations can never be finally achieved' (Sibley 1995a: 14). Sibley goes on to examine the relationship between these processes and organization of space, social values, and power relations. He shows how cultural and social values in Western society construct particular groups (such as gypsies, lesbians and gay men, ethnic minorities, and so on) as 'dirty' or polluting. Using examples, Sibley (1995a) demonstrates how people often respond to these 'abject others' with hatred, attempting to create social and spatial boundaries to exclude or expel them. Through this literal mapping of power relations and rejection Sibley argues that particular exclusionary landscapes are developed in different times and places (see also Chapter 6).

■ **Summary**

- The body is a surface upon which cultural values, morality and institutional regimes are inscribed.

- It is constituted within discourse. Different discursive regimes produce different bodies.

- The body is also a sensuous being, the material basis for our connection with, and experience of, our environment.

- Some geographers have used psychoanalytical theory to try to understand the relationship between the individual and the external world.

■ **2.4 The body as a project**

■ 2.4.1 The individual body

In the late twentieth and early twenty-first centuries affluent Western societies have been marked by a growth in mass consumption, the democratization of culture, a decline in religious morality and a postindustrial emphasis on hedonism and pleasure (Turner 1992). In this context, the body has emerged in consumer culture as an important bearer of symbolic value (see section 2.3.1.2) and as constitutive of our self-identities. As Chris Shilling (1993: 3) explains: 'For those who have lost their faith in religious authorities and grand political narratives, and are no longer provided with a clear world view or self-identity by these trans-personal meaning structures, at least the body initially appears to provide a firm foundation on which to reconstruct a reliable sense of the self in the modern world.' In particular, the exterior surface or appearance of the body is seen to symbolize the self.

Developments in everything from plastic surgery and laser treatments to sports science mean that we each have increasing possibilities to control and (re)construct our bodies (although this is not to say that we all have the interest, desire or resources to do so) in terms of size, appearance, shape, and so on. The body is malleable and dynamic rather than static (Lupton 1996). Therefore '[i]n the affluent West there is a tendency for the body to be seen as an entity which is in the process of becoming; a *project* which should be worked at and accomplished as part of an *individual's* self-identity' (Shilling 1993: 5). This emphasis on our individual responsibility for our bodies is evident in government health campaigns. The onset of all sorts of illnesses (such as cancer or diabetes) and even the ageing process, we are told, can be avoided or at least delayed, if we look after our bodies properly, eat the right foods, watch our weight, exercise, do not smoke, drink in moderation and so on. The media, advertising, fashion industries, medicine and consumer culture all provide a framework of discourses within which we can each locate, evaluate and understand our own bodies (Featherstone 1991).

In this context then, where the body is a symbol of the self, an object of public display and the responsibility of its 'owner', we are expected to be vigilant about our size, shape and appearance and to discipline ourselves in order to produce our bodies in culturally desirable ways (Valentine 1999a, 1999b). Drawing on Foucault's notion of the surveillant gaze (see section 2.3.1.3), writers have argued that disciplinary power is most effective when it is not external but is exercised by, and against, the self. The failure to maintain a slim, youthful, heterosexually desirable body is thus commonly seen not only as a social, but also a personal and moral failing (Synott 1993). Anthony Synott (1993) suggests that our bodily identities are central to our life chances. People who are overweight are often stereotyped as self-indulgent, lazy, untrustworthy and non-conforming (see Chapter 5). There are many examples of organizations as diverse as The Disney Corporation, the New York City Traffic Department and City of London banks who have established their own corporate bodily norms, or fired employees for being overweight (Bell and Valentine 1997, Valentine 1999a).

There are other pay-offs too for those who treat their bodies as a project. According to Shilling (1993: 7), 'Investing in the body provides people with a means of self-expression and a way of potentially feeling good and increasing the control they have over their bodies. If one feels unable to exert influence over an increasingly complex society, at least one can have some effect on the size, shape and appearance of one's body.' More extreme forms of control – which may develop as extensions of mundane everyday bodily practices such as dieting, exercise and dress – include plastic surgery, liposuction, body building (Johnston 1998) and body modification (Bell and Valentine 1995b).

It is also important to remember that this sort of identity construction is not an option within reach of everyone, however. Mike Featherstone (1999: 5) points out, for example, that racialized bodies 'cannot be so easily reconstituted and made into a project'.

■ 2.4.1.1 Body building

'The body builder . . . is involved in actively reinscribing the body's skeletal frame through the inscription of muscles (the calculated tearing and rebuilding of selected muscles according to the exercise chosen) and of posture and internal organs' (Grosz 1994: 143). A muscular physique has been associated in the Western tradition with the male sex (Johnston 1998). Elizabeth Grosz (1994: 224) argues that male body building can be read as an attempt to render the whole body into the phallus, 'creating the male body as hard, impenetrable, pure muscle'. Interestingly, though, Lynda Johnston (1998) points out that the narcissistic practices of waxing, oiling and posing which this investment in the body can involve might be seen as a feminine activity.

In a paper titled 'Creating the perfect body' Lee Monaghan (1999) highlights the extent to which male body building can be a narcissistic act. He identifies a range of muscular bodies that men can potentially produce through body building, their obsessions with their ethnophysiology (separation between muscles, muscle defini-tion, body fat, and so on) and evaluation and appreciation of other men's bodies in body building magazines and competitions. Drawing on interview material, Monaghan (1999) exposes the anxieties and dismay some men experience when they spend con-siderable amounts of time building bodies which they think will have sexual cap-ital only to find that women laugh at them when they parade on the beach. One, for example, describes the horror of working his body to produce a 'rhino' or 'frog' shape (big back and shoulders, small waist, big thighs) only to find out that he should have gone for a more athletic or sleeker look, such as the 'tiger' or 'puma'. Others fret that wearing clothes over bulked up muscles makes them look fat rather than fit.

The female body builder both reinforces and destabilizes traditional notions of femininity. By exercising self-surveillance, constantly monitoring their diets and shape, and employing plenty of feminine signifiers such as lipstick and earrings, some female body builders accentuate their femininity (Aoki 1996, Johnston 1998). Others, who develop strong, muscular bodies, understand body building as a transgressive act which destabilizes a feminine bodily identity and questions women's difference from, and otherness to men (see section 2.2.2). As a consequence of diet and train-ing women can reduce their body fat to such an extent that their breasts are reduced in size or even eliminated and menstruation can stop (Johnston 1998). By develop-ing attributes which are usually associated with men – strength, stamina, muscular-ity and control – some women feel empowered. Likewise, pumping iron in itself is also a corporeal and potentially erotic sensation. The quotes in Box 2.2 capture the pleasure and pain of working out and watching the pump transform the body.

■ 2.4.1.2 Body modification

Body modifications in the form of tattooing, multiple piercing, branding, cutting, binding, scarification and inserting implants are increasingly visible aspects of many contemporary sub-cultural styles in Western societies (Sanders 1989, Mascia-Lees and Sharpe 1992, Myers 1992). '[T]oday tattoos and body piercing have become increasingly stylish; even fashion models get delicate piercing, and modern bohe-mians sport pierced lips, cheeks, nipples, tongues and genitals' (Steele 1996: 160).

Many accounts by body modifiers emphasize it as a way of taking control of the body, possessing it, and expressing a self-identity. David Curry (1993: 69) explains:

Box 2.2: Body building

Pain

Fiona: 'It's not just the way you end up looking, um, it's the feel of lifting the heavy weights, I really enjoy that. There's something about it. Like this morning – I did chest this morning – just chest alone and I left and I felt all tight across here [points to her chest]. It's a really good feeling and also you get to *like* [original emphasis] the pain you get the next day. For legs you usually get sort [of] two days of pain ...I'm still feeling it from the other day on me.'

Pleasure

Fiona: 'Yeah I feel good when I'm walking along and even covered up, um, there is some sense of feeling good and holding your head up, walking square shoulders and knowing underneath you've got this changed body that you're working on.'

Sarah: 'I used to think *no way* [original emphasis]. I'm not going to be a woman with muscles because again I envisaged this big huge thing and until you start to push a little bit of weights and notice the change in your body you start to appreciate that, you know, it looks nice.'

Johnston 1998: 255–6

'Body decoration lies at the interface between the private and the public. The skin is the actual membrane between what, on one side, is inside me and, on the other side, is outside me. It is superficially me and at the same time a surface onto which I can both consciously and unconsciously project that which is more deeply me.' In this respect, body modifications are regarded as a way of articulating individuality, of resisting the superficiality of consumer culture and of making a statement of difference. People often go to great lengths to design their own unique tattoos. One of Paul Sweetman's (1999: 68) interviewees explains '. . . it makes you feel individual . . . you know like, everyone's born with roughly the same bodies, but you've created yours in your own image [in line with] what your imagination wants your body to look like. It's like someone's given you something, and then you've made it your own, so you're not like everyone else any more.' While a man with a large-scale back tattoo feels 'I am a different person now, and I realise that in many ways, I am not the average guy on the street. On a more public level, my tattoo affirms that difference. It visually sets me apart from the masses' (Klesse 1999: 20).

In contrast to other ways of articulating a self-identity many body modifications have the advantage of permanence (although this varies as piercings and some tattoos can be removed). As one woman with a tattoo explains, 'Before, I could express myself in clothes and things like that, now it's actually something that's permanent

Plate 2.1 Body modification can be kept private by clothing or displayed at will

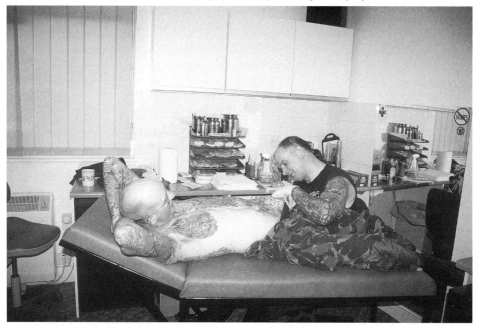

and that's definitely me' (Sweetman 1999: 58). In this way, body modifications can be a stronger, definitive statement of the self, 'an attempt to fix and anchor the self by permanently marking the body' (Featherstone 1999: 4). Indeed, some use them as a way of recording defining moments in their lives (e.g. a wedding) onto their bodies. As such, body modifications are a 'permanent diary that no one can take off you' (Sweetman 1999: 69). Consequently, the American tattooist Don Ed Hardy argues that it is misleading to suggest that tattoos and piercings are just fashion. He says, 'It is on your body; it's permanent; you have to live with it and it hurts' (quoted in Sweetman 1999: 72).

Although body modifications are on the exterior of the body, levels of visibility vary widely, from whole-body tattoos or piercings on the face, to discreet butterflies hidden on the buttocks (see Plate 2.1). The fact that body modifications can be kept private by clothing and displayed publicly at the will of the bearer offers one of the transgressive pleasures of adornment: the knowledge that under the smart suit is a pierced nipple (Bell and Valentine 1995b).

In contrast to those who use body modification to express individuality, others use it as a group marker. These non-mainstream body modifiers include 'Modern Primitives', a movement which originated in California, USA in the 1970s. These people claim to have primal urges to alter their bodies. Among their practices, which they term 'body play', are contortion, including foot-binding and stretching of body parts; bondage and various forms of body constriction; deprivation, including being caged in boxes, bags, and body suits; wearing 'iron' (manacles, heavy bracelets, etc); burning, branding and shock treatment; penetration including piercing, skewering, tattooing; and various forms of suspension from different body parts (Klesse 1999). Modern Primitives seek their inspiration, and appropriate what they regard as 'primitive rituals' and body modification techniques, from indigenous traditions of Polynesia and elsewhere. Their movement emphasizes spirituality and community (see also Chapter 8, section 8.5.2). Through the creation of 'new tribes' they claim to give people a sense of belonging which they feel they have lost in contemporary society. It is a movement, however, which is widely criticized for fetishizing other cultures – most notably idealizing and essentializing primitive cultures – and for repro- ducing repressive gender and racialized stereotypes (Klesse 1999).

■   2.4.2 The connected body

The popular and academic focus on the body as a project (2.4.1 above) reflects what Chris Shilling (1993: 1) describes as 'the unprecedented *individualisation* of the body' in contemporary affluent societies. Yet, this perhaps overstates the degree of control individuals actually have over their own bodies. Writing about places, Doreen Massey (1993: 66) argues that they 'can be imagined as articulated moments in networks of social relations and understandings' (see also Chapter 1). In the same way, it is possible to think of bodies not as bounded and discrete entit- ies but as relational 'things', as the product of interactions, as constituted by con- stellations of other social relations. This means that rather than thinking about the individualization of bodily practices, geographers need to recognize the *connected- ness* of bodies to other places (see also Nast and Pile 1998), and specifically, the ways in which bodies are inflected by material practices, representations, social rela- tions and structures of political-economic power in wider locations such the home, workplace, community, state and so on.

For example, the extent to which individuals can produce the space of their bod- ies in accordance with their own individual desires depends upon the extent to which they feel able to demarcate ownership or control over their own bodies. Within spa- tial settings such as the home and the workplace our bodies can be 'open' locations that are subject to control and regulation by others. David Morgan (1996: 132) uses the term 'bodily density' to describe the ways that close proximity to a person over a period of time 'can result in knowledge, control and care of each other's bodies in numerous repeated and often unacknowledged ways'. Nowhere is this more appar- ent than in sexual relationships and in shared households where bodies can become contested terrain between partners, housemates, parents and children (see Box 2.3). Likewise, the role of workplaces and communities in defining how 'members' should dress and manage their bodies (Green 1991, McDowell 1995, Valentine 1999a) and the role of the state in regulating bodily behaviours, from sado-masochistic sex- ual practices (Bell 1995a, 1995b), to assisted suicide and abortion, further expose the limits of individuals' corporeal freedoms.

Box 2.3: Bodily density

Carol: 'I mean there are times when I say "Right I'm going to be good for the next month, cut down, you know. I know I've put on weight", and then two days later I'll be eating something and Mike [her husband] will say to me "What are you eating that for? I thought you were trying to lose some weight" and, er, it gets me mad.'

Valentine 1999a: 333

Mike: 'I worked in buying and buying was, you were always taken out for lunch by sales people who came to visit you and, er, so you know, I don't know, possibly three or four times a week really you'd be taken out for meals and go to restaurants or pubs or whatever . . . I mean five pints at lunch time was common, you know in buying . . . I started to put, well I started to put a lot of weight on because I was just eating all the time. I remember coming up to Christmas and I'd been out for 18 or 19 Christmas dinners . . . I just got, you know, a big stomach and, er, my face was fat.'

Valentine 1999a: 346

Tasmin: 'I've got good intentions but they never work. I mean, I like to go swimming but it's, like, finding the time when I can go swimming when I haven't got the children, or I haven't got to go and pick children up from nursery and things . . . just doesn't work.'

Valentine 1999a: 341

These wider locations can also frustrate individuals' attempts to discipline and manage their bodies in chosen ways. Although the bodies of workers in many professional occupations are constituted within discourses which elide slimness with success, productivity and professionalism (see Chapter 5), the cultural practice of the 'business lunch' can actually drive 'professional' bodies out of shape by making them indulge in excess consumption (see Box 2.3). Likewise, women often find it hard to diet and exercise when their corporeal freedom is limited by the need to provide meals to suit the tastes of other family members and by the time–space constraints of household routines (see Box 2.3).

The erosion or permeability of individuals' body boundaries in these ways demonstrate how little corporeal freedom we can actually have to shape our bodies according to our individual desires. Contrary to popular (and some academic) discourses about the body as an individual project, activities such as healthy eating, dress, and exercise, are not necessarily individualized practices. Rather, a focus on the 'situated interdependence' (Thrift 1996: 9) of everyday life, reveals that our bodies are better understood as porous locations which are inflected by wider socio-spatial relations (Valentine 1999a).

There is also a tendency within popular discourses and the academic literature on the body to emphasize discipline and control. Yet, there are other meanings around bodily practices such as eating and exercise which stress their pleasurable and hedonistic dimensions as well as the physical sensations and emotions such as comfort, release of stress, and happiness, which they can produce (Lupton 1996). It is these sensual pleasures that create ambivalences and tensions for individuals who, on the one hand, want to manage their bodies in line with dominant discourses around self-discipline, yet who, on the other hand, enjoy the physical sensations of eating, are sceptical about medical advice or who take a fatalistic attitude towards their body shape and appearance (Lupton 1996, Valentine 1999a).

Indeed, individuals may sometimes experience competing understandings of how they should be producing the space of their bodies in different locations in which their body is sometimes constructed as a 'public' location and sometimes a 'private location' (Valentine 1999a) – for example, encountering pressure from employers and colleagues at work to diet, while at home taking pleasure from cooking for, and dining with others, or relaxing and comfort eating alone. Consequently, individuals' bodies are rarely completely disciplined or, in Foucault's term, 'docile' (see section 2.3.1.3) yet neither are they completely free from the shadow of ascetic discourses about self-discipline and individual responsibility. As Deborah Lupton (1996: 153) writes, 'In consumer culture there is, therefore, a continual dialectic between the pleasures of consumption and the ethic of asceticism as means of constructing the self.'

■ Summary

- The body has emerged in contemporary consumer culture as a project which should be worked at and accomplished as part of an individual's self-identity.

- Individuals are expected to take responsibility for their own bodies by exercising vigilance and self-discipline.

- The popular and academic emphasis on self-discipline can overstate the degree of corporeal freedom or control individuals may have.

- Bodies are inflected by material practices, representations, social relations and structures of political-economic power in wider locations.

- In addition to ascetic discourses there are other meanings around bodily practices which stress their pleasurable and hedonistic dimensions.

- In consumer culture there is a continual tension between the bodily pleasures of consumption and the ethic of asceticism as means of constituting self-identities.

- Our bodies are rarely completely disciplined, yet neither are they completely free from the shadow of these ascetic discourses.

■ 2.5 Bodies taking up space

This section examines how we physically occupy space, connect with surrounding spatial fields and take up space through our size and appearance (2.5.1) and bodily comportment (2.5.2). Both examples emphasize how women's bodies are expected to occupy less space than men's.

■ 2.5.1 Size and appearance

Research suggests that the citizens of contemporary Western societies are getting fatter, yet slimness is the aesthetic ideal promoted by the media, fashion and consumer industries. Being thin is equated with health and sexual attractiveness in what Deborah Lupton (1996: 137) terms the food/health/beauty triplex. Although there is some evidence that men are increasingly coming under pressure to watch their weight and take care of their bodies (Bocock 1993), it is women who have traditionally been expected to pursue what Susan Bordo (1993: 166) describes as the 'ever changing, homogenizing, elusive ideal of femininity'. She writes: 'Through the exacting and normalising disciplines of diet, make-up and dress – central organising principles of time and space in the day of many women – we are rendered less socially oriented and more centripetally focused on self-modification. Through these disciplines, we continue to memorise on our bodies the feel and conviction of lack, of insufficiency, of never being good enough' (Bordo 1993: 166).

While not all women pursue this commodified ideal, and some women deliberately reject it (Orbach 1988), the 1980s onwards have witnessed a growth in eating disorders such as anorexia and bulimia. A number of feminist writers have argued that eating disorders are an indictment of women's position in society. They suggest that many women willingly accept the bodily 'norms' they are encouraged to aspire to because, through dieting, women are able to experience a sense of control and power (traditionally coded as male) and independence which they do not have in other aspects of their lives and in return they receive admiration (both for their self-control and for their shape) from a world where they often feel excluded and undervalued (Orbach 1988). For some women these rewards become so addictive that they take the issue of controlling their body's demands for food to the extreme where they seek to kill off its needs altogether. Jenefer Shute (1993), in her fictional account of a woman's battle with anorexia, *Life Size* describes how her character, Josie, struggles to lose weight and so minimize the space she takes up, while sneering at those she thinks are too fat and are occupying too much space.

Bordo (1993) suggests that dieting to the point where a woman loses her feminine curves and develops a 'boyish' body represents for some women a way of escaping the vulnerability (both social and sexual) which is associated with having a female body and of entering a male world. However, she argues that the sense of power or control this may induce is illusionary. For feminist writers such as Bordo (1993: 175) anorexia is not a product of individual psychopathy or dysfunction but rather is a form of embodied protest against women's social and cultural position. She writes: 'It is unconscious, inchoate, and counterproductive *protest* without an effective language, voice, or politics, but protest nonetheless.'

In contrast, those who are overweight are often accused of taking up too much space. Schwartz (1986: 328) observes how everyday environments are designed to accommodate only certain body shapes and sizes: 'Airplane seats, subway turnstiles, steering wheels in cars are designed to make fat people uncomfortable.' In the face of unaccommodating environments, the hostile gaze of slim bodies, and sometimes even overt discrimination, overweight bodies can feel pressurized into self-concealment and be inhibited in everyday spaces from the restaurant to the beach (Cline 1990).

Not surprisingly, the body (fat and thin) has also become a site of resistance. In response to fat discrimination in the USA the National Association to Aid Fat Americans held a Fat-In in New York in 1967. Fat is also, as Orbach (1988) famously declared, a feminist issue. In the 1970s and 1980s feminists argued that bodily practices such as dieting, wearing make-up and shaving legs were aimed at pleasing men. Women were encouraged to give up all forms of body maintenance and concerns about their appearance (Green 1991). At the turn of the twentieth and twenty-first centuries, however, there has been a backlash against what became seen as oppressive attempts by feminism to police women's bodies. Rather the emphasis is now on re-engaging with femininity on new terms (Bell *et al.* 1994).

■ 2.5.2 Bodily comportment

In an essay titled 'Throwing like a girl' Iris Marion Young (1990) argues that women are alienated from their bodies and, as a result, occupy and use space in an inhibited way compared to men. She begins her analysis by drawing on the observations of the writer Erwin Strauss about the different way that boys and girls throw a ball. Whereas boys use their whole bodies to throw, leaning back, twisting and reaching forward, girls, Strauss noted, tend to be relatively stiff and immobile, only using their arms to produce a throwing action. Young (1990) argues that women demonstrate similar restricted body movements and inhibited comportment in other physical activities too. For example, women tend to sit with their legs crossed and their arms across themselves, whereas men tend to sit with their legs open and using their hands in gestures. In other words, Young (1990) claims that women do not make full use of their bodies' spatial potentialities. This is not because women are inherently weaker than men but rather it is to do with the different way that men and women approach tasks. Women think they are incapable of throwing, lifting, pushing and so on, and so when they try these sorts of activities they are inhibited and do not put their whole bodies into the task with the same ease as men (for example, only using their arms to throw). Young (1990: 148) describes this as 'inhibited intentionality'. It is a bodily comportment which is learned. A number of writers have argued, for example, that teenage girls give up sport and leisure activities in order to spend time with boys (Griffin 1985), whereas schools promote physicality amongst boys through sport (Mac an Ghail 1996).

Not only do women underestimate their physical abilities and lack self-confidence, but they also fear getting hurt. Describing women as experiencing their bodies as a 'fragile encumbrance', Young (1990: 147) writes that 'she often lives her body as a burden which must be dragged and prodded along and at the same time protected'. Women also experience their bodies as fragile in another sense too, in that their bodies are the object of the male gaze. Young (1990) suggests that it is acceptable for men to look at, comment on or touch women's bodies in public space and that, as a result, women are fearful that their body space may be invaded by men in the form of wolf whistles, minor sexual harassment or even rape (see also Chapter 6). As part of a defence against this fear of invasion, women experience their bodies as enclosed and disconnected from the outlaying spatial field. Young (1990: 146) writes, 'For many women as they move . . . a space surrounds us in imagination that we are not free to move beyond; the space available to our movement is a constricted space.' It is important to note, however, that Young (1990) does point out that her observations apply to the way women typically move but not to all women or all of the time.

In contrast to women, men learn to experience a connectedness between their bodies and their surrounding spatial field and to view the world as constituted by their own intentions. Bob Connell (1983), for example, argues that whereas women are valued for their appearance, men are expected to demonstrate bodily skill in terms of their competence to operate on space, or the objects in it, and to be a bodily force in terms of their ability to occupy space. This competence is developed through cults of physicality, sport (formal and informal), drinking, fighting, work and so on. For instance, certain forms of manual labour like lifting, digging, carrying are closely linked to some sense of bodily force in masculinity. Although economic restructuring means the stress on pure labouring has declined, the social meanings and relations of physical labour and bodily capacity to masculinity have not (Connell 1995). According to Young (1990), men live their bodies in an open way. They feel about to move out and master the world. Connell (1983: 19) explains: 'To be an adult male is distinctly to occupy space, to have a physical presence in the world. Walking down the street, I square my shoulders and covertly measure myself against other men. Walking past a group of punk youths late at night, I wonder if I look formidable enough.'

The difference in the meanings of men and women's physicality is evident in relation to naked bodies. Whereas a male stranger's naked body is seen as a sign of aggression and as frightening or threatening to women, a female stranger's body is not read in the same way by men. Men do not feel assaulted or threatened by seeing an unknown woman naked. It is assumed that men want to look at the nude bodies of women because they are an opportunity for pleasure. Consequently, in the eyes of the law women cannot commit the crime of 'flashing' because, in contrast to men, their naked bodies are regarded as non-aggressive and not sexually threatening, being read instead as entertaining. The only time a woman can be arrested for indecent exposure is if her actions are understood to be an offence against public sensibilities (Kirby 1995).

■ **Summary**

- We physically occupy space, connect with surrounding spatial fields and take up space through our bodily size and appearance and bodily comportment.
- Women's bodies are expected to occupy less space than men's.
- Fat and thin bodies have become political issues and sites of resistance.

■ 2.6 Bodies in space

Adrienne Rich (1986) observes that our material bodies are the basis of our experience of everyday spaces. Our bodies are what people react to; we read into them stories of people's age, lifestyle, politics, identity, and so on. They connect us with other people and places but they also serve to mark us out as different from other people and as 'out of place' (Cresswell 1996). Reflecting on how people have viewed and treated her because she is white and female, Rich (1986: 216) writes that to locate herself in her own body means recognizing 'the places it [her body] has taken me, the places it has not let me go'.

'[O]ur bodies make a difference to our experience of places: whether we are young or old, able-bodied or disabled, Black or White in appearance does, at least partly, determine collective responses to our bodies . . .' (Laws 1997a: 49). Corporeal differences are the basis of prejudice, discrimination, social oppression and cultural imperialism (Young 1990b). These exclusionary geographies operate at every scale from the individual to the nation, discursively defining what different bodies can and cannot do, dividing conceptual space and operating materially to structure physical and institutional spaces (Young 1990b). 'Bodily differences open and close spaces of opportunity: because their bodies are sexed female and thereby subject to the threat of violence, many women will not travel alone at night; because they are old, some women will avoid certain parts of town; because of their skin color, some people find it difficult (if not impossible) to join certain clubs' (Laws 1997a: 49).

Within geography there is a significant amount of work on 'bodies in space' which explores how different bodies, most notably those of pregnant women (Longhurst 1996), lesbians, gay men and bisexuals (Bell and Valentine 1995a), ethnic minorities (Davis 1990), children (Philo 1992, Valentine 1996b), 'Gypsies' (Sibley 1995a), the sick (Moss and Dyck 1996), the disabled (Butler and Parr 1999) and the mentally ill (Parr 1997) are defined as 'other' and are marginalized and excluded within a range of spatial contexts (see, for example, section 2.3.1 above, Chapter 6 and Chapter 8). Such oppressions can be produced through formal laws or policy, but more often are the product of informal everyday talk, evaluation, judgements, jokes, stereotyping, and so on.

■ 2.6.1 The disabled body

Understandings of disability have, until recently, focused on the physical body. In biomedical models of disability individual bodies are defined, usually by medical institutions, as 'disabled' because they do not meet clinically defined 'norms' of form, mobility or ability. The medical and social significance which is accorded to bodily 'normality' means that 'disabled' individuals are further categorized as socially inferior and as a 'problem' for society. (A more detailed outline of the medical model and its critique is found in Parr and Butler 1999.) Bodily and sensory technologies (such as hearing aids, wheelchairs, and so on) are regarded as the solution to the problem of the 'deviant body'.

Disability theorists, however, have challenged this biomedical model of disability (examples of such work within geography include Brown 1995, Chouniard 1997, Dorn and Laws 1994, Gleeson 1996, Hahn 1986, Laws 1994, etc). In the social model of disability, they shift the focus from the physical body to emphasize the role of society in creating disability (see Parr and Butler 1999). While illness and accidents cause bodily impairments it is everyday socio-spatial environments which dis-able people by marginalizing them economically, socially and politically (Chouniard 1997, Dyck 1995, Laws 1994) (see Box 2.4). Disability theorists are critical of the economy for excluding and devaluing bodies that cannot meet the demands of capitalist work regimes (see, for example, Hall 1999). Likewise, urban planning is blamed for constructing environments that are designed for, and prioritize the needs and abilities of the able-bodied and so restrict the mobility of those with physical impairments (Imrie 1996). Commenting on a wheelchair user's description of Los Angeles as 'a vast desert containing a few oases', Hahn (1986: 280) describes the city as having an 'impenetrable geography'.

In the past, because the problems disabled people experienced in finding paid employment and navigating environmental obstacles were regarded as a product of their individual functional limitations (i.e., their unspoken biological inferiority), rather than as a shared or political problem, many disabled individuals withdrew from public life or were confined to relatively barrier-free environments (Hahn 1986). In this way, they were further marginalized. Paterson (1998) describes the impaired body as 'dys-appearing' in everyday environments because it does not fit in either functionally or aesthetically. Nevertheless, disability activists are now challenging the laws and policies which produce dis-abling landscapes. They imagine a city where all spatial barriers are eliminated and where there are residential environments which

Box 2.4: The dis-abled body

'In 1984, after walking on crutches for more than thirty years, I finally made the transition to a wheelchair. Many friends obviously felt that this decision implied a reduction in status, but I regarded it initially as a liberating experience. Small pleasures such as having lunch, which I had previously passed, were suddenly open to me. And yet, as I ventured into the major thoroughfares of Los Angeles, which are less accessible than Pennsylvania Avenue in Washington but more accessible than Broadway in New York, or Michigan Avenue in Chicago, or most of the major streets in European or third world countries, I gained an enhanced appreciation of the importance of simple measures such as curbcuts to freedom of movement. Frequently, I found that I simply could not "get there from here". The chair did have some compensating advantages. I discovered again the pleasures of washing my hands, which was nearly impossible on crutches. But the search for accessible rest-rooms, the frustration of encountering steps in front of buildings that I wanted to enter, and similar barriers gradually curtailed my sense of adventure ... Often I travelled countless blocks in a futile search for an accessible route to my destination. And the inability to maintain eye contact while seated in my chair places me at a serious disadvantage in personal conversations. Sometimes I have found paths formed by skateboards, baby carriages, and shopping carts. These modifications suggest that others also may have difficulty in moving through the environment. But I suspect that, until they learn about the benefits of having their own "scooter chairs", the effort to change the environment is going to entail a major struggle.'

Hahn 1985: 3

can suit individual notions of independent living, arguing that environments which enable expanded contact between the able-bodied and the disabled may also help to reduce or eliminate discrimination and prejudice. (However, there are some debates amongst disabled groups about whether to seek integration in able-bodied communities or whether to aim for separate residential communities where members might provide mutual support and protection for each other – see, for example, De Jong 1983.)

Campaigners in both the USA and the UK have taken non-violent direct action, such as handcuffing themselves to buses and trains, blocking roads and occupying public offices with wheelchairs, in order to disrupt able-bodied space and to draw attention to the dis-abling effects of the environment. In 1990 the *Americans with Disabilities Act* (ADA) which makes it mandatory for public buildings to be accessible to disabled people came into effect and in 1995 the *Disability Discrimination Act* (DDA) was passed in the UK.

Figure 2.2   The disability movement has campaigned against discrimination (© Angela Martin)

Accessible environments are, however, only a small part of the concerns of the disabled. It is not enough to change the built environment if discriminatory attitudes persist. Contemporary Western cultural values emphasize corporeal perfectibility (Hahn 1986). Yet physical impairments are often regarded as 'ugly' or may produce involuntary bodily movements or noises that are considered socially inappropriate (Butler 1998). As a result, Tom Shakespeare (1994: 296) argues, disability is often seen to represent 'the physicality and animality of human existence'. He goes on to suggest that the disabled are often objectified through the gaze of the able-bodied in a similar manner to the way that women are objectified by the gaze of men. As Morris explains, 'It is not only physical limitations that restrict us to our homes and those whom we know. It is the knowledge that each entry into the public world will be dominated by stares, by condescension, by pity and by hostility' (Morris 1991: 25, quoted in Butler and Bowlby 1997: 411).

The disability movement has therefore not only campaigned to change the physical environment but to challenge the hostility, patronizing behaviour, misunderstandings and discrimination experienced by disabled people in everyday spaces (see Figure 2.2). It has sought to encourage people not to be ashamed of their impairments, criticized representations of the disabled as dependent and in need of help, and mobilized people with impairments to recognize oppression and to fight for their civil rights (Butler and Bowlby 1997, Parr and Butler 1999). In turn, however, the disability movement has itself been criticized for being the domain of white, heterosexual men, for failing to acknowledge the heterogeneity of different forms of impairment (e.g. deafness, spinal injuries, and so on) and the way that other axes of difference such as gender and class intersect with a disabled identity (Butler and Bowlby 1997, Parr and Butler 1999). Writing about the experiences of disabled lesbians and gay men, Ruth Butler (1999) observes how they encounter 'ableism' in the lesbian and gay 'community' and homophobia amongst the disabled 'community', while Vera Chouniard (1999) highlights the efforts of disabled lesbian and heterosexual women activists in Canada to challenge their political invisibility.

Recently, disability theorists have argued that illness should be considered along-side impairment because those who are sick also experience the dis-abling effects of physical environments and encounter the sort of social attitudes described above (see, for example, Moss and Dyck 1996, 1999). Indeed, Hester Parr and Ruth Butler (1999) argue that categories of health/illness and ability/disability are leaky and unstable. They point out that the healthy majority may only temporarily occupy able bodies. At different times and in different spaces we may each experience different states of physical and mental illness/wellbeing (for example as a result of ageing, pregnancy, accidents, and so on). Thus, the problems disabled and ill bodies encounter in every-day spaces are not just a concern for so-called 'deviant bodies' but are potentially an issue for everyone. In this way, the body is being seen as an important site of resistance and emancipatory politics (Brown 1997a, Dorn and Laws 1994, Dyck 1995).

■ **Summary**

- Our bodies make a difference to our experience of places. They – at least partly – determine collective responses to us.

- Corporeal differences are the basis of prejudice, discrimination and oppression. Geographical work examines why/how particular bodies are defined as 'other' and marginalized.

- In biomedical models of disability individual bodies are defined by medical institutions as 'disabled' because they do not meet clinically defined 'norms' of form, mobility or ability.

- Contemporary theorists challenge the biomedical model. In the social model of disability, they emphasize the role of society in creating disability.

■ 2.7 The body and time

Age is important to understanding our social world. The medical model of ageing, like the medical model of disability, is being challenged by those who recognize that age, like gender, race and disability, is a social rather than a biological category (Featherstone and Wernick 1995) (see section 2.6.1). Bodies are marked by social norms and expectations which shape what we think they can and cannot do at different ages, what they should or should not be doing, where they should or should not be going and how they should or should not be dressed (Harper and Laws 1995). For example, the menopause is supposed to mark the age at which a woman is too old to cope with the responsibility of looking after a dependent child, yet women of this age are routinely expected to care full-time for dependent elderly relatives (Laws 1997a). Expectations about young–old bodies, then, are not predicated on biology but are actively socially constructed in discursive practices which vary across space and time. They have important consequences because they shape individual opportunities, structure collective experiences, and have spatial ramifications (see sections 2.7.1 and 2.7.2 below) (Harper and Laws 1995, Katz and Monk 1993).

People are often expected to follow a linear sequence through the lifecourse, spending their childhood and youth in education, adulthood in paid employment and old age in retirement. Yet this model is increasingly being challenged as people live more individualized and diverse lifestyles, in which the body becomes a project to be worked at, and identities are defined more by consumption than production (see section 2.4.1). For example, older people may be retiring from work but are then taking up part-time work or returning to education (Harper and Laws 1995, Featherstone and Wernick 1995). These complexities and multiple understandings of what it means to have a body of a different age are beginning to challenge some of the stereotypical images and forms of discrimination which are based on assumptions about the inherent bodily characteristics of different age groups (which bear striking similarities to assumptions made about female bodies and black bodies, see section 2.2.2): notably the young and the old.

■ 2.7.1 Young bodies

Western thought has imagined the child's body in two ways: as 'innocent' and as 'evil' (Valentine 1996c, Holloway and Valentine 2000). In the romantic images of poets and artists such as Blake and Wordsworth children are imagined as innately good. The eighteenth-century writer Rousseau, in particular, contributed to this 'Apollonian' (Jenks 1996) view of childhood in which children are imagined as 'pure' and 'innocent' in contrast to the ugliness and violence of the adult world from which they must be protected. This discourse of the 'innocent child' which emerged in the eighteenth and nineteenth centuries laid the foundations for child-centred education, concerns about children's safety and vulnerability and the belief that children should be the concern of everyone because they represent the future (James, Jenks and Prout 1998). Allison James, Chris Jenks and Alan Prout (1998: 152) note that this imagining of childhood explains the horror and outrage child abuse engenders in adults. They write, 'Children's bodies are to be preserved at all costs, any violation signifying a transgressive act of almost unimaginable dimensions. To strike a child is to attack the repository of social sentiment and the very embodiment of "goodness".'

It is important to remember, however, that the Apollonian view of childhood is an imagining of childhood rather than the reality experienced by most children. The experience of childhood has never been universal; rather, what it means to be a particular age intersects with other identities so that experiences of poverty, ill health, disability, having to care for a sick parent, or being taken into care have all denied many children this idealized time of innocence and dependence (see, for example, Stables and Smith's 1999 account of the lives of young carers).

In contrast to the Apollonian view of childhood, the Dionysian (Jenks 1996) imagining of childhood stems from the notion of original sin, in which 'evil, corruption and baseness are primary elements in the constitution of "the child"' (James, Jenks and Prout 1998: 10). The evil child threatens the stability of the adult world and is in need of education and discipline in order to develop sufficient bodily control to be civilized into membership of the human race (see also Chapter 5). It is an imagining of childhood evident in books such as *Lord of the Flies* and in contemporary constructions of teenagers as troublesome and dangerous (see Chapter 6).

While both the Apollonian and Dionysian understandings of childhood are always present in children's complex and diverse experiences, at different historical moments one or the other of these binary conceptualizations often appears to dominate the popular imagination. At these times, other meanings of childhood can be overlooked or forgotten, before being periodically rediscovered (Stainton Rogers and Stainton Rogers 1992).

Both these understandings of childhood have been used to justify children's exclusion from public space. The Apollonian understanding constructs children as less knowledgeable, less competent and less able than adults and therefore as vulnerable and in need of protection from adults and the adult world (see Chapter 6), while the Dionysian understanding constructs children as dangerous, unruly and potentially out of control in adultist public space (Valentine 1996b, 1996c).

The child's body has not just been understood in terms of these discursive constructions, but has also been the subject of ethnographic studies in both sociology and geography which have sought to understand the body as entity experienced by children within the spaces of their own social worlds and cultures (see, for example, James, Jenks and Prout 1998, Holloway and Valentine 2000, Valentine 2000). In a study of the body in children's everyday lives Allison James (1993) points out that 'children's perceptions of their own and Other bodies constitute an important source of their identity and personhood'. She argues that a ruthlessly patterned hierarchy characterizes children's cultures. While there is no necessary relation between physical difference and marginality or outsiderhood, different bodily forms are given significance in terms of social identity by other children. James (1993) identifies five aspects of the body that have significance for children's identities: *height* (specifically, the importance of physical development – where size marks social independence and 'titch' is a form of abuse), *shape, appearance, gender* (shape, appearance and gender are all based on adult, heterosexual notions of desirability and issues of morality), and *performance* (this includes dynamic aspects such as gracefulness of bodily movement and sporting prowess or ability). While some differences may be temporary – a growth spurt might can help to shake off the nickname 'titch' – others produce more permanent stigmatized identities.

James (1993) further explores how children have to negotiate their changing bodies within the context of changing institutional settings in which the meanings of their bodies change drastically. Summarizing her work, she and two co-authors explain (James, Jenks and Prout 1998: 156):

> in the later stages of nursery school children came to think of themselves as 'big'; their apprehension of the difference between themselves and children just entering the nursery plus the significance of the impending transition to primary school signalled their identity. But once they had made the transition and were at the beginning of their career in primary school, they were catapulted back into being 'small' once again. This relativity produced a fluidity in their understanding of the relationship between size and status, generating what James identifies as a typical 'edginess' among children about body meanings. The body in childhood is a crucial resource for making and breaking identity precisely because of its unstable materiality.

In the adult world children's bodies are socially and spatially segregated from grownups through the schooling system (see Chapter 5) and through laws, curfews and informal regulations which bar them from certain public spaces at specific times (e.g. bars, cinemas, the street, shops) (see also Chapter 6). It is worth noting, though, that children can and do (re)negotiate the meanings of their biological age (see Chapter 3, section 3.6). David Sibley (1995a) notes how, further up the age scale, the bodies of youth are ambiguously wedged between childhood and adulthood because the legal classifications of where childhood ends and adulthood begins are notoriously vague. The age at which young people can drink alcohol, learn to drive, earn money, consent to sex or join the armed forces varies widely in different spatial and historical contexts. This point is also made by Cindi Katz and Jan Monk (1993), whose edited book *Full Circles: Geographies of the Lifecourse* demonstrates how young people's spatial freedoms vary considerably according to their age, gender, environmental setting and household type.

■   2.7.2 Older bodies

Like the meanings of childhood, the meanings of older bodies are not static but change over time and space. In some cultures elderly people are respected for their life

Box 2.5: When I am an old woman

Warning
When I am an old woman I shall wear purple
With a red hat which doesn't go, and doesn't suit me,
And I shall spend my pension on brandy and summer gloves
And satin sandals, and say we've no money for butter.
I shall sit down on the pavement when I'm tired
And gobble up samples in shops and press alarm bells
And run my stick along the public railings
And make up for the sobriety of my youth.
I shall go out in my slippers in the rain
And pick the flowers in other people's gardens
And learn to spit.

You can wear terrible shirts and grow more fat
And eat three pounds of sausages at a go
Or only bread and pickle for a week
And hoard pens and pencils and beermats and things in boxes.

But now we must have clothes that keep us dry
And pay our rent and not swear in the street
And set a good example for the children.
We must have friends to dinner and read the papers.

But maybe I ought to practise a little now?
So people who know me are not too shocked and surprised
When suddenly I am old, and start to wear purple.

Joseph 1996

experience and wisdom. However, in contemporary Western societies older bodies are often not valued because, as non-participants in the labour market, and as past the age of reproduction (for women at least), they are deemed to have no economic or social worth and are stigmatized and disempowered (Featherstone and Wernick 1995, Hugman 1999). The state creates structured dependency among older people by establishing retirement ages and defining resources and services for them. Yet, while 60–65 years old is commonly defined as retirement age in the affluent West, this definition of 'old age' does not necessarily accord with specific biological markers of ageing, nor with the diverse experiences of this age group. While some elderly people are frail, housebound and relatively dependent on others socially and economically, others are fit, active, mobile and possibly still working or travelling (see Box 2.5, Plate 2.2). Richard Hugman (1999) claims that less than one-third of the population

Plate 2.2 'Gray power'

aged over 65 are actually frail or unwell enough to need professional care or support in their everyday lives. Graham Rowles (1983) makes a distinction between the 'old-old' who are physically frail and isolated and the 'young-old' who are more mobile and adept at drawing on support beyond their immediate community.

In both the USA and Australia some retirees have begun to emerge as a significant social force, dubbed 'gray power'. As a group, despite no longer being in paid employment, some older people have a higher than average income (from pensions, savings and investments) and have begun to establish homogeneous, relatively small communities, such as 'Sun City' (in California, Arizona, Nevada and Florida, USA) based on single-family houses and condo developments in accessible locations such as the edge of cities or highways (Laws 1995, Laws 1997b). These are usually based around a golf course and often have recreation or arts and craft centres. They are landscapes of consumption rather than production and are promoted as ideal and harmonious 'communities' away from the problems of city life. In effect, they are quite self-contained, security-conscious and sometimes gated communities in which outsiders, particularly those aged under 50, both stand out and are not welcome (see also Chapter 4 and Chapter 6). Glenda Laws (1995: 26) argues that the emergence of these types of age-segregated retirement communities – which she describes as 'embodied built environments' – reflects the inhospitable nature of cities for older bodies who may be relatively immobile and frail, and the extent to which these bodies are socially marginalized in everyday spaces. Laws (1995: 26) further argues that, in turn, these embodied built environments contribute to the further social and spatial differentiation of society according to age, while what she terms the 'emplaced bodies' within them also contribute to the construction of particular social identities.

These sorts of communities stand in stark contrast to institutional forms of residential care for the elderly, which are variously labelled 'nursing homes', 'old people's homes', 'rest homes', 'aged-care hostels', etc (Hugman 1999). Such institutions have been critiqued for spatially segregating the elderly from the rest of society in such a way that they may also lose contact with family and friends, for depriving individuals of the right to determine their own use of space and time, and for (re)producing the older body as dependent and docile. In this sense, and because residents may not have chosen to be there or feel 'at home' there (see Chapter 3), residential care institutions are often compared with other total institutions such as prisons and asylums (Hugman 1994, 1999, Laws 1997b; see also Chapter 5).

As our bodies age we accumulate different experiences that shape our perceptions of space and place (Laws 1995). In two studies of elderly people, one group of whom had lived in an old industrial town in New England, USA (given the pseudonym 'Lancaster') for more than 30 years, and the other group of whom lived in a rural town in Appalachia, USA (given the pseudonym 'Colton'), Rowles (1978, 1983) found that as the respondents' bodies had aged, so their investment in the places in which they lived had changed. As they became more physically immobile and their activities were inhibited, they invested more emotionally and psychologically in their homes. What he terms the 'surveillance zone' – the field of vision around their home – became an important source of support in which certain supportive, if transitory, relationships developed. For example, he explains, 'observing the daily routine of neighbours, chastising the children for overly zealous play in the path outside and watching those who regularly pass by, provides support through a sense of ongoing social involvement' (Rowles 1983: 120).

His rural participants claimed to have a strong sense of 'insiderness' in Colton because of their familiarity with the place. Rowles (1983: 114) explains: 'Repeated use of the same route in the journey to the store, or to church over a period of several decades means that these paths become ingrained within the participants' inherent awareness of the setting . . . The old person comes to wear the setting like a glove.' For both the urban and rural elderly, the places where they lived had become a landscape of memories which provided them with a sense of identity: where a child was born, a husband was met or a first job was obtained (Rowles 1983). In this sense, Rowles (1983) suggests that, for some elderly people, it is still as if rooms or the neighbourhood are inhabited by missing people (the dead, their children who were now grown up and had left home) because they can still visualize them there. For these people, retreating in their imaginations into the past becomes an escape from bodies that are relatively confined in space by physical limitations. Thus, he argues, their lives become lived in their minds rather than their bodies.

■ **Summary**

- Age is a social rather than a biological category.

- Expectations about young–old bodies are socially constructed in discursive practices which vary across space and time.

- These social norms matter because they shape individual opportunities, structure collective experiences, and have spatial ramifications.

- Multiple understandings of what it means to be a particular age are beginning to challenge stereotypical images and forms of discrimination which are based on assumptions about the inherent bodily characteristics of particular ages.

2.8 Future bodies?

Some commentators argue that we are on the edge of undergoing a third technological revolution. The first involved *transport* (steam engine, car, aircraft), the second *transmission* (radio, television, Internet) and the third will involve *miniaturization* (the transplant revolution). Both the second and third of these revolutions have the potential to transform the human body. The emergence of new information and communication technologies (ICT) and the development of new possibilities for medical interventions in the body (biotechnologies such as transplants, in vitro fertilization and plastic surgery and nanotechnologies and miniaturization) are thus producing new uncertainties about what the body is, what is 'natural', and which scientific or medical bodily interventions should be allowed. All these are alleged to be leading to a crisis in the body's meaning. First, ICT, which is regarded as a *disembodied* form of communication, offers a vision of the future in which we can do away with the body altogether. Second, there is speculation about the extent to which the body is being merged with technology: *the cyborg.*

2.8.1 Disembodiment

Dubbed 'meat' or 'wetware trash' by cyberenthusiasts, the body is regarded as a nuisance. Morse (1994: 86, in Lupton 1995: 100) writes: 'For couch potatoes, video games addicts and surrogate travellers of cyberspace alike, an organic body just gets in the way.' Its demands to be fed, to be washed, to sleep and so on, interrupt cyber-pursuits and interactions. Writers such as Barlow (1990) imagine a utopian vision of a future *post-biological world* where we will be able to escape the 'meat', for example, by transferring our memories into computers or robots (Kitchin 1998). 'The idealized virtual body does not eat, drink, urinate or defecate; it does not get tired; it does not become ill; it does not die' (Lupton 1995: 100). In this sense, the desire of cyberutopians to have 'your everything amputated' (Barlow 1990: 42) harks back to the Cartesian division and subordination of the body to the mind (see section 2.2.1), in which the mind is considered to be the 'authentic' self.

While technology does not yet offer us the possibility of achieving the desire of the utopians to be completely liberated from our bodies, ICT is claimed by some writers to offer disembodied forms of communication that enable us to be freed from some of the constraints of our bodies. Most famously, it is argued that because of the meanings which are read off from our physical bodies and the judgements which are made about particular bodily characteristics such as age, health, race and gender (see section 2.6), the body can act as a social barrier to some relationships (Van Gelder 1996). ICT therefore offers a cloak of anonymity. As Turkle's informant (1996: 158) argues, 'You don't have to worry about the slots other people put you in as much. It's easier to change the way people perceive you because all they've got is what you show them. They don't look at your body and make assumptions. They don't hear your accent and make assumptions. All they see is your words.' The advantage of ICT, then, is that it potentially enables individuals to construct one or multiple on-line identities for themselves which may bear no relation to their physicality. There are many famous examples of men claiming to be women on-line, the able-bodied to be disabled, and so on (Slouka 1996).

This practice of creating multiple alternative identities which are played with and then forgotten is termed 'cycling through' (Turkle 1996). It is often promoted as appealing to those – such as the young, the old and the disabled – who want to escape off-line bodies which are commonly regarded as incompetent and undesirable (Valentine and Holloway 2000).

Yet, this discourse of disembodiment has been subject to critique. A number of writers have argued that on-line textual persona cannot be separated from the off-line physical person who constructs them. Rather, disembodied identities and conversations are commonly based on embodied off-line identities. For example, in Box 2.6 Francesca describes how she uses her off-line interests as a basis of her on-line identity and friendships. Even when individuals make up new personas, these on-line identities are often chosen because of off-line experiences – in other words, because they are a way of resisting or overcoming the problems of off-line bodies. In Box 2.6 Andy and Steve describe how they try to 'pass' as adults by pretending to be bouncers or to go to pubs (Valentine and Holloway 2000). In this way their on-line identities are a product of their dis-identification with their off-line bodies.

Not surprisingly, many of the practices and structures which shape off-line lives also mediate on-line interactions (Kitchin 1998). Most notably, cyberspace has been critiqued as the domain of white males (although there is a significant feminist and lesbian on-line culture). There are many well-known accounts of women being harassed and even 'raped' on-line (Gilbert 1996).

Box 2.6: Disembodied identities

Francine: I just have a couple of handles that I use from books that I've read that I like, people's names and stuff. I think it's kind of fun, but I don't have an alter ego or anything, you know, I just go on there and talk about stuff that I'm, I, me actually I'm interested in. I know you get people on there who pretend they're models or whatever but I don't really see it like that.

Interviewer: How do you kind of represent yourself on screen [in relation to a conversation about a Teen Chat Room]? You just give yourself a nickname of something like that, is that how it works?
Andy: Well you just give yourself a name, just make something up and then just describe yourself or whatever.
Interviewer: And so you can just pretend to be somebody else?
Andy: Yeah
Interviewer: Do, have you done that, have you pretended to be?
Andy: Yeah
Interviewer: What have you done?
Andy: I posed to be a bouncer. [laughs]
Interviewer: A bouncer?
Andy: Yeah.
Interviewer: Why was that?
Andy: Oh, I don't know. It was just that I was in this erm, the Teen Chat one, and there was this, there's girls on it, so, so I pretended to be a bouncer of 22.

Steve: Oh me and a friend acted we were 17 years old.
Interviewer: 17?
Steve: Yeah and it was like we was drunk and we kept writing different things, like strange things.
Interviewer: You were drunk or were pretending to be drunk?
Steve: Pretending . . .
[edit]
Interviewer: Why was it, why did you enjoy pretending to be older than you are?
Steve: Because you can write about all different things and just normal games. You can write about sort of like going to the pub and all that.

Valentine and Holloway 2000: 10

For Vivenne Sobchack the impossibility of escaping her material body on-line was brought home to her when she had to have major surgery for cancer. Referring to her experience of physical pain while reading about discourses of disembodiment, she writes, 'there is nothing like a little pain to bring us back to our senses, nothing like a real (not imagined) mark or wound to counter the romanticism and fantasies of techno-sexual transcendence that characterise so much of the current discourse on the techno-body that is thought to occupy the cyberspaces of post-modernity' (Sobchack 1995: 207). Sobchack's experience is an important reminder that – certainly at the current level of technology – the discourse of disembodiment is something of a misnomer. Cyberspace is always entered into and interacted with from the site of the body, the body is always present. '[W]e do not just have bodies, but . . . we are our bodies. Bodies cannot be transcended; rather, they are a fundamental constituent of us, of being' (Kitchin 1998: 83).

■   ### 2.8.2 The cyborg

In a 1985 (republished in 1990) paper titled *A Manifesto for Cyborgs* the feminist historian of science Donna Haraway argued that the body and technology are merging or coalescing into what she termed 'the cyborg': a 'hybrid of machine and organism' (Haraway 1990: 191) where technology is a substitute or supplement for the flesh.

Contemporary science fiction contains many imaginings of creatures who are part-human or animal, part-machine (see, for example, the writing of William Gibson, Octavia Butler, Vonda McIntyre, and so on). The cyborg has also been popularized in films such as *Blade Runner*, *Robocop*, and *Terminator*, where it is represented as stronger than the more fallible and vulnerable human body, immune to pain and able to repair any injury or damage to itself and so cheat 'death' (Lupton 1995). In these movies '[T]he cyborg body is constituted of a hard endoskeleton covered by soft flesh, the inverse of the human body, in which the skin is a vulnerable and easily broken barrier between 'inside' and 'outside' (Jones 1993: 84, quoted in Lupton 1995: 101).

Yet the cyborg is not only fiction or image but also partly fact. Modern medicine and biotechnologies are increasingly transforming and recreating the human body. Technology is already built into human bodies, from assistive technologies to replace or supplement organs, limbs and sense (such as plastic heart values, prostheses and cochlea implants) to gene therapy, trans-species transplants and trans-genetic organisms. In Haraway's (1990, 1991, 1997) terms therefore we are all cyborgs – fabricated hybrids of machine and organism. All predictions suggest that the future will only bring a further merging of bodies and technologies.

Nanotechnology, for example, offers possibilities of inner body interventions with 'molecular machines roving the bloodstream to search and destroy viruses' (Featherstone 1999: 2) and smart pills that will be able to transmit information about our nerves or our blood to external monitors. Other visions, of a future in which we will be able to plug our bodies into cybersystems and machines so that they will function more powerfully and precisely, further modify the horizons of what the body can be. The cybercritic Paul Virilio (1997: 53) imagines a nightmare future scenario of a body colonized by machines – a 'citizen terminal' in which the body becomes not a site but a collection of parts:

this citizen-terminal soon to be decked out to the eyeballs with interactive prostheses based on the pathological model of the 'spastic' [sic]; wired to control his/her domestic environment without having physically to stir: the catastrophic figure of an individual who has lost the capacity for immediate intervention along with natural moticity and who abandons himself [sic] for want of anything better, to the capabilities of captors, sensors and other remote control scanners that turn him into a being controlled by the machine with which they say, he talks.

In writing about the cyborg Haraway argues that technology is breaching three rather leaky boundaries:

- *The boundary between humans and animals*. Trans-genetic organisms such as Onco mouse, and trans-species, such as pig–human, organ transplants (Haraway 1997) suggest that '[f]ar from signalling a walling off of people from other living things, cyborgs signal disturbingly and pleasurably tight couplings' (Haraway 1990: 193).

- *The boundary between organism (animal-human) and machine*. Distinctions between what aspects of the body are 'natural' or 'artificial' and what is self-developing rather than externally designed (e.g. through gene therapy) are becoming more ambiguous so that it is increasingly difficult to be certain about what counts as 'nature' or 'natural'.

- *The boundary between physical and non-physical*. High technology and scientific culture are producing machines which are increasingly fluid, portable, mobile, and even opaque or invisible. 'Our best machines . . . are nothing but signals, electromagnetic waves, a section of a spectrum' (Haraway 1990: 195).

The cyborg is unsettling because it disrupts our sense of these boundaries, particularly the distinction between humans and other living and non-living things. As Haraway (1990: 219) says, it is not clear 'who makes and who is made in the relation between human and machine'. Yet, she argues that we should take pleasure in the confusion or transgression of these boundaries and the potent fusions and possibilities they offer. The world, she suggests, is being restructured through social relations of science and technology; communication technologies and biotechnologies are creating new moments for recrafting bodies and therefore opening up possibilities for new social relations. She sees cyborg imagery as important because it challenges universal totalizing theory and assumptions about purity and identity, and it offers a way out of dualistic Western thought (human–animal; mind–body; male–female; internal–external, nature–culture, self–other, etc) by producing and destroying identities, categories, relationships, spaces, and so on.

This has appealed to some feminists because, by making cultural categories and bodily boundaries indeterminate and fluid, the cyborg offers us possibilities for escaping the limitations of gender and other stereotypes. Female embodiment is traditionally seen as organic (see section 2.2.2), but because cyborg embodiment is dynamic, fluid and partial rather than given or waiting to be reinscribed, women have the chance to re-code the self and the body, to replace patriarchal dualisms and to re-appropriate and contest new social relations (Lupton 1995). 'The cyborg is Haraway's figuration of a possible feminist and posthumanist subjectivity' (Prins 1995: 360).

However, some time after writing the *Cyborg Manifesto*, Haraway warned against the liberatory cyborg she promoted, pointing out that it denies mortality, that 'we really do die, that we really do wound each other, that the earth really is finite, that there aren't any other planets out there that we know we can live on, that escape-velocity is a deadly fantasy' (Penley and Ross 1991: 20). She has also been criticized because the cyborg life-forms she identifies are essentially organic rather than mechanical. She therefore tends to 'find kinship with animals, not machines, as well as lodge identity in creatures, not apparatuses' (Luke 1997: 1370).

■ **Summary**

• The development of technology is producing new uncertainties about what the body is, what is 'natural', and is bringing about a crisis in the body's meaning.

• Disembodied forms of communication offer a vision of the future in which we can do away with the body altogether.

• At the current level of technology the discourse of disembodiment is something of a misnomer. Cyberspace is always entered into, and interacted with, from the site of the body.

• The body and technology are merging or coalescing into what is termed the cyborg: a 'hybrid of machine and organism'.

• Cyborg imagery challenges universal, totalizing theory, and offers a way out of dualistic Western thought (body/mind; human/animal; nature/culture, etc).

■ **Further Reading**

• Useful overviews of geography and the body are provided by Longhurst's (2000) single-authored text *Bodies: explaining fluid boundaries*, Routledge, London; and a number of edited collections, including: Ainley, R. (1998) (ed.) *New Frontiers of Space, Bodies and Gender*, Routledge, London; Duncan, N. (ed.) (1996) *Bodyspace: Destabilizing Geographies of Gender and Sexuality*, Routledge, London; Nast, H. and Pile, S. (eds) (1998) *Places Through the Body*, Routledge, London; Pile, S. and Thrift, N. (eds) (1995) *Mapping the Subject*, Routledge, London, and Teather, E.K. (ed.) (1999) *Embodied Geographies: Spaces, Bodies and Rites of Passage*, Routledge, London.

- Beyond geography's porous boundaries good overviews of the body and social theory are found in: Featherstone, M., Hepworth, M. and Turner, B.S. (eds) (1991) *The Body*, Sage, London; McNay, L. (1994) *Foucault: a Critical Introduction*, Blackwell, Oxford; Shilling, C. (1993) *The Body and Social Theory*, Sage, London; Synott, A. (1993) *The Body Social: Symbolism, Self and Society*, Routledge, London, and Turner, B. (1996) *The Body and Society*, Sage, London.

- Influential feminist writings on the body include: Bordo, S. (1993) *Unbearable Weight: Feminism, Western Culture and the Body*, University of California Press, Los Angeles; Grosz, E. (1994) *Volatile Bodies: Towards a Corporeal Feminism*, Indiana University Press, Bloomington and Indianapolis, and Young, I.M. (1990b) *Throwing Like a Girl and Other Essays in Feminist Philosophy and Social Theory*, Indiana University Press, Bloomington and Indianapolis.

- Questions of embodied identities and/or socio-spatial exclusion are the subject of a number of key books, notably: Bell, D. and Valentine, G. (1995a) *Mapping Desire: Geographies of Sexualities*, Routledge, London; Blunt, A. and Wills, J. (2000) *Dissident Geographies*, Prentice Hall, Harlow; Bonnett, A. (2000) *White Identities*, Prentice Hall, Harlow; Butler, R. and Parr, H. (eds) (1999) *Mind and Body Spaces: Geographies of Illness, Impairment and Disability*, Routledge, London; Gleeson, B. (1999) *Geographies of Disability*, Routledge, London; Jackson, P. and Penrose, J. (eds) (1993) *Constructions of Race, Place and Nation*, UCL Press, London; James, A. (1993) *Childhood Identities*, Edinburgh University Press, Edinburgh, London; and Sibley, D. (1995a) *Geographies of Exclusion: Society and Difference in the West*, Routledge, London. These issues also feature heavily in articles in journals such as *Body & Society, Environment and Planning D: Society and Space, Gender, Place and Culture* and *Social and Cultural Geography*.

- The best edited collection on technological embodiment is Featherstone, M. and Burrows, R. (eds) (1995) *Cyberspace, Cyberbodies and Cyberpunk: Cultures of Technological Embodiment*, Sage, London. A useful summary of the cyberbodies literature is found in Kitchin, R. (1998) *Cyberspace*, John Wiley & Sons, Chichester; while a more difficult, theoretical, but extremely influential book is: Haraway, D. (1991) *Simians, Cyborgs and Women: The Reinvention of Nature*, Free Association Books, London.

- Jenefer Shute's (1993) novel *Life Size* (Mandarin paperbacks) contains reflections on the issue of taking up space in her account of anorexia; Nancy Mairs (1995) *Remembering the Bone House* (Beacon Press, Boston, MA) writes powerfully about the experience of MS; while Carol Shields *Stone Diaries* (1994, Fourth Estate, New York) explores the process of ageing. William Gibson's popular and influential science fiction is worth reading because of the way he reconfigures bodies and represents geographies of cyberspace (e.g. *Neuromancer*, 1984, Grafton, London).

■ Exercises

1. Keep a diary for a few days about your everyday movements and encounters. Use this to reflect on how you perform your own identity in different places, and how you think your identity is read by others. Are there places that your body will or won't let you go or where you feel 'out of place'? Are you aware of altering your dress, behaviour, and so on because of the disciplining gaze of friends, lecturers, employers, etc or of exercising self-surveillance in different spaces?

2. Drawing on the work of Seamon (1979, see section 2.3.2), with a small group of friends reflect on and discuss the everyday ways your own bodies move through and occupy everyday spaces, and the relationship of your bodies to their surroundings. To what extent do your experiences parallel those of Seamon's students? How might you criticize his work?

3. Collect a range of different women's and men's magazines. How are different bodies represented in these texts? What discourses can you identify? What are the similarities and differences in the way that men and women are encouraged to locate, evaluate and manage their bodies?

4. Spend some time in on-line chatrooms. Write a field diary about your observations as you would if you were conducting participant observation in an off-line space. Think about how people represent themselves, how you choose to construct your own identity, the nature of the social relations in these disembodied spaces, and the relationship between your on-line and off-line worlds.

■ Essay Titles

1. Critically evaluate Adrienne Rich's (1986) claim that the body is 'the geography closest in'.

2. Using examples, critically assess the role of the mind/body dualism in shaping the way geographical knowledge has been produced.

3. To what extent do you agree with Shilling's (1993) claim that in contemporary affluent societies we have witnessed the unprecedented individualization of the body?

4. Critically evaluate Law's concept of 'embodied built environments'.

The home

■ 3.1 The home

The home is not just a three-dimensional structure, a shelter, but it is also a matrix of social relations (being particularly valorized as the site of heterosexual family relationships) and has wider symbolic and ideological meanings (for example, during the Second World War symbolic images of home and homeland were used to mobilize support for the war effort). Traditionally, the home has been constructed as a private space in opposition to the public space of the world of work: an understanding articulated in the construction of postwar suburban housing estates in North America and Europe. As such, it is commonly regarded as a safe, loving and positive space.

It is women who traditionally have been charged with the responsibility of making and maintaining the home, hence its characterization as a 'woman's place'. Feminists have challenged the idealization of the home, pointing to some women's experiences of domestic work, violence and oppression within familial homes. In contrast, black feminist writers have extolled the virtues of the home as a site of resistance in the face of white hegemony.

The home is an important site of consumption. When goods are purchased they are domesticated within the specific social relations of individual households who construct personal economies of meaning. Such processes, however, can be the source of domestic conflict. The home is an important site where spatial and temporal boundaries in relation to both domestic space and public space are negotiated and contested between household members. (Even within the home multiple different temporalities and spatialities are produced: as a crude example, a child's bedroom can be produced as a very different space from the 'family' living room or the adults' study.)

All estimates suggest that homelessness is a growing social problem, which has been attributed to global processes of economic restructuring, national welfare reforms and individualization. However, there are many different definitions of homelessness, some of which recognize that there is a continuum between being homed and homeless. Empirical work suggests that the so-called 'homeless' create relationships, social networks and appropriate spaces which take on many of the meanings of home which the homed attribute to conventional forms of housing.

This chapter explores each of these themes. It begins by evaluating feminist work which argued that housing designs articulate assumptions about gender roles and relations. It then goes onto explore the literature on the meanings of home, which tends to emphasize it as an ideal location. This is contrasted with diverse accounts of actual experiences of the home as a site of work, as a site of violence, as a site of resistance and as a site of non-heterosexual relationships. The focus then switches to the moral economy of the household and to the way that multiple spatialities and temporalities are negotiated within and beyond the home by household members. The final section explores the causes and experiences of homelessness.

In addition to the understandings of home outlined in this chapter, there are other conceptualizations of home which are addressed in different chapters. Institutional 'homes' such as prisons and asylums are the subject of Chapter 5. The notion of being 'at home', which captures a sense of belonging, identity and rootedness, is not only experienced at the scale of the individual home but also at the scale of the nation. The homeland often has particular importance for diasporic communities, who maintain emotional and physical connections with their countries of origin. This notion, and the sense of displacement and alienation which is experienced by some refugees and migrants who are forced to leave their homes as a result of violence and persecution, are explored in Chapter 9.

■ 3.2 Housing design

We tend to take houses for granted, yet they are not merely neutral containers for our social relationships. They are designed and built by people and are thus the outcome of the society that produced them. In the 1970s and 1980s feminist architects, planners and geographers (Werkele *et al.* 1980, Hayden 1984, Matrix 1984, Women and Geography Study Group 1984) argued that the physical form of housing could literally be read as a map of the social structures and values which produced it. As Jos Boys (1984: 25) wrote at this time, 'Architecture seems to make a physical representation of social relations in the way it organises people in space. It does this both symbolically – through imagery and "appropriateness of place" for a particular activity [and person] – and in reality – through physical boundaries and the spatial relationships made between activities.'

In particular, feminist writers observed that housing designs contained ideas about the proper 'place' of women. They therefore suggested that because the built environment could embody meanings – for example, about what is 'private' and what is 'public' activity and for whom, and what behaviour is appropriate for men and women in particular locations – it could be restrictive, especially for women. In this way, feminist academics argued that the built environment contributed towards perpetuating the **patriarchal** society that produced it.

The following subsections look at some of this work and the criticisms made of it. Although it has been attacked for being simplistic and environmentally deterministic, questions about the relationship between society and space are still important because the buildings and cities we live in are instrumental in shaping our everyday experiences.

■ 3.2.1 Gender and the built environment

Before the development of industrial capitalism in Europe commodity **production** (e.g. weaving, baking, farming) and reproduction (child rearing) took place in the same sphere. There was not the same level of separation of activities into different spaces – home and work – as there is today. Rather, people lived more communal lifestyles. An average household would include not only immediate kin such as parents and children but also other relatives, servants and apprentices. According to a group of feminist designers known as Matrix (1984), this form of social relations was evident in the design and layout of houses at this time. These were basic structures that accommodated reproductive activities such as eating, sleeping, and cooking as well as tasks associated with production and trading.

With the development of industrial capitalism there was a separation of activities, with production increasingly taking place in large-scale factories, and reproduction being removed from the communal sphere of the village and relegated to the private sphere of the home. At the same time the meanings attached to family and home, and to men's and women's roles, also changed (McDowell 1983). Families in pre-industrial societies were not very child-oriented; instead, most acted like small businesses with all members, including children, working in order to contribute to the household economy (England 1991). When production moved out of the home into the workplace the house became a private place for the family – in other words, a home. Whereas before, cooking, childcare, cleaning, and so on had been done on a collective basis, this communal style of living broke down and families became emotionally and physically more enclosed or privatized. This definition of 'home' as a place separate from employment devalued the unpaid work done within it, precisely because it was not paid.

Women's roles also changed. In the sixteenth and seventeenth centuries women had participated in commercial life; with the privatization of the family and the separation of home and work a new **ideology** of gender difference emerged. A key element of this was the 'cult of true womanhood'. Women were perceived as having the sort of emotional qualities necessary to nurture families and run the house (i.e. gentle, mild, passive), whereas men were seen as fiery, active, aggressive and so more suited to the public world of work. Soon the idea that a mother/wife was necessary for the healthy functioning of the family home became an accepted 'norm' (England 1991). Women were regarded as responsible for the upkeep of the house, the emotional wellbeing of the family and reproducing the paid labour force (Bowlby *et al.* 1982). This quote comes from a mid-nineteenth-century engineer: 'Among the working class the wife makes the home . . . The working man's wife is also his housekeeper, cook and several other single domestics rolled into one; and on her being a managing or mismanaging woman depends whether a dwelling will be a home proper, or a house which is not a home' (engineer 1868, cited in Cockburn 1983: 34).

By the end of the nineteenth and early twentieth century, this privatization of family life and women was articulated in changes in the built environment. On a city scale residential areas developed along road and railway lines, allowing men to travel into the city to the workplace, leaving women and children in residential suburbs in the urban fringe. In other words, the built environment became characterized by a divide between specialist areas of reproduction: the suburbs, and production in the centre of cities. During this time having a non-working wife at home became a hallmark of respectability in upper- and middle-class households. While many low-income, working-class and immigrant women have always been engaged in paid employment outside the home, such households also began to aspire to replicate upper- and middle-class gender ideology (England 1991). The Second World War brought a breakdown in many class divisions and a collapse of divisions between middle- and working-class women (e.g. with the decline in numbers of domestic servants, middle-class women became responsible for the domestic tasks previously carried out by them), leading to the emergence of the classless 'housewife' (Ravetz 1989).

Assumptions about gender roles were articulated in the design and layout of houses built immediately after the Second World (Roberts 1991). One of the major ideals of wartime social policy had been to preserve and protect the sanctity of family life. After the war nuclear families became prioritized over all other household types. Planners in both the UK and the USA took what was called a *pro-natalist* approach to housing design. They were concerned about falling birth rates and argued that improved family housing would persuade more women to have children and remove temptations for them to work outside the home. Lewis Mumford wrote (1945: 9):

> . . . the first consideration of town planning must be to provide an urban environment and an urban mode of life which will not be hostile to biological survival: rather to create one in which processes of life and growth will be so normal to that life, so visible, that by sympathetic magic it will encourage in women of child-bearing age the impulse to bear and rear children, as an essential attribute of their humanness, quite as interesting in all its possibilities as the most glamorous success in an office or factory.

With rising standards of housing after the war also came rising standards of house-work. Consumer durables such as washing machines and vacuum cleaners became commonplace. In particular, the development of open-plan houses began to erode traditional divisions between formal and informal space so that women were expected to keep a much larger space clean and tidy for 'show' (Matrix 1984). Domestic ideology was such that housework became understood not just as a set of chores but as a moral undertaking. A woman's moral status could be read from the way she managed her house and, by implication, her family too (Roberts 1991). A dirty home was equated with slovenliness, while cleanliness was equated with goodness. This ideology placed a heavy burden on women, as this statement from Mackintosh illustrates. 'The house is inseparable from the housewife. If it becomes dilapidated it becomes the wheel on which the housewife is broken' (Mackintosh 1952: 110, cited in Roberts 1991: 93).

Despite significant social changes in subsequent decades, the text of the UK hous-ing manual *Housing the Family*, with which all state and most private houses were supposed to comply in the 1970s and early 1980s, continued to reflect and repro-duce this ideology about men's and women's roles, and the relationship of space and design to these roles. As the time-space diagram of Mr and Mrs Average and their children (see Figure 3.1) taken from the manual shows, the design guide con-tinued to presume a separation of productive and reproductive activities. It was taken for granted that men spent most of the day in paid employment outside the home: for them the home was understood to be a place of comfort and rest, while for women the home was still assumed to be a place of work in the form of childcare and do-mestic chores (Matrix 1984).

Similar conservative assumptions about family structure and roles for women were codified and enforced in the designs for the planned suburbs called Greenbelt Towns which were built in the USA from the 1930s onwards (Wagner 1984). The planners assumed that the husband would commute to work, while illustrations in the Resettlement Administration pamphlets showed women engaged in domestic tasks. One caption read: 'Housewives in Greenbelt towns will enjoy complete, airy kitchens, fitted with modern and durable but inexpensive equipment. In the nearby allotment gardens, the housewife can raise her own fruits and vegetables if she wishes' (cited in Wagner 1984: 37). Some of the leases even went so far as to prohibit women from working outside the home and to ban women from using the home for any trade or work without the written permission of the government (Wagner 1984).

In the late twentieth and early twenty-first centuries the growth in the numbers of women engaged in paid work outside the home, and the emergence of more female-headed households, have led to a contradiction between the urban spatial form and contemporary gender roles and relations (England 1991). Studies suggest that when women are in paid employment they continue to do the lion's share of domestic work and childcare. In trying to juggle these dual roles, woman confront spatial constraints in terms of a lack of affordable childcare in accessible locations and poor public transport, which inhibit their employment opportunities outside the home (Tivers

Figure 3.1 Mr/Mrs Average and their children. Redrawn from *Matrix* (1984), Pluto Press, London

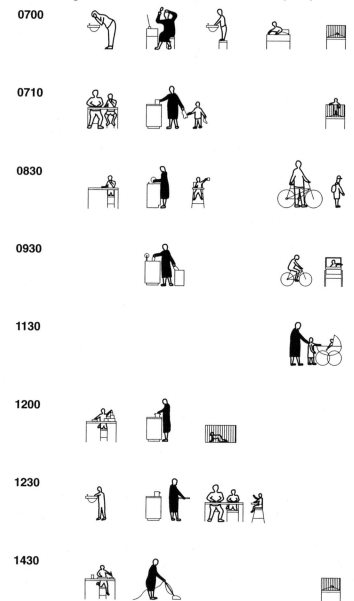

1985). As a result, feminist geographers have argued, the built environment with its dichotomous assumptions about home and work has become dysfunctional for suburban women with multiple roles, and indeed that it never worked for those who were single, lone parents, elderly, or who wanted to live in other household arrangements (MacKenzie 1984, England 1991). It is a critique shared by the disabled, who point to the 'ableist' assumptions of the design professions and the consequent lack of affordable housing that is built to meet the needs of people with physical impairments (see Imrie 1996, Gathorne-Hardy 1999).

The emphasis within housing design on the stereotype of the nuclear family – which physically represents and reinforces the cultural norm of the reproductive, monogamous heterosexual unit – assumes and reproduces a privatized form of family life in which all tasks such as cooking, eating and childcare are contained within the home. There is a rich history, however, of academics speculating or theorizing about alternative ways of living and of organizing society that transcend the traditional divisions and limitations of home and workplace. For feminist planners and architects this has taken the form of considering what non-sexist housing and non-sexist cities might look like.

Dolores Hayden (1980) makes an argument for designing and building private housing grouped around collective spaces and activity centres (including kitchens, food co-operatives, allotments, childcare and home help services) that would unite housing, services and jobs in one environment and enable food preparation and childcare to be undertaken collectively. She calls for the establishment of a campaign: Homemakers Organisation for a More Egalitarian Society (HOMES) to transform housing and residential neighbourhoods. Her programme to achieve change is based on involving men and women in the unpaid labour associated with housekeeping and childcare on an equal basis; eliminating residential segregation by class, age and race; eliminating all state programmes and laws that offer implicit or explicit reinforcement of the unpaid role of the female homemaker; minimizing unpaid domestic labour and maximizing choices for households concerning recreation and sociability.

While Hayden's (1980) design is an idealized vision, others have tried to put their ideas into practice. In the 1980s a group of ten professionals in Sweden, who named themselves *Bo i gemenskap* (BIG) (which translates as Live In Community), set out to create a housing model for themselves that would combine privacy and community while also being both practical and desirable. They established what they termed co-housing (rather than collective housing), in which residents have their own private flats but share some common facilities. The intention was that, by sharing some space, equipment and tasks, the residents would be able to maximize their personal, economic and spatial resources and so get 'more for less'. The first BIG project, a converted multi-storey house called Stacken ('ant hill') just outside Gothenberg, was followed by 40–50 similar houses built across Sweden (Sangregorio 1998).

■ 3.2.2 Critiques of work on gender and housing

Studies of gender, housing design and the spatial structure of the city played an important part in the development of feminist geography in the 1980s. In particular, by highlighting the social construction of the public as male and the private as female, and by demonstrating how the home has been regarded as the primary space of women's identification and work as the primary space of men's identification, feminist work has shown how these binary categories have played a part in defining women as secondary and as other in relation to men. In so doing, this feminist writing laid the groundwork for the challenging of dualistic (male/female, public/private, home/work) ways of thinking and the collapsing of these boundaries. However, these studies of gender, housing design and the spatial structure of the city have also been criticized for a number of reasons:

- Much of this writing was based on the misconception that the built environment is a simple metaphor for the society that produced it. Jos Boys (1998) recognizes this failing in her own early work and acknowledges that feminists in the early 1980s often failed to think closely enough about who has access to, and control over how meanings about gender and the built environment are made, and about the mechanisms of translation through which society is articulated in space.

- The work generally assumed uniform approaches to housing design, ignoring the fact that architectural knowledges, and positions within and between architects and builders about 'appropriate' socio-spatial concepts, are contested (Boys 1998).

- It often came close to being environmentally determinist in the way that it cast women as the passive victims of housing designs and urban spatial structures produced by architects, planners, property developers and the State (England 1991). In doing so, it oversimplified complex relationships between society and space, failing to recognize that material environments are physically realized in different ways by different residents or 'consumers' and that confusions about designs, unintended uses, and transgressions therefore often arise. It also underplayed the role of women in actively contesting and transforming housing and the spatial structure of the city. For example, women have set up neighbourhood self-help networks (Genovese 1981 and Stamp 1980) and developed alternative housing and ways of living (Ettore 1978, Holcomb 1986).

- In focusing on male–female and public–private dichotomies, feminist work has also treated men and women as homogeneous groups, ignoring the ways that gender identities are cross-cut by other social identities. Most notably, the research tended to focus on traditional heterosexual, white nuclear family households, thereby overlooking the fact that other social groups (such as lesbians and gay men, women gentrifiers, low-income female-headed households, and so on) may have had different living arrangements, experiences of, and relationships with, their spatial settings.

Despite these criticisms, geographers should not abandon thinking about the relationship between society, housing design and the spatial structure of the urban environment altogether. As Boys (1998: 217) explains, 'while architecture does not "reflect" society, and is only partially shaped by our continuing and contested struggles for identity, the buildings and cities we inhabit remain deeply implicated in shaping our everyday experiences'.

■ **Summary**

- In the 1970s–1980s some academics claimed that the physical form of housing could literally be read as a map of the social structures and values which had produced it.

- Feminist writers argued that housing designs and the spatial structure of the city contained ideas about the proper 'place' of women and therefore that the built environment contributed towards perpetuating the patriarchal society that produced it.

- This work has been credited with exposing how binary categories – male/female, public/private, home/work – have played a part in defining women as secondary and as other in relation to men, and with challenging these dualistic ways of thinking.

- There is a rich history of academics speculating about alternative ways of living and what a non-sexist city might look like.

- However, studies of gender, housing design and the spatial structure of the city have also been criticized as simplistic and deterministic.

■ **3.3 The meanings of home**

The home is not only a physical location but also a matrix of social relations. It is the location where our routine everyday lives are played out. Not surprisingly, our homes – perhaps more than any other geographical locations – have strong claims on our time, resources and emotions. Stretton (1976: 183, cited in Saunders 1989: 177) explains:

> In affluent societies (as in most others) much more than half of all waking time is spent at home or near it. More than a third of capital is invested there. More than one third of all work is done there. Depending on what you choose to count as goods, some high proportion of all goods are produced there and even more are enjoyed there. More than three quarters of all sustenance, social life, leisure and recreation happen there. Above all, people are produced there and endowed with the values and capacities which will determine most of the quality of their social life and government away from home.

Box 3.1: Meanings of home

'I can dress how I like and do what I like. The kids always brought home who they liked. It's not like other people's places where you have to take your shoes off when you go in.'

Retired man

'To feel at home – you never call it your house, it's your home. It's the way you build it up and the people who live in it...It's the house you build together. A home isn't the bricks and mortar, it's the love that's in it.'

Female home owner

'If I bought my own house I think I'd feel it actually belongs to me; whereas with a council house you walk along a row of little boxes until you come to a little box that's yours.'

Male semi-skilled manual worker, council tenant

'When you own you can do a lot more with it, make it more homely because its yours. You don't bother too much when it's rented.'

Female, semi-skilled manual worker, council tenant

'Home relates to happiness, not bricks and mortar. As a house this is very plain – a box – but we've had so much happiness here. And sentiment. It's familiarity with the surroundings I suppose.'

Female, retired clerical worker, home owner

'The things around you become part of you – you grow with them...Home to me is a sanctuary – somewhere you can shut the world out.'

Male, skilled manual worker, owner occupier

Cited in Saunders 1989: 180–7

The home is endowed with powerful meanings, of which Peter Sommerville (1992) identifies seven (see Box 3.1):

- *Shelter*: home is a material structure that provides physical security and protection from the elements.

- *Hearth*: home provides a sense of warmth, relaxation, comfort and a welcoming atmosphere for visitors.

- *Heart*: home is based on relations of mutual support and affection – it is a site of love, emotion, happiness and stability.

- *Privacy*: 'Being in a private place is a central part of what it means to be "at home"' (Allan and Crow 1989: 4). This involves the power to control who can enter the house. Fences, hedges, spyholes, gates, and alarms all create boundaries between the home and the outside world. Legal statues also govern rights of entry and exclusion. Who an occupier allows into the house, and which rooms they are invited into is a signal of the closeness of their relationship with the visitor. Some people are only invited in as part of planned visits, whereas close friends may 'pop in' spontaneously. While some houses may be quite open locations (with friends and neighbours welcome to come and go), others are more spatially closed off or privatized locations (Valentine 1999a). Linked to the notion of privacy is also the concept of privatism. Saunders and Williams (1988) define this as home-centredness. They argue that increased domestic consumption and the commodification of the private sphere (television, the Internet, home shopping, takeaway foods, etc) are leading people to withdraw from public life into the home and that one consequence of this home-centredness is that the streets are becoming more dangerous (see Chapter 6).

- *Roots*: home can be a source of identity and meaningfulness. It helps to reduce our sense of alienation from the mass of society. It is somewhere we feel we belong, and to which we return. Indeed, the home often becomes a symbol of the self. Domestic decor and conspicuous consumption are just some of the ways we can articulate and communicate our sense of identity (see Plate 3.1).

- *Abode*: this meaning can refer to anywhere you happen to stay. It does not have to be a house as we conventionally define it, but could be a tent or a park bench. In this sense, home is nothing but a spatial location, one that does not even have to be fixed. It is merely a place for sleep and rest.

- *Paradise*: this is the idealization of all the positive features of home fused together (a sort of spiritual bliss). In this sense, the home is an ideological construct created from people's emotionally charged experiences of where they happen to live, have lived or want to live again. Indeed, people who are homeless may value this meaning of home, despite the fact that the material underpinnings for it have disappeared, finding new ways of defining 'home' on the streets. In contrast, some people who have a physical house do not consider it to be 'home'. Sommerville (1992) therefore argues that rooflessness rather than homelessness is a more accurate way of describing the reality of having no physical house.

3.3.1 Tenure, lifecourse and the meanings of home

In some respects the meanings of home are independent of tenure but in others they are predicated upon it. Peter Saunders (1989) argues that tenants are more likely to live in one neighbourhood for longer than owner occupiers and so their sense of

Plate 3.1   Domestic decor and conspicuous consumption are some of the ways we articulate our sense of identity within the home (© Becky Ellis)

'home' is often bound up with the neighbourhood 'community'. In contrast, owner occupiers' perceptions of 'home' are often unrelated to their sense of belonging in the locality because there is a much higher turnover of privately owned houses, particularly on suburban estates, as people move up the housing chain (see Chapter 4).

Owner occupiers are more likely to equate the 'home' with the personal possessions which fill it and to see the home as a place where they can go and relax and enjoy home comforts. The greater powers they have than tenants to choose where they live and shape the design, layout and decor of the home means that home owners more readily associate it as a place where they can relax and be themselves; they are more strongly attached than tenants to the houses in which they live; and they get satisfaction and pleasure from working on their homes. In other words, home ownership contributes to a sense of personal autonomy and ontological security. However, this can sometimes be offset by fears about keeping up mortgage payments and about the costs of maintaining a property and its interior (Saunders 1989).

Although tenants can be more carefree about how they treat a house because they do not own it, Saunders (1989) found that local authority tenants often voiced sentiments of alienation from their houses by denying responsibility for them or by complaining that any improvements they made would only benefit the owners (in this case the local authority) rather than themselves.

At different points in the lifecourse the meaning of home can change. People redefine their own space differently as their relationships, family, work and interests develop and they accumulate more personal possessions. Notably, the importance of 'home' increases as people get older and become more home-centred or, in some cases, 'housebound'. Older people tend to be more concerned about the use or quality of their house rather than its monetary value, regarding it as permanent home, whereas for young people the home can be just a commodity, a trading-up point in a housing career trajectory. Saunders' (1989) work suggests that people's desire and willingness to move house declines drastically after the age of 45. His survey found that about 50 per cent of those aged under 45 said they planned to move in the near future, compared with only 15 per cent of those over 65. Over 80 per cent of those of retirement age said that they felt attached to their house and did not want to leave it. For older people the home often becomes a place where they have put down roots, it is an embodiment of the past, somewhere in which they have invested emotionally as well as financially. For example, particular rooms or features in the garden have meanings because they are associated with memories of children who have now grown up and left the family home (see also Chapter 2, section 2.7.2).

In view of this, the elderly often strongly resist having to enter institutional residential care, fearing that they will lose their sense of roots, identity, control and privacy: ' "being in a home", where they may feel "at home" but not "in *my own* home" ' (Hugman 1999: 198, emphasis in the source). Instead, the preferred option is often what is known as 'home care' in countries such as France and Canada or 'community' care in the UK and New Zealand (Hugman 1999), where family members and/or or professionals provide support services within the home. While the recipients of this form of care may still lose some degree of control over, and privacy within, the space of the home, in contrast to those who live in institutions (see Chapter 5), they are at least still able to retain a sense of individuality, normality, belonging and identity by remaining 'at home' (in both senses of the word: location and emotion) (Hugman 1999).

■ **Summary**

- The home is more than a physical location; it is also a location which is imbued with meaning.

- The home has multiple meanings. These involve variable distinctions between ideals and reality.

- The meanings of home are (re)produced and experienced differently according to tenure and age.

■ 3.4 Experiences of home

A common theme underlying many of the meanings of home outlined above is the notion that it is a positive place – a sentiment epitomized by popular sayings that the home is 'a haven' or 'a castle'. However, people's actual experiences of home do not always accord with its idealized meanings. The home is not only an enabling social and physical location, it can also be a constraining one; thus, the meanings of home may vary between individuals within homes and between different households.

■ 3.4.1 Home as a site of work

After the Second World War, British and North American governments were concerned about falling birth rates and about the problem of finding work for returning soldiers. One solution was to try to persuade women, many of whom had played an active part in the war filling traditionally male roles such as working in munitions factories and on farms, to return to the home and have children (see also section 3.2.1). This domestic ideology was actively promoted through television, radio and the print media (see Figure 3.2). Cover lines from 1950s American women's magazines included: *Femininity Begins at Home*; *Have Babies While You are Young*; *How to Snare a Male*; *Training Your Daughter to be a Wife*; and *Cooking to Me is Poetry*. Inside, the magazines featured stories of women renouncing their jobs to become homemakers.

In 1963, Betty Friedan wrote a ground-breaking book that became an international bestseller and inspired what is known as the second wave of feminism (the first wave being the suffragette movement of the turn of the century). Her book begins:

> If a woman had a problem in the 1950s and 1960s she knew that something must be wrong with her marriage or herself. Other women were satisfied with their lives, she thought. What kind of woman was she, if she did not feel this mysterious fulfilment waxing the kitchen floor? She was so ashamed to admit her dissatisfaction that she never knew how many other women shared it. But on an April morning in 1959 I heard a mother of four having coffee with four other mothers in a suburban development, say in a tone of quiet desperation: the problem (Friedan 1963: 1)

'The problem' which Friedan (1963) outlined was the sense of alienation, loneliness, boredom and oppression that she identified women as experiencing at home. She quotes one women who explains:

> I'm desperate. I begin to feel I have no personality. I'm a server of food and a putter-on of pants and a bedmaker, somebody who can be called on when you want something. But who am I?

Friedan's book ruptured the 1950s and 1960s imagery of the 'happy housewife'. In the late 1970s and early 1980s white Anglo-American feminists challenged the

Figure 3.2   Domestic ideologies were evident in advertising and the media

Artwork from The Advertising Archives

notion of home as a haven or hearth (Davidoff *et al.* 1979), arguing that 'for women in their domestic role, the ideal single-family home has always been primarily a work-place for their reproductive work, and often a very oppressive and isolating one at that' (MacKenzie and Rose 1983: 159). In particular, socialist feminists argued that, by giving birth to future workers and feeding, clothing and servicing the labour force, women maintained and reproduced the capitalist system both ideologically and materially (Dalla Costa and James 1975). They criticized the domestic ideologies that said 'women belonged at home', arguing that the value of women's invisible labour should be recognized by paying women a wage for their housework.

Other feminists saw this 'wages for housework campaign' as merely reinforcing women's role as domestic workers. They claimed that if this campaign was successful, although working in the home would become a paid occupation, women would still be seen as 'naturally' domestic and home-loving. In their view, a wage for house-work would not in itself challenge the ideology that constructed women as belonging at home and as being the primary homemaker. Instead, it was argued that a wage for housework would merely lock women into this domestic role. Rather, the problem was understood to lie with the ideal of the single-family home as a location divorced from the workplace (see section 3.2.1). Much emphasis was put on getting women into paid employment outside the home and in looking at radical ways of redesigning housing so that domestic labour could be shared by men and women on a collective basis (see section 3.2.1).

Contrary to this work by feminists, a number of studies have suggested that many women are untroubled by domestic inequalities, do not feel any more tied to the home than men, and identify it in equally positive terms (Saunders 1989, Baxter and Western 1998). This may be because the emotional meanings people attach to the home are divorced from what they actually do in the home. It also reflects the expansion in the participation of women in the paid labour market in the late twentieth and early twenty-first centuries (Holloway 1999). As a result, 'the home is no longer the primary space identified with women but rather one space amongst many, a situation which has contributed to the multiple, frequently contradictory nature of women's identifications' (Gregson and Lowe 1995: 227).

Domestic tasks in middle-class households are increasingly being performed by a combination of waged and unwaged labour. For example, middle-class households (especially where both partners are in full-time paid employment) can afford to buy in help with domestic chores such as cleaning, the laundry and childcare (Gregson and Lowe 1995). However, although middle-class women may no longer have do the domestic chores themselves, this work is still being done in the home by other women.

The absence of affordable collective day-care facilities for the under-fives in the UK means that parents usually have a choice between a nanny who comes into the parental home to care for the child, or a child minder who looks after children within her own home. Nicky Gregson and Michelle Lowe's (1995) research found that, given this choice, most mothers preferred to employ a nanny because she would work in the parental home. They explain that the home was repeatedly cited as 'the *best place* to care for young children whilst the nanny (by virtue of her location within the parental home) was presented . . . as the closet approximation to maternal care and therefore as the *best substitute* for this maternal care' (Gregson and Lowe 1995: 230).

Thus, despite changes in women's economic position, the parental home is still regarded as a safe haven for children, and the home remains unchallenged as the space for mothering. Rather than rejecting the role of 'homemaker' or 'mother' in favour of that of 'professional worker', women in paid employment are attempting to juggle these identities. For the middle classes in contemporary Britain, traditional cultures of domesticity (such as maintaining a clean and tidy home) and motherhood are still important and part and parcel of homemaking; they merely recast and reproduce these ideologies by employing waged domestic labour to do this work for them (Gregson and Lowe 1995).

This leads on to a further criticism of those meanings of home which valorize it as place of relaxation and comfort and which implicitly conceptualize it as a separate and distinct time-space from the workplace, for as the Gregson and Lowe paper implies, the home may be a site of paid work as well as of unpaid domestic labour. In a study of Appalachia, West Virginia, USA, Anne Oberhauser (1995) observes that the loss of traditional male jobs in manufacturing and mining has led to a growth not only in female employment in low-skilled, part-time work but also in home-work. This involves the home-based production of goods and services (such as sewing piece-work, knitting, weaving, telephone sales, catalogue marketing, hairdressing, secretarial work, pet care, and so on) that are sold in both the formal and informal sectors.

The use of the home as a workplace can cause conflict between family members over the control and use of domestic space (see also section 3.5). Oberhauser (1995) points out that home-work is often not regarded by other people as 'real work' because it is located in the home and often draws on domestic skills (such as sewing). As such, it is generally seen as an extension of domestic work rather than as paid work. Family and friends also tend to think that because the worker is 'at home' they can interrupt them in a way they would not disturb someone who was working in an office or a factory.

Even where paid work is performed outside the home, work can still invade the porous space of the home (see also Chapter 5). This is demonstrated by Doreen Massey's (1998a) study of men employed in the high-technology sector in Cambridge, UK. The men she interviewed were all dedicated to their work. They placed a strong emphasis on responding to their customers' needs, even if this meant working late or taking telephone calls at home in the middle of the night. To succeed in a labour market where the workplace culture glorified long hours, the employees recognized the need to work long and flexible hours and many of them were happy to do because they loved their jobs. This commitment and one-way temporal flexibility meant that the priorities and pre-occupations of the workplace intruded into their homes far more than the other way round. PCs, modems, faxes, and so on all allowed work to literally spatially invade the home. The men's partners and children also complained that even when they were physically present in the home their minds were often 'at work'. The men too cited examples of getting up in the middle of the night to work on a problem and being unable to switch off their minds, even in the bath. Although some individual employees did try to resist the way work polluted the home – for example, by staying late in the office rather than taking work into the domestic environment – they were also aware this might hinder their careers.

Men who participated in Glendon Smith and Hilary Winchester's (1998) research in Newcastle, Australia, also commented on how the demands of the workplace in terms of long hours and stress affected home and family life. While they attempted to share domestic work and childcare, it was their partners who bore the burden of these responsibilities. Again, there was a great asymmetry of power between the two spaces of home and work. All the negotiation and compromise around the tensions between competing commitments and temporal flexibility took place in domestic space rather than the workspace.

However, this is not to suggest that the workplace is always the dominant location. It is often assumed that it is paid employment which determines where individuals/households decide to live. Studies of the job search process, for example, often presuppose that individuals will move home in the pursuit of suitable employment. Yet Susan Hanson and Gerry Pratt's (1988) research shows that the home can affect individuals' decisions about whether to work outside the home, the hours they work and the type of work they do. For many people the home is regarded as the fixed point from which they look for work. This immobility can affect what employment they seek and where they work. Likewise, employees' private lives can also intrude into the workplace and contribute to shaping workplace culture and identities (see Chapter 5).

To summarize this section, some of the positive meanings of home (such as hearth) are predicated on an assumption that home and work are separate spaces in which the home is conceptualized as a space of relaxation and pleasure that is located in opposition to the responsibilities and stress of paid employment in the workplace. However, the range of examples cited in this section, including women's unpaid domestic work, waged domestic labour, home-working, paid work invading the home, and the impact of residential location on paid work all demonstrate that home and work are not distinct time-spaces but rather are mutually constituted. These examples challenge binary ways of thinking (work/home, public/private, paid/unpaid, responsibility/relaxation, stress/pleasure) and emphasize the need for geographers to collapse these boundaries and to think more in terms of home/work interdependencies.

■ 3.4.2 Home as a site of violence

Privacy is often identified as a positive meaning of home, yet privacy is not always experienced positively. The elderly can experience it as loneliness, non-working mothers often suffer from lack of contact, and the veil of privacy can also hide abusive domestic relations within households. While the home is generally regarded as a safe haven in a dangerous and heartless world, for those who experience domestic violence this is a paradox. As Elizabeth Wilson (1983) says, 'the place to which most people run to get away from fear and violence can be, for women, the context of "the most frightening violence of all" '.

Although men are occasionally subject to domestic abuse (and it occurs in same-sex relationships too), most statistics show that it is usually women who are the victims and men the perpetrators. Research in the UK suggests that between one and two women are murdered by their male partners each week and that more than a quarter of all violent crime reported to the police is domestic violence. This makes domestic abuse the second most common violent crime in the UK. In London alone, around 100,000 women a year seek medical treatment for violent injuries sustained

---

Box 3.2: Three experiences of domestic violence

'I have had ten stitches, three stitches, five stitches, seven stitches where he has cut me. I have had a knife stuck through my stomach; I have had a poker put through my face; I have no teeth where he knocked them all out; I have been burnt with red hot pokers; I have had red hot coals slung over me; I have been sprayed with petrol and stood there while he has flicked lighted matches at me. But I had to stay there because I could not get out. He has told me to get out. Yet if I had stood up I know what would have happened to me. I would have gotten knocked down again.'

'I don't know. I kept thinking he was changing, you know, change for the better … He's bound to change. Then I used to think it's my blame and I used to lie awake at night wondering if it is my blame – You know, I used to blame myself all the time.'

'I went to my parents and of course he came – I left him because of his hitting and kicking me – I went home to them, but he came there and I had to go. I went back really to keep the peace because my parents weren't able to cope with it.'

Cited in Stanko 1985: 51, 55–6, 58

---

at home (URL 1). A similar pattern is evident in the USA. The New York City Police Department receives over 200,000 calls per annum about family disputes. Surveys suggest that 7 per cent of US women who are married or living with a partner are subject to physical abuse, 37 per cent are verbally and emotionally abused, and as many as one in three Americans have witnessed an incident of domestic violence (URL 2).

But despite these attempts to quantify domestic violence, the real extent of abuse is largely unknown. As Smith (1989: 6) argues, 'By its intrinsic nature, domestic violence is an elusive research topic: it takes place behind closed doors; is concealed from the public eye; and is often unknown to anyone outside the immediate family.' The general understanding amongst academics and the police is that figures for domestic abuse are almost certainly underestimates. It is thought that the figure for domestic violence may be greater than that for any other crime (Smith 1989).

Qualitative research shows that this violence includes a broad range of forms of abuse which vary in their nature and severity from bruising and cuts, to rape and broken bones (see Box 3.2). In addition to physical injuries, it is common for women, and children witnessing the abuse, to suffer psychological damage, including depression and panic attacks (Stanko 1985).

Dobash and Dobash (1992) argue that two of the main sources of conflict leading to violent attacks are men's expectations concerning women's domestic work in the home and men's belief in their right to punish women for perceived wrongdoing. This logic is also sometimes apparent in the responses of the police and courts to cases of domestic violence where women have effectively been put on trial over their ability to look after the house and produce a good home (Edwards 1987). This shows the strength and persistence of ideologies about women's place in the home and domesticity. The ideology that only 'bad' or failed women get hurt is, according to Elizabeth Stanko (1985), one of the reasons why many women are to ashamed to report domestic violence to the police, as well as their concern that the police and judicial system will not be sympathetic to them or able to act effectively.

The traditional binary distinction between public and private space has also contributed to the reluctance of the police to respond to domestic incidents. The public world has been seen as the legitimate arena for intervention by the police, while the home has been regarded as the domain of the family. The relationships within it have therefore been understood to be private, voluntary and non-compulsory and so to be beyond the legitimate control of the state. Hence the popular saying 'a man's home is his castle'. Indeed, when women are subject to abuse most turn in the first instance to their extended family for help rather than to the police (Smith 1989). One tactic employed by Women's AID (an urban social movement established to support women who experience domestic violence) in response to the ineffectiveness of the state at dealing with domestic violence has been to establish alternative 'private' spaces in the form of refuges and safe houses to help women escape from dangerous homes.

■ 3.4.3 Home as a site of resistance

> When I was a young girl the journey across town to my grandmother's house was one of the most intriguing experiences . . . I remember this journey not just because of the stories I would hear. It was a movement away from the segregated blackness of our community into a poor white neighbourhood. I remember the fear, being scared to walk to our grandmother's house because we would have had to pass that terrifying whiteness – those white faces on the porches staring down with hate. Even when empty or vacant, those porches seemed to say 'danger', 'you do not belong here', 'you are not safe'. Oh! that feeling of safety, of arrival, of homecoming when we finally reached the edges of her yard, when we could see the soot black face of our grandfather, Daddy Gus, sitting in his chair on the porch, smell his cigar, and rest on his lap. Such a contrast, that feeling of arrival, of homecoming, this sweetness and the bitterness of that journey, that constant reminder of white power and control (hooks 1991: 41).

bell hooks uses this recollection of her childhood as a springboard to describe how black women have constructed the home as a space of care and nurture in the face of the brutal reality of racist oppression. She claims that 'historically, African-American people believed that the construction of a homeplace, however fragile and tenuous (the slave hut, the wooden shack), had a radical political dimension' (hooks 1991: 42). It was the one site where black people could achieve the dignity and strength which they were refused in the public realm.

For black women, making a homeplace was not just a matter of creating a comfortable domestic environment, it was about the construction of a safe place where black people could affirm their identity, a space where they could be free from white racism and a site for forming political solidarity and organizing resistance. hooks writes (1991: 42): 'We could not learn to love or respect ourselves in the culture of white supremacy, on the outside; it was there on the inside, in that "homeplace" most often created and kept by black women, that we had the opportunity to grow and develop to nurture our spirits.' She further claims that, without the space to construct a homeplace, black men and women cannot build a 'meaningful community of resistance' (hooks 1991: 47).

Her argument is supported by other black writers. Patricia Hill Collins (1990) points out that the terms 'public' and 'private' have different meanings to African Americans than to white Americans. Private is not always just equivalent to the domestic home, it can also refer to black community spaces that are beyond the reach of white people. For example, mothering in a black community might often involve 'other mothers' in the community as well as blood mothers. In this way, private space can be a resource for black women rather than a burden. Thus Hill Collins (1990) argues that the notion of everyday domestic oppression suffered by 'housewives' isolated in the home (see section 3.4.1) is a problem specific to white women.

In this way black writers have exposed the universalism of 1970s feminism and its assumptions that white middle-class women's experiences of the home could stand in for all women's experiences of this space. hooks, for example, takes white feminists to task for focusing on the home as a site of patriarchal domination and oppression. She suggests that by seducing black women into being concerned about domesticity and the domestic division of labour, white feminism encouraged black women to forget the importance of the home as a site of subversion and resistance and that, as a consequence, black liberatory struggles have been undermined. hooks (1991: 48) calls for black people to focus on a revolutionary vision of black liberation built on the foundations of the home. She writes:

> Drawing on past legacies, contemporary black women can begin to reconceptualise ideas of homeplace, once again considering the primacy of domesticity as a site for subversion and resistance. When we renew our concern with homeplace, we can address political issues that most affect our daily lives.

■ 3.4.4 Home as a site of non-heterosexual relationships

The nuclear family is the traditional living arrangement experienced by many people in contemporary Western societies at some point in their lives. Yet we spend a relatively short span of our total life in this household form. As a result of separation, divorce and the growth of other household forms, the nuclear family is becoming a much less common form of living arrangement. Recent figures suggest that more than two in five marriages end in divorce and three in ten children's parents separate before they are 16 (URL 3). Even within heterosexual family units there may be intermittent disruptions in household membership because, for example, one parent or partner may be absent from home for periods of time because they work in another city or country or because of temporary work commitments. Indeed, Judith Stacey (1990: 269) argues that the family 'is an ideological concept that imposes a mythical homogeneity on the diverse means by which people organise their intimate relationships [such that] "the family" distorts and devalues the rich variety of kinship stories'.

Despite its declining statistical significance, the nuclear heterosexual family is often presented as being synonymous with 'the home'. Yet, the home can take on very different and contradictory meanings for people living in household forms which are not based on kinship, such as single people in shared housing or elderly people in various forms of sheltered accommodation (Valentine 1999b). This section focuses particularly on the experiences of lesbian and gay men who either share a house with heterosexual family members or create their own home space (Johnston and Valentine 1995, Elwood 2000).

Lesbians living in the 'family' house who have not 'come out' to their parents can find that a lack of privacy from the parental gaze constrains their freedom to perform a 'lesbian' identity 'at home' and to have friends and partners to stay (Valentine 1993, Johnston and Valentine 1995, Elwood 2000). In this sense the public/private dichotomy is much more complex than a simple duality. Having privacy from the outside world is not the same thing as having privacy within the home. The distinction between public and private is often hard to draw within households, especially in families where personal boundaries are often blurred by pressures to share activities, socialize with each other, co-ordinate schedules, and so on. Often individual privacy is only recognized at certain times in certain rooms, if at all (Allan and Crow 1989). Although walls and doors can provide some visual privacy in different rooms, aural privacy from noises generated within and between households can be less easy to achieve.

The home is supposed to be a place where family members participate in communal activities, socialize and share their feelings. These basic patterns of social relations are often underlain with a heterosexual ethos. At the kitchen table or round the television the heterosexual family can serve up a relentless diet of heterosexism and homophobia that can alienate gay members of the family and undermine for them many of the positive meanings which are popularly assumed to be associated with the family home (such as roots and hearth).

Although the home is supposed to be a medium for the expression of individual identity, a site of creativity or a symbol of the self, in practice this can mean that family homes express a heterosexual identity in everything from pictures and photographs, to furnishings and record collections, while the lesbian or gay identities of individual household members are submerged or concealed. Because of this, many sexual dissidents can feel 'out of place' and that they do not belong 'at home'. A survey of 239 lesbians and gay men aged 16–26 in Scotland, UK found that one-third had actually been forced to leave their family residence as a result of coming out (Scottish Federation of Housing Associates 1997). The freedom to perform a lesbian identity, to relax, be in control, and to enjoy the ontological security of being at home therefore appears to be best met when lesbians and gay men can create and manage their own home spaces.

Section 3.2.1 pointed out that housing is often designed with the presumption that it will be occupied by nuclear families. Lesbians and gay men often subvert this, for example, by making structural changes to the house to express a non-heterosexual identity or lifestyle, or by decorating the home in appropriate ways (Elwood 2000). However, this can cause a problem when heterosexual relatives or work colleagues call round, and can also provoke hostility from neighbours or property owners (Egerton 1990). The Scottish survey revealed that one in three of those questioned had been harassed because people suspected them of being gay.

Despite these moments, the lesbian home can still be a sanctuary from the heterosexuality of the public world. Indeed, in provincial towns and rural areas (see Chapter 8) where there are no lesbian or gay venues such as bars and clubs, individuals' homes can take on much broader roles as social venues and political meeting places (Valentine 1993, Rothenberg 1995, Kennedy and Davis 1993). As such, the home can also become a site of resistance and a place to build a politics of resistance for sexual dissidents.

■ Summary

- People's actual experiences of home do not always accord with its idealized meanings, and some valued meanings are not always experienced positively.

- The home can be a site of oppression and danger rather than a haven, and a place of work rather than a space of relaxation.

- The home can be a site of subversion and resistance and a place from which it is possible to build a politics of resistance.

- How the home is experienced depends on who has the power to determine how the space of the home and social relations within it are produced.

■ 3.5 The moral economy of the household

The home is 'the focus and pivot of consumption . . . [it] is not only itself an object of consumption, but it is also the container within which much consumption takes place' (Saunders 1989: 177). Roger Silverstone and Eric Hirsch (1992: 6) describe the household as a moral economy which they define as 'a social, cultural and economic unit actively engaged in the consumption of objects and meanings'. They argue that objects and technologies such as televisions, computers, food and so on are crucial to the production of households' identities, integrity and security, while also embedding households within the wider public sphere of the formal economy (see for example Morley 1986, Livingstone 1992, Valentine 1999b).

When household members purchase domestic goods these objects cross over from the public sphere, where they have been produced, into the home, where they become part of the everyday dynamics of household life. Here they are domesticated within the context of the specific social relations of individual households who draw on their own biographies, histories, politics and beliefs to construct a personal economy of meaning. Roger Silverstone, Eric Hirsch and David Morley (1992) identify four processes through which this occurs: appropriation, incorporation, objectification and conversion.

First, objects are *appropriated*, that is, they are taken possession of by an individual or household. This can cause disputes between different household members. The meanings given to objects in the home in this way may differ from those ascribed to them in the public sphere.

Second, they are *incorporated* into daily routines through the different ways that they are used. These may be ways that were not intended either by the manufacturer or the purchaser. For example, parents often buy home computers as an educational tool for their children but in practice they emerge as glorified games machines (Valentine, Holloway and Bingham 2000). Where objects are located within the microgeography of the home, and when or how they are used or consumed by different individuals can also form the basis of processes of differentiation and identification both within, and between, households. For instance, what constitutes a 'proper meal' and where it may be eaten can be a source of identification and conflict along the axes of gender, class, age, etc (see Valentine 1999b).

Third, artefacts are *objectified* through usage and through the different ways that they are spatially displayed within the home. The way that we decorate rooms and organize and arrange the objects within them provides an objectification of the values, aesthetics and meanings of the household or those who identify with them. Examples of studies that have looked at such domestic spatial arrangements include Daniel Miller's (1988) work on kitchens and Tim Putnam and Charles Newtons' (1990) work on living rooms.

Fourth, the process of *conversion* defines the relationship between the household and the outside world. For example, television programmes, computer games, or the purchase of goods such as microwaves and washing machines, are all the source of everyday conversations outside the home. Participation in these exchanges can signal individuals' membership of, and competence within, wider local, national and global cultures. For example, children can use conversations about computer games as a way of being accepted into local peer groups. At the same time the discourses they draw on are part of a wider global cultural language (Holloway and Valentine 2000, see also Chapter 5). In this way, objects may also increase the permeability of domestic boundaries by linking household members into wider networks at school, at work, and so on. In turn, these connections may threaten or change what individuals or households do, or take for granted, in terms of domestic routines. Through the process of conversion 'the boundary of the moral economy of the household is extended into and blended into the public economy' (Silverstone, Hirsch and Morley 1992: 26).

■ 3.5.1 The television

In a study of television in the USA in the 1950s, Lynn Spiegel (1992) examines how the television set was seen as a kind of household glue that would bond together the fragmented lives of families who had been separated during the Second World War. This was also a time when many families had moved out to the new suburbs, leaving friends and extended family members behind in the city. The television was expected to strengthen these isolated households as a family unit, as well as having the added benefit of keeping children off the streets. Having a family room in which the set could be located was identified as important in producing this sense of togetherness. The television replaced the fireplace or piano as the focal point around which seating and other furnishings were arranged. It was, claims Spiegel (1992: 39), a 'cultural symbol par excellence of family life'.

At the same time there were fears in the 1950s that the television might dominate the household. The upper middle classes regarded the set as an eyesore and its public display as the epitome of bad taste. These sorts of shameful meanings led companies to promote furniture designs that would camouflage it.

Even in the 1950s it was children who influenced the purchase of televisions, and, as the most enthusiastic viewers, often appropriated them. This triggered popular concerns that children were becoming addicted to watching television, that this was undermining their education, manners and even causing physical disorders, and that the 'innocent child' (see Chapter 2, section 2.7.1) might be corrupted by inappropriate programmes. These concerns have subsequently been replicated in other generations, and in relation to other domestic objects such as the home personal computer (Buckingham 1997, Oswell 1998, Valentine, Holloway and Bingham 2000). In the 1950s parents were advised to incorporate television into household life in a controlled manner by using the location of the set, and rewards and punishments to discipline children's viewing habits and to encourage them to watch morally uplifting and educational programmes such as documentaries. In this way, national moral panics, and their local interpretations, were expressed in everyday household decisions about who could watch what, when, and with whom. At this time the television was also seen not only as a threat to childhood but to the masculine position of power within the home. In the competition for cultural authority between fathers and the television, the television was assumed to win.

These examples which Spiegel cites from the 1950s demonstrate that, while objects may be given meanings by their owners, objects can also transform their users. The television is a good example of the importance of the relationships which people enter into with objects. When the two come together in practice particular identities or social relationships between different household members may be produced. These are not static; although they may be continually reproduced they can also be undone or changed.

The increase in the number of television sets within the average contemporary home is leading to suggestions that the television (and other technologies too), rather than unifying the modern family, as was claimed in the 1950s, may actually be encouraging the dispersal and fragmentation of the household. There are multiple spatialities and temporalities within homes; for example, living areas have traditionally been 'public' rooms where visitors are entertained and where household members participate in activities such as watching television, listening to music, eating and using the telephone. In contrast, bedrooms have traditionally been more privatized spaces, which have primarily been used for (un)dressing, sex and sleep. However, the expansion of technologies (such as TVs, stereos, computers, telephones) into the bedroom is facilitating the ability of individuals to lead a separate existence from other members of the household, accentuating the privatization of the bedroom as a space. This process is being fuelled by the discordant lifestyles that many contemporary household members lead. The increase in women working in paid employment outside the home; the number of people working in part-time, shift work or demanding jobs; and a growth in after-school institutionalized play and leisure opportunities for children are just some of the factors which are producing complex and diverse life patterns within postmodern families. Thus, although different members of a family may live under the same roof, they may occupy very different time-spaces within the home.

■ **Summary**

- The household is a moral economy defined as a social, cultural and economic unit actively engaged in the consumption of objects and meanings.

- Households domesticate 'public' goods to construct a personal economy of meaning through four processes: appropriation, incorporation, objectification and conversion.

- Domestic objects are implicated in the production of households' identities and embed households within the wider public sphere of the formal economy.

■ 3.6 Home rules: negotiating space and time

As section 3.5 hinted, the home is an important site where spatial and temporal boundaries, in relation to both domestic space and public space, are negotiated and contested between household members.

In a study of the living room of 435 Cutler Street, the home of Denis and Ingrid Wood and their two sons Randall (then aged 11) and Chandler (then aged 9), Denis Wood and Robert Beck (1994) identify 223 rules governing the room and the 70 objects it contains (such as the couch, walls, table, sofa, plants and fireplace). In a book appropriately titled *Home Rules* they present a photograph and description of each object in turn and a list of the rules relating to it (see Box 3.3). While the rules are assumed to be self-evident to other adults and are rarely explicitly stated

---

Box 3.3: Home rules

rules: the glass in the sidelights
30: Don't breathe on them
31: Don't touch the windows
32: Don't let your friends put their noses on the windows
33: Don't wipe off the windows with your hands. You only smear them
34: Don't pound on them
35: Don't spit on them

rules: the white couch
129: Don't throw yourself onto the couch – sit down on it
130: Don't sit on the arms
131: Don't sit on the back
132: Don't hang off the back of the couch
133: Don't mark back with shoes
134: Don't climb over the couch
135: No fighting on the couch
136: No shoes on the couch
137: Don't put food on it
138: Don't throw up on it – go outdoors or in the bathroom
139: Don't get the couch dirty
140: Don't hide things under or behind the pillows
141: Don't take up the pillows and sit underneath them
142: Don't leave the pillows not spiffed up
143: Don't leave things under the couch
144: Don't leave your shoes under the couch

Wood and Beck 1994: 37, 158–9

---

to them, children are presumed to be less civilized (see also Chapter 2 and Chapter 5) and less able or willing to read or understand the rules in the same way. When children are in a room adults constantly repeat a battery of home rules: 'Mind the vase', 'Close the door', 'Don't smudge the windows', etc. These enable parents to maintain their values by protecting the room from the children, while at the same time reproducing these values by instilling them into their children.

In the analysis accompanying the restrictions Wood and Beck (1994) argue that home rules articulate the values and meanings of objects but they are also part of a wider belief system. For example, while some rules are specific to individual households, others are more universal within contemporary western societies. As they point out, 'You wouldn't do that at home, would you?' is a popular adult refrain heard in public places such as schools or restaurants. In the discussion of universal rules they observe that different rules apply in different rooms, for instance the rules relating to artefacts in the bathroom and living room are not the same. They conclude that the arrangement of a room's parts and objects physically stores or represents how the household members interact together. Thus, because home rules are in part derived from adults' memories of their own childhood homes, the rooms within a house are, in effect, an 'instantiation of a kind of collective memory' (Wood and Beck 1994: xv).

Wood and Beck's (1994) book is a case study of only one family, whereas David Sibley and Geoff Lowe (1992) reflect on the boundaries of acceptable behaviour, which include limits on use of space and time, within different households. They draw on Bernstein's (1971) distinction between positional and personalizing families. Positional families are those in which the parents are authoritarian and establish unambiguous boundaries in terms of adult space and time within the home. In contrast, personalizing families are less constrained by these sorts of arbitrary rules. Rather, children are given a greater voice in negotiating domestic spatial and temporal boundaries. However, in using this distinction Sibley and Lowe (1992) also recognize that 'normal' behaviour is often characterized by a mixture of personalizing and positional tendencies.

In subsequent solo work David Sibley (1995b) draws on data about middle-class childhoods in the UK from a Mass Observation archive which shows how children, when given their own bedrooms, appropriate, transform and secure the boundaries of this space to make it their own. He contrasts these descriptions with those of other spaces in the home where children have less freedom, and are more subject to parental restrictions. In some households, whether a room is a child space or an adult space can change with the time of day. Sibley (1995b) cites the case of a living room which was shared by a mother and the children during the day, but was transformed into an 'adult space' when the father returned home from work. In other cases the timing of activities and division of space can create more liminal zones. He suggests that domestic tensions around home rules and the use of different rooms represent conflicts between adults' desire to establish order, regularity and strong domestic boundaries, and children's preferences for disorder and weak boundaries.

It is not only children's use of domestic space which is the subject of home rules, their spatial range outside the home is also a constant source of negotiation between parents and their offspring within households. Children do not passively adhere to adults' definitions of what they are capable of doing or where they are capable of going at specific biological ages. Rather, they play an active part in negotiating the meanings ascribed to their 'age' within the household (see Chapter 2, section 2.7.1). One tactic some children adopt is to perform or demonstrate competence in one aspect of their lives and then to use this as evidence of their maturity to try to negotiate more independence in other spaces. For instance, a child may use the fact that they are competent enough to use kitchen appliances as a lever to win more autonomy in public space. Definitions of children's competence are not always allowed to spill over from one space into another, however. Understandings of children's maturity at home or in the street can therefore stand in awkward contradiction to each other. In particular, children are often more closely supervised in public space than private space, which illustrates the extent to which contemporary parents consider traffic and stranger dangers to be more important threats to children's safety than domestic hazards (Valentine 1999c).

Mothers and fathers often take on different parenting roles and discipline their children in different ways – although they may try to present a united front at home to their offspring. Children can exhibit a sophisticated knowledge of the gender division of parenting and power dynamics in the family, playing mothers and fathers off against each other (Valentine 1997a). This is particularly the case for children living in reconstituted families with a biological parent and a social parent while maintaining contact with the other biological parent who is now living in a different household, because they can exploit the fact that their welfare is often a source of major tensions between their mothers, their fathers and their parents' new partners.

Household negotiations which determine children's spatial boundaries are also framed in relation to popular public discourses within the law, childcare professions, and the media, about what it is appropriate for children to do at different ages. Conversations between parents within their 'communities' play an important role in informing the way that individuals interpret these public discourses within the home, and produce common-sense definitions of what it means to be a 'good' mother or father (Dyck 1990, Holloway 1998, Valentine 1997a, see also Chapter 4). Parents whose home rules or spatial boundaries are out of line with these local parenting cultures can find themselves stigmatized by other adults. The consequences of individualized domestic practices can also spill over into other spaces such as the school, where any form of 'difference' can be grounds for children to be marginalized or excluded by their peers (see Chapter 5).

Parents sometimes sidestep children's resistance to their spatial boundaries by disguising their attempts to restrict the children's use of space. This is done by subtly structuring children's leisure time, for instance by taking them to after-school clubs, institutional play schemes or sports clubs (Valentine and McKendrick 1997, Smith and Barker 2000, McKendrick *et al.* 2000), so they do not have the opportunity to be in public space independently (see also Chapter 5). The effect Sibley (1995b: 16) claims is that 'For children in the most highly developed societies, the house is increasingly becoming a haven. At the same time the outside becomes more threatening, populated by potential monsters and abductors so the boundary between the home (safe) and the locality (threatening) is more strongly drawn.' In this way, parental fears and the associated restrictions placed on children's use of space are being blamed for robbing the young of the opportunity for independent environmental exploration and for destroying children's street cultures (see also Chapter 6).

■ Summary

- The home is a key site where spatial and temporal boundaries in relation to domestic and public space are negotiated between household members.

- Home rules represent conflicts between adults' desire to establish order and strong boundaries, and children's preferences for disorder and weak boundaries.

- Children actively negotiate the meanings ascribed to their 'age' within the household.

- Household negotiations which determine children's spatial boundaries are framed in relation to public discourses and local cultures of parenting.

- Parental restrictions are being blamed for making children's lives home-centred and for destroying their street cultures.

■ 3.7 Homelessness

Early academic definitions of homelessness emphasized the absence of a place to sleep and receive mail as well as what was termed 'social disaffiliation'. More recently, the focus has switched to recognizing that there is a continuum between being homed and being homeless, with people living in temporary, insecure or physically substandard accommodation falling somewhere between these two extremes (Veness 1993). In most cases homelessness is, in any event, a temporary circumstance rather than a permanent condition.

The *official homeless* are those people who are designated homeless by the state and who have applied for housing. The *single homeless* are those with no statutory rights to housing who end up living on the streets or in shelters (Warrington 1996). A further category are the '*hidden homeless*'. These are people who fall outside official definitions of homelessness and are not actually living on the streets because they are in precarious or unstable housing situations, such as staying with family and friends, or in secure units. The term *protohomeless* has also been coined to describe those households who are at risk of homelessness (Warrington 1996).

Homelessness is often assumed to be an urban problem because homeless people are more numerous and more visible in cities, yet it is a problem in rural areas too. The idyllization of rural lifestyles in the UK and the glorification of the struggles and hardship endured by the pioneers in the USA both contribute to obscuring the extent of rural poverty and social problems in the countryside (see Chapter 8). The poor quality of the rural housing stock and the lack of shelters or transitional units in rural areas produce particularly high levels of hidden homelessness. This includes people who live in cars, camper vans, tents, caves, with relatives, or in inadequate and dilapidated facilities (URL 4).

Figures for the homeless at any given time are difficult to calculate because of the range of definitions of homelessness and the lack of information about the length of time for which people are homeless and the proportion of people who experience repeated episodes of homelessness. Projected estimates for the USA in 1999 suggested that over 700,000 people would be homeless on any given night and up to 2 million people would experience homelessness during one year (National Law Center on Homelessness and Poverty 1999). Other studies suggest that over 12 million US residents have been homeless at some time during their life (URL 5).

There is a geography to these statistics as well. The communities that produce homelessness are often not those that end up caring for homeless people. In particular, people experiencing economic, social or housing difficulties often seek work or accommodation in large cities. Many remain there even when they cannot get the work, social support or housing which they were seeking, rather than return to their home town. For example, young people and lesbians and gay men often migrate from small towns and rural areas to major cities to escape tight-knit communities and to forge their own identities in the anonymity of the city (see Chapter 8).

There are also wide variations between places in terms of the service provision for those in need because authorities adopt varying approaches to the homelessness crisis. This reflects differing demand for help but also variations in the institutional and financial capacities of different local government and voluntary organizations, as well as the political will of community leaders in particular cities (Dear 1992, Wolch and Dear 1993, Takahashi 1996, 1998).

The homeless are stigmatized through association with alcoholism, drugs, crime and mental ill-health, and so are often blamed for becoming or remaining homeless (Takahashi 1996, 1998). Indeed, some homeless people, such as the young, are regarded as more worthy of support than others. Those with mental ill-health are the least tolerated. In the USA some authorities attempt to foist their homeless population onto neighbouring cities, for example, by using the police to escort or transport homeless people to other jurisdictions, by enforcing nuisance laws and anti-bum ordinances, by watering public parks to deter homeless people from congregating there, and so on. Neighbourhood communities also help to (re)produce these patterns of social exclusion by opposing the siting of facilities for the homeless in their midst (so-called NIMBYISM – Not-In-My-Back-Yard, see Chapter 4). Residents and businesses (particularly in affluent areas) commonly claim that the presence of the homeless in the neighbourhoods disrupts the moral order of the streets and undermines their personal safety, property values and quality of life (Dear and Taylor 1982, Dear and Gleeson 1991, Dear 1992, see also Chapter 6). The result is what Takahashi (1996: 297) terms a 'rising tide of rejection' which results in the concentration of services and support for the homeless in relatively few locations.

▓ 3.7.1 Causes of homelessness

There is no single pathway to homelessness – different people end up on the streets for different combinations of reasons. However, the late twentieth and early twenty-first centuries have been marked by a number of significant economic and social changes which have contributed to a rise in the homeless population.

In the 1970s and 1980s a fundamental restructuring of international and domestic economies occurred. 'This restructuring process has involved fundamental alterations in patterns of international finance and trade, shifts in the mix of industrial outputs and changes in the organisation and geographical location of production systems. Its effect has been to create a transition from *Fordism* (traditional assembly-line, mass-production industry) to a qualitatively different mode of economic organisation, *post-Fordism*' (Wolch and Dear 1993: 3). These changes have resulted in deindustrialization and the retreat of manufacturing in North America and Europe to cheap-labour locations in less developed countries. In rural regions a parallel decline occurred in extractive industries such as mining, timber and fishing (Wolch and Dear 1993).

At the same time new jobs emerged but these were primarily in finance, property, insurance and other professional services. As a result, the service sector has surpassed manufacturing as the mainstay of contemporary Western economies. Post-Fordism also involved flexible specialization: the ability to be sensitive to, and respond rapidly to, changes in consumer preferences or the market. To increase their flexibility and reduce overheads, firms in manufacturing and services restructured their employment practices, hiring part-time and temporary workers – a process which particularly drew women into the labour force. In turn, dual-income households began to employ childcare and other domestic service workers as substitutes for the unpaid work women previously did at home. This trend further promoted the growth of jobs in personal service and the retail trade (Wolch and Dear 1993).

These employment shifts have seen a spiral in long-term unemployment, especially among traditional blue-collar workers, and less-educated, low-skilled workers. Consequently, the gap between those in high-wage jobs in the service sector and those who are unemployed or on low wages has widened drastically, with record numbers of people being cast into poverty and homelessness. In this way, individuals on the streets of New York, London or Paris are regarded as a local problem for the city, but their presence is a product of global economic processes.

Most contemporary Western societies have a commitment, to a greater or lesser extent, to provide for those citizens who unable to support themselves. After the Second World War most expanded their welfare expenditure and services. However, when economies in countries like the USA and the UK were restructured in the 1980s, welfare provision was also revised and cut back. One aspect of this contraction was a policy of deinstitutionalization which moved psychiatric patients out of state and local mental hospitals into community-based service settings (see Chapter 5). For example, there were over half a million people in US asylums in the 1950s, yet within two decades this population had been cut to under 100,000 (Wolch and Dear 1993). This policy of deinstitutionalization was subsequently also extended to include people with disabilities, the dependent elderly, and people on probation (Takahashi 1996, Kearns and Smith 1993). Yet, though the expenditure on state institutions was slashed, there was no corresponding provision of funding to support local community services. At the same time there was also a general reduction in levels of benefits and definitions of who was entitled to claim support. These cutbacks particularly affected the young. In the absence of support many people ended up drifting into inner-city neighbourhoods where cheap accommodation was available, and from these unstable housing arrangements onto the streets. The stereotype of the homeless used to be middle-aged male alcoholics fallen on hard times, but now homelessness is being democratized, with more women, young people and those from ethnic minority groups joining its ranks.

In effect deinstitutionalization and cuts in welfare expenditure – which were widely adopted across North American and European societies – served to shift responsibility for vulnerable groups from national governments to voluntary agencies, and from the state (i.e. public provision outside the home) to local communities, and especially to families. In other words, there was movement away from public provision outside the home towards an expectation that support would be provided within the wider family home.

These problems were compounded by a lack of affordable housing. Parallel cuts were made in expenditure on subsidized and public housing, marking a general transfer of responsibility for housing provision from national governments to local agencies and charities. This squeeze on supply occurred at a time when a number of demographic shifts were creating a rise in demand for housing, producing a resultant shortfall in provision. These included a drastic increase in single-person households facilitated by the growth in female employment rates which has enabled more women to live independently than ever before; a rise in female-headed households as a result of a decline in marriage and higher divorce rates which particularly fuelled the demand for smaller affordable houses; and the greying of the population (it is estimated that by 2040 there will be 87 million Americans over 65 and 24 million over 85), which also prompted a demand for small housing units, often with special facilities (Wolch and Dear 1993).

These economic and demographic changes in turn also altered the structure of urban housing markets, stimulating **gentrification** (see Chapter 7 and Chapter 8) and exacerbating the homeless problem. Historically, affluent groups have preferred to live in the suburbs but in the 1980s those profiting from economic restructuring began to move back into the central city, a trend which has continued into the twenty-first century. Here rents and prices of properties tended to be low because they were often in a poor condition, but the potential rents were high because of the desirable location and so the properties could be redeveloped at a profit. The effect of gentrification (by mainly young, white professionals) has been to drive up rent and property prices and to increase the competitiveness of the housing market, displacing those on low incomes (mainly blue-collar workers, the elderly and ethnic minorities). Not surprisingly, the poorest and most vulnerable invariably end up homeless.

A final cause of homelessness is personal crisis. The late twentieth and early twenty-first centuries are a period which have been marked by a process of **individualization** (Beck 1992). This is a shift in ways of thinking about how individuals relate to society which has been characterized by a decline in pressures to follow conventional lifepaths and to adhere to traditional institutions such as the church, and the development of new social forms and types of commitment. Biographies are increasingly reflexive in that people can choose between different lifestyles, subcultures and identities (see also Chapter 2, section 2.4.1), but these opportunities have also brought increased risks that individuals may end up on the margins of society as a consequence of their own choices.

As a result of this shift, a number of positive social trends have emerged which have also had negative consequences for some individuals. First, the growing number of women in paid employment has enabled more women to live independently and to leave unsuccessful or abusive relationships. However, because women generally work in lower-paid, lower-skilled jobs than men and maintain responsibility for any children from a relationship, there has been a feminization of poverty. Second, there has been a decline in the nuclear family and increased tolerance of alternative lifestyles. This has given more lesbians and gay men the confidence to 'come out', but, as a consequence, some individuals face discrimination and find themselves evicted from parental homes or rented accommodation. Third, the post-industrial emphasis on hedonism, pleasure and high levels of conspicuous consumption has produced a rise in drug abuse, alcoholism, HIV and individual indebtedness.

Because of this, more individuals have found themselves in situations of personal crisis that have resulted in homelessness. In many cases this is not the outcome of one incident, like the loss of a job, but is the outcome of an extended period of cumulative difficulties in a person's life in which there is a gradual drift or slide into homelessness. The process often involves a series of transitions in which a person loses their house, moves to rented accommodation, then stays with friends or parents before ending up on the street.

■ 3.7.2 Experiences of homelessness

Veness (1993) argues that the traditional dichotomous distinction which is made between the homed and the homeless does not fit the messy and complicated reality of poor people's lives. In a study of three groups of people living in poverty in Delaware, USA she showed that, while badly housed or homeless by society's standards, these people still created homes within homeless shelters and a diverse range of other locations such as cardboard cities, derelict buildings, camper vans, caves and tunnels (see Plates 3.2 and 3.3). Veness argues that these people are neither homed nor homeless, rather she describes them as un-homed, arguing that this is both a valid social category and a lived environment. Indeed, Rowe and Wolch's (1990) study of homeless women on Skid Row in Los Angeles, USA suggests that the social

Plate 3.2   Alternative 'home' spaces (© April Veness)

UNDERGROUND RAILWAY THEATER PRESENTS
Home Is Where

Plate 3.3　Alternative 'home' spaces (© April Veness)

relations (in terms of gender, perceptions of community, and so on) of the homed and some of their meanings of home are also replicated by people living on the street.

While the homed often stigmatize the homeless as dangerous and a threat to their safety (see Chapter 6), people living on the street are attacked or robbed of their possessions by passers-by and vigilantes. Yet they have little recourse to the police or criminal justice system. Rowe and Wolch (1990) argue that women, in particular, are at much greater risk of assault and feel more vulnerable on Skid Row than men. There are far fewer shelters provided for women than for men and, as a consequence, many women on Skid Row see sexual relationships as a means of gaining some protection. These relationships have other benefits too, because a partner is a person with whom possessions and messages can be left and with whom resources can be pooled. There can even be a quasi-domestic division of labour on the street. In this sense partners can take on the role of 'home base' for each other. Rowe and Wolch (1990) illustrate this with the example of Rita and Paul. Paul would often stay with their possessions in the park where the couple had spent the night, allowing Rita to go and panhandle for money or find food. His location in the park functioned as a home for her: it was a place of continuity, where her day began and ended (see also Box 3.4).

Box 3.4: Life on the streets

Relationships
Pam: 'We didn't have no way, we didn't have nothin' to eat and there ain't no missions or nothin' out there in Santa Monica at all . . . [H]e was tryin' to make a game of it, you know. And you know singing and "A little bit further". Jokin' and laughin' and stuff.'

Being moved on
Lisa: '. . . Otis, Sue, Roger, all these people that lived on our side of the street had been there for all those months. They had been together for years. They were used to this being moved from one place to another. Linda, who had been on the street for seven years, hey, this was nothing new to them. We're just getting moved again. They'd gone from one parking lot to another. This was nothing, to say hey, you got to pack up and go.'

The street encampment
Lisa: '. . . so many people had donated so many things to us. For instance, Thrifty's with all the health supplies, first aid. We had the grills that people had donated. Those beautiful grills that the church people donated. The tents that the church people donated. Fred Jordan's gave me mine.'

Panhandling
Lisa: 'My preference is I'm going to go panhandle, I'll make more money doing that. And so for a couple of days, I made forty dollars. Ah hey, I'm good. Then all of a sudden it went from forty dollars to almost ten, twelve dollars. It was a boom, a real drop.'

Cited in Rowe and Wolch 1990: 194, 195, 196, 197

Such relationships also provide a sense of identity with people even establishing traditional family units. Like the homed, however, the homeless can also become involved in abusive relationships. The parallels continue because, like homed women who experience domestic violence, homeless women also often stay with violent partners because the relationship serves the logistical and material functions of a home base. With no alternative forms of support, many women would rather put up with predictable patterns of abuse than face the unknown dangers of the street alone.

While for some homeless people a relationship becomes a surrogate home, for others groups or networks of other people perform the same role. Rowe and Wolch (1990) found that on Skid Row some homeless people organized informal communities in the form of street camps. These had names, such as Justiceville and Love Camp, and were used as places where possessions and messages could be left and as social gathering points where news and information could be passed on. Although these camps were not proper physical structures with walls and a roof, Rowe and Wolch (1990) argue the homeless women often preferred them to hostels, shelters and other short-term accommodation provided by the authorities because they provided a sense of identity and belonging and some degree of privacy and control for their 'residents'. In contrast, most shelters and hostels have strict rules to make homeless people conform to ideals of appropriate home life. In a study of homeless policy in the late nineteenth and early twentieth centuries in the State of New York, USA, Veness (1994) describes how shelters attempted to reproduce middle-class ideologies of cleanliness, order, privacy and material comforts through the imposition of rules about personal hygiene, furnishings, appropriate clothing and behaviour.

Street encampments vary in size and most are temporary because they are dispersed by the police. Ruddick's (1996, 1998) work on young people in Los Angeles, USA, provides a good example of how, in such ways, homeless people are marginalized not only in space but also through space. Her study focused on about 200 punk squatters who rejected the shelters and services provided by the authorities for homeless youth and instead created their own oppositional space. They did this by occupying two derelict residential estates in Hollywood Hills known as Doheny Manor and Errol Flynn Manor, which became sleeping places, and by tactically appropriating and subverting bars on Hollywood's sunset strip as places for gigs, and marginal spaces such as cemeteries as places to gather. In doing so, Ruddick (1996, 1998) shows how the homeless youths were able to transform the meanings both of themselves, and of the spaces to which they were limited, through their use of these environments. However, the authorities eventually demolished the buildings they were occupying. The loss of this strategic space which was controlled by the young people led eventually to the demise of the squatting movement and its substitution by shelters and services defined and controlled by the authorities.

Veness, Rowe and Wolch and Ruddick's work clearly demonstrates that just because the so-called homeless do not have a house does not mean that they do not create homes. Rather, the homeless develop their own matrices of social relations and alternative spaces which provide many of the meanings of home such as continuity, identity, privacy and control (see section 3.3), which the homed attribute to conventional forms of housing.

■ Summary

- There are many different definitions of homelessness, some of which recognize that there is a continuum between being homed and being homeless.

- Problems of defining and recording homelessness mean that it is hard to calculate how many people are without a home. Estimates suggest that it is a growing social problem.

- There are wide variations between places in terms of the service provision for those in need because authorities and local communities adopt varying approaches to homelessness.

- The rise in homelessness has been attributed to global processes of economic restructuring, national welfare reforms, and individualization.

- Homeless people create relationships, social networks and appropriate spaces which take on many of the meanings of home (e.g. abode, identity, roots), which the homed attribute to conventional forms of housing.

Further Reading

- There are a number of books focusing on gender and housing design, notably: Hayden, D. (1984) *Redesigning the American Dream: The Future of Housing, Work and Family Life*, Norton, New York; Matrix (1984) *Making Space: Women and the Man Made Environment*, Pluto Press, London, and Roberts, M. (1991) *Living in a Man-Made World: Gender Assumptions in Modern Housing Design*, Routledge, London. These issues are also nicely summarized in overview papers by Hayden, D. (1980) 'What would a non-sexist city be like? Speculations on housing, urban design and human work', in *SIGNS: Journal of Women in Culture and Society*, 5, 170–87; Boys, J. (1984) 'Is there a feminist analysis of architecture?' *Built Environment*, 10, 25–34, and England, K. (1991) 'Gender relations and the spatial structure of the city', *Geoforum*, 22, 135–47. It is important to remember, however, that this work has been subject to a number of criticisms which are nicely outlined by Boys (1998) herself in 'Beyond maps and metaphors? Rethinking the relationships between architecture and gender', in Ainley, R. (ed.) *New Frontiers of Space, Bodies and Gender*, Routledge, London.

- The meanings of home are the subject of three key papers: Saunders, P. (1989) 'The meaning of "home" in contemporary English culture', *Housing Studies*, 4, 177–92; Saunders, P. and Williams, P. (1988) 'The constitution of the home: towards a research agenda', *Housing Studies*, 3, 81–93; Sommerville, P. (1992) 'Homelessness and the meaning of home: rooflessness or rootlessness', *International Journal of Urban and Regional Research*, 16, 529–39. These should be read in conjunction with work that explores actual experiences of home, for example: Dobash, R. and Dobash, R. (1980) *Violence Against*

Wives, Open Books, Shepton Mallet; hooks, b. (1992) *Yearning: Race, Gender and Cultural Politics*, Turnaround, London; and Elwood, S. (2000) 'Lesbian living spaces: multiple meanings of home', *Journal of Lesbian Studies*, 4, 11–28. The interdependencies of the home and the workplace are evident in Hanson, S. and Pratt, G. (1988) 'Reconceptualising the links between home and work in urban geography', *Economic Geography*, 64, 299–321; Massey, D. (1998a) 'Blurring the binaries? High tech in Cambridge', in Ainley, R. (ed.) *New Frontiers of Space, Bodies and Gender*, Routledge, London, and Smith, G.D. and Winchester, H.P.M. (1998) 'Negotiating space: alternative masculinities at the work/home boundary', *Australian Geographer*, 29, 327–39.

- The concept of the moral economy is explored in chapters in Silverstone, R. and Hirsch, E. (1992) *Consuming Technologies: Media and Information in Domestic Spaces*, Routledge, London. A specific example of the domestication of the television within the home is provided by Spiegel, L. (1992) *Make Room for TV: Television and the Family in Postwar America*, University of Chicago Press, Chicago.

- The negotiation of household spatialities, temporalities and social relations is captured in a number of studies, such as Wood, D. and Beck, R.J. (1994) *Home Rules*, Johns Hopkins University Press, Baltimore; Sibley, D. (1995b) 'Families and domestic routines: constructing the boundaries of childhood', in Pile, S. and Thrift, N. (eds) *Mapping the Subject: Cultural Geographies of Transformation*, Routledge, London, and Valentine, G. (1999c) ' "Oh please, Mum. Oh please, Dad": Negotiating children's spatial boundaries', in McKie, L., Bowlby, S. and Gregory, S. (eds) *Gender, Power and the Household*, Macmillan, Basingstoke. An overview of contemporary changes in the 'family' is provided by Stacey, J. (1990) *Brave New Families: Stories of Domestic Upheaval in Late Twentieth Century America*, Basic Books, New York.

- There are lots of geographical studies of homelessness. Some of the most influential work has been produced by three key authors: Michael Dear, Jennifer Wolch and April Veness. Examples of their many books and papers include Dear, M. and Wolch, J. (1987) *Landscapes of Despair From Deinstitutionalisation to Homelessness*, Princeton University, Princeton, NJ; Wolch J. and Dear, M. (1993) *Malign Neglect: Homelessness in an American City*, Jossey-Bass Publishers, San Francisco, CA; Rowe, S. and Wolch J. (1990) 'Social networks in time and space: homeless women in Skid Row, Los Angeles', *Annals of the Association of American Geographers*, 80, 184–204 and Veness, A. (1993) 'Neither homed nor homeless: contested definitions and the personal worlds of the poor', *Political Geography*, 12, 319–40. A more specific (and excellent) study of homeless youth is found in Ruddick, S. (1996) *Young and Homeless in Hollywood*, Routledge, New York.

■ Exercises

1. Make a list of the ways in which you think conventional housing serves, and does not serve, the needs of employed women and their families, and the disabled. Bearing these in mind, now try to draw a non-sexist and non-ableist housing design. What are the principles behind your design and what might its limitations be? How realistic or idealistic are such utopian housing projects?

2. In a small group, imagine that you live in a small country, Noma. There has just been a revolution and the new government wants to create a more equal and harmonious country. Prior to the revolution, Noma was organized around nuclear family units living in individual and privatized homes (like the contemporary Western society ideal). The new government is concerned that this living arrangement may create unequal social relations and lead to unhappiness and oppression. Hold a public inquiry to decide whether the family home should be abolished as the main social unit. Choose four or five key witnesses (e.g. a homemaker, a homeless person, etc) to prepare different cases for/against the 'traditional home'. They should prepare and make brief arguments (5–10 minutes) outlining their position, based on academic reading but embellished with the character's personality and experiences. When each witness has presented their case, the audience may cross-examine them. Then take a vote on whether the traditional home should or should not be abolished.

3. Choose one object from your own home, or your childhood home (e.g. the television, the computer, the ice-maker). Write a brief account of how this is/was domesticated within the context of your household's social relations.

4. Make a list of all the 'home rules' you remember as a child. How do these compare with the home you live in now? What do these rules tell you about the social relations, and the spatialities and temporalities in your household? How did you as a child negotiate and contest such rules, and how did you appropriate, transform and secure the boundaries of rooms (or parts of rooms) to make particular spaces and times your own?

■ Essay Titles

1. Critically evaluate the assertion that the plan of a house tells us a lot about how women are expected to organize their lives.

2. Critically assess the popular saying that 'a man's home is his castle'.

3. Critically examine the use of the term 'private' in relation to the home.

4. Using examples, account for the contemporary rise in homelessness.

4 Community

■ 4.1 Community

The concept of 'community' has a very long and complex history within the social sciences, being defined, researched and theorized in very diverse and contradictory ways at different times and by different academics.

At the beginning of the twentieth century, drawing on the work of plant ecologists, a group of sociologists from Chicago developed a theory of how 'natural' communities emerge in cities. This work, although subject to frequent criticisms, for many years inspired empirical studies to locate, map and measure the cohesion of neighbourhood communities.

Other writers, however, have disputed not only this definition of community but also the longevity of neighbourhood communities, arguing, for a whole host of different reasons, that if they ever existed at all they are certainly now in decline. Indeed, the complexity and contradictory nature of these debates has led some writers to argue that community is a meaningless and over-romanticized concept.

Since the late 1980s the notion of community has been retheorized as a structure of meaning or imagining. This marks a major shift in understandings of community. This work recognizes that imagined communities are fluid and contested but are still important to their 'members' and have wider political meanings.

The many positive meanings associated with community (such as solidarity, support, etc) mean that at different moments this ideal has been mobilized in different ways by political parties, social movements and groups of citizens for their own political purposes. At the same time, however, writers such as Iris Marion Young have challenged community as an ideal, claiming that it privileges unity over difference, that it generates social exclusions and that it is an unrealistic vision.

This chapter attempts to capture all these conflicting understandings and evaluations of the worth of the 'community'. It begins by outlining the work of the Chicago School of Human Ecology and its critics and then looks at geographers' and urban sociologists' efforts to define and understand place-based or neighbourhood community. The chapter then outlines the arguments of those who have theorized the decline of neighbourhood community, and looks at why some writers argue that it is a concept with no analytical value. From here the chapter focuses on 'imagined communities', drawing particularly on the empirical work of Gillian Rose and Claire Dwyer. The penultimate section turns to community politics, examining the different ways 'communities' mobilize around different issues by considering community service and the state, communitarianism, and community-based activism (including NIMBYISM and the actions of stretched-out communities). The chapter concludes by questioning the desirability of community. The issues raised in this chapter are also important to major sections within Chapter 7 and Chapter 8.

4.2 'Natural communities'

At the beginning of the twentieth century a group of sociologists at Chicago University carried out detailed studies of where different kinds of people lived within the city. They became known as the Chicago School of Human Ecology (CSHE) because they used an analogy with plant communities to interpret the residential patterns of Chicago and to develop a theory of how 'natural communities' emerge within cities. Three men – Robert Park, Ernest Burgess and Roderick McKenzie – were responsible for advancing this school of work. Robert Park was originally a journalist and it was while tramping the streets of Chicago for stories that he developed the idea of the city as a social organism. He went to Chicago University in 1914, and, influenced by the work of Charles Darwin and plant ecologists, began to carry out research and to explore his idea that an analogy could be drawn between plant and human communities. Park sent his students out to complete detailed mapping exercises of the social and economic characteristics of the city. His research findings were published in a book, *Introduction to the Science of Society*, co-authored by Burgess. In turn, this influenced and inspired the work of a number of other authors and led in 1925 to the publication of a book called *The City* (1967, 4th edition) This was a collection of the work of the people who became known as the CSHE.

According to Darwin's web of life there is an intimate interrelationship between different biological organisms and their environments. Park believed that, because humans are also organic creatures, we must also be subject to the same general laws of the organic world as plants and animals, although he tempered this analogy with a recognition that, unlike plants and other creatures, humans are driven not only by a survival instinct, but also by social and cultural needs. Park's conceptual framework of '*human ecology*' therefore distinguished between two aspects of human life: community and society. Community, he argued, was the product of biotic activity and could be understood through ecology, while the study of society (culture, social processes of communication and so on) should be based on social psychology.

Park's theory of community was underpinned by three key concepts derived from his analogy with plants: *competition, ecological dominance* and *invasion and succession*.

- *Competition*: Park believed that, just like plants and animals, humans have an urge to survive, and in order to do this they must compete for the best places in which to live and to set up businesses. This competition, he claimed, takes place in the form of pricing mechanisms which determine land values. As a result of the operation of this market, Park argued, businesses and people are segregated into different areas of the city according to their ability to buy property or to pay rent, so that similar types of people live in similar parts of the city. This explains the development of Central Business Districts (CBD) in the heart of cities and why people with what he termed 'low economic competency' live in the least desirable parts of the city.

- *Ecological dominance*: Within different types of plant associations one species will often have a dominant influence. This means that it can affect the environmental conditions, either encouraging or restricting the development of other species. For example, the height and thickness of a tree determines how much light will get through to the area underneath it and so what types of plants can grow there. In the same way, Park argued that, within the city, the Central Business District (CBD) is dominant because it is competition between businesses who want to locate in this area which drives up land prices in the city centre and this, in turn, determines who can afford to live there and so explains the segregation of people in the surrounding residential areas. He also pointed out that some groups of people can exert dominance over others. For example, high-status neighbourhoods are dominated by wealthy people who can afford to pay high property prices or rents and so people on lower incomes are unable to move into their neighbourhoods.

- *Invasion and succession*: Plants can change the micro environments in which they live and so can generate conditions in which other plants can grow. These species can then invade the environment, eventually becoming dominant. Using this analogy, Park pointed out that invasion and succession also occurs in human communities. For example, residential areas may become invaded by some businesses, people living there then begin to move out and gradually the area becomes a commercial rather than a residential area; likewise, lower-income neighbourhoods may be invaded by middle-class groups (see Chapter 7) or an area dominated by one ethnic community be invaded by another.

Following these acts of invasion and succession, communities were then assumed to stabilize again and to maintain similar boundaries, albeit predicated on a different way of living.

Park's theory was developed by Burgess and McKenzie, who made a further comparison between the plant world and the city of Chicago. They argued that, just as evolutionary growth in the plant world causes increased differentiation of an organism's component parts and the concentration of control at the point of dominance, so too this was found in the city. They noted that where the gradient of land values fell from a peak in the centre, other social factors, such as rates of crime and mental illness, mirrored this pattern, declining outwards from the centre on the same gradient. It was also Burgess and McKenzie who introduced the concept of 'natural areas' to describe the homogenous neighbourhoods which they identified as subsystems within the larger community. They described these as '[a] mosaic of little worlds which touch but do not interpenetrate' (Park et al. 1967: 40). Burgess brought all these ideas together in his famous model of city growth: the concentric ring model (Figure 4.1).

■　4.2.1 Criticisms of the Chicago School of Human Ecologists

The work of the CSHE has been criticized on both empirical and theoretical grounds. The research inspired numerous empirical studies of other cities. When the CSHE's theory was applied to other places it did not always fit very well, communication routes and the physical landscape often distorting the concentric ring model. For example, one study of 20 US cities found that it was communication routes rather than ecological processes which influenced the emergence of communities (Davie 1937). Notably, this work suggested that industry located near railways or waterways and that, in turn, low-grade housing would develop in these areas. Other empirical studies have found that urban land values vary more within than between concentric zones (Hoyt 1939), and that, despite the homogeneity and stability of some residential groups, residential solidarity does not necessarily occur or develops in only a patchy way.

The CSHE's work has been criticized on theoretical grounds as conceptually flawed. The distinction between biotic community and cultural society was challenged by academics who argued that it is impossible to distinguish between types of human behaviour in this way (Firey 1947). Competition is not free and unconstrained and people do not necessarily make rational economic decisions about where to live. Rather, factors like culture, taste and emotional attachments can shape residential choices. This theoretical point was demonstrated through empirical examples which showed that ecological processes were distorted because particular places have symbolic or sentimental connotations which counter the rational allocation of land use. Walter Firey (1945) used the exclusive area of Beacon Hill in Boston to prove this point.

Figure 4.1 The concentric ring model. Redrawn from Park, Burgess and McKenzie *The City* (1967).

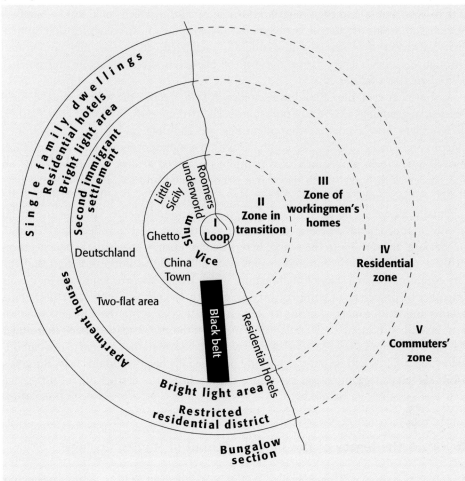

He observed that this was an inner high-status area, which, according to ecological theory, should have suffered decay, invasion and loss of status over time. He argued that it had remained an exclusive area for over half a century because it was an important historic neighbourhood with literary and familial traditions.

Other writers (e.g. Sjoberg 1960) criticized Park and Burgess for developing what was presented as a universal theory applicable to any city, from research which was actually time–space specific, being based on particular US cities at a particular moment in time. For David Harvey (1973) the work of the CSHE is problematic because of the way that it legitimates capitalist processes of economic competition.

More fundamentally, the ways in which the CSHE conceptualized social difference and urban spatiality have subsequently been overturned and are now an anathema to contemporary thinking on identity and difference (see, for example, Jacobs and Fincher 1998). Burgess and his colleagues used social distinctions like 'race' as if they were 'natural', stable, pre-given, essential categories which could just be mapped. Such assumptions are deeply problematic. Geographers now recognize that subjectivity, whether it is sexualized, gendered, racialized, etc, is socially constituted through discourse. There is no 'natural' or pre-given subject (Jacobs and Fincher 1998, see also Chapter 2).

In a similiar vein, the CSHE has been heavily criticized for the way its work defined residential groupings in terms of only one social attribute such as class or race. Rather, geographers now recognize that identities are multiple and variably positioned (Jacobs and Fincher 1998). The emphasis within contemporary social geography has moved away from naive, uncritical mappings of social groups towards a focus on the processes through which difference is constituted and the ways in which particular subjectivities are privileged or marginalized (see Chapter 1 and Chapter 7).

Finally, Burgess's model contained implicit moral judgements. Some of the social groups mapped, such as black people and prostitutes, were labelled 'problematic' or 'deviant' in contrast to the white, heterosexual 'norm'. The spatial concentration of these negatively differentiated groups was, according to Burgess, a sign that the city organism was in a state of malfunction, which would be corrected by the inevitable outcome of processes of competition and growth (Jacobs and Fincher 1998). Jane Jacobs and Ruth Fincher (1998: 6) suggest that 'In this sense the Chicago sociologists actually produced an assimilationist model of the city. It was assumed that under the right conditions these 'deviant' groups would conform to the characteristics embodied by the suburbs and then the city would become fully healthy.' In this way, Burgess's work served to legitimize and naturalize the essentialist constructions of difference that were implicit in his model.

■   **4.2.2 The legacy of the Chicago School of Human Ecology**

Despite the many criticisms levelled at the work of the CSHE, it is important not to downplay the influence it had within geography, sociology and urban studies, particularly in the 1960s and 1970s. During this time it influenced thinking about 'community' in a number of ways (Suttles 1972):

• The term 'natural community' and the analogy between plant communities and processes of competition and segregation appealed to academics, particularly in the USA, because it appeared to fit with a general belief in the importance of unregulated economic competition. Competition, it was argued, produced differences between people and their lifestyles. These differences, in turn, produced residential segregation according to lifestyle, income and ethnicity.

• The theory appeared to show that urban neighbourhoods were not planned or artificially contrived, but emerged out of ecological processes. The term 'natural' areas was interpreted to mean that universal residential solidarity or a sense of community would develop because these areas were a grassroots phenomenon based on differences such as income and ethnicity, rather than the artificial creation of administrators. Suttles (1972: 14–15) writes: 'All these elements were in tune with one another and seemed to conjure up an almost inevitable reality in which sentiment, nature, primordialism, universality and unplanned clannishness combined in a particular social configuration which we could scarcely doubt ever existed.'

- The work of CSHE promoted a belief that communities were homogeneous and stable, albeit punctuated by periods of invasion and succession; and for a long time it inspired many empirical studies to locate, map and measure the cohesion of neighbourhood communities (see section 4.3). Indeed, Peter Jackson (1984) claims that the methodology of participant observation can be linked to the techniques practised by Park and his students.

■ **Summary**

- The CSHE used an analogy with plant communities to interpret the residential patterns of Chicago and to develop a theory of how 'natural communities' emerge within cities.

- This theory was underpinned by three key concepts derived from the analogy with plants: *competition*, *ecological dominance* and *invasion and succession*.

- The work of the CSHE has been criticized on empirical grounds because research in other cities has not always replicated the CSHE model; and on theoretical grounds, as conceptually flawed.

- The legacy of the CSHE was to inspire geographers and urban sociologists to locate, map and measure the cohesion of neighbourhood communities.

■ 4.3 Neighbourhood community

The notion of community as a positive social relationship embracing a sense of shared identity and mutually caring social relationships dates, according to Yeo and Yeo (1988: 231), back to the Latin term *communitatem* (fellowship) and has been in popular use in this way since the sixteenth century. Selznick (1992: 361–4, quoted in Smith 1999: 21) suggests that a fully realized community should include *historicity* (interpersonal bonds developed through a shared history and culture), *identity* (evident through loyalty), *mutuality* (interdependence and reciprocity between members), *plurality* (members taking part in intermediate associations or group attachments), *autonomy* (the flourishing of unique and responsible persons), *participation* (in different roles and aspects of society) and *integration* (via political, legal and cultural institutions).

Geographers have tended to add a spatial dimension to these sorts of definition. According to the *Dictionary of Human Geography*, community is 'a social network of interacting individuals, usually concentrated in a defined territory' (Johnston *et al.* 2000: 101). The scale at which this definition is evoked to represent socio-spatial relations ranges from the neighbourhood to the nation and even the globe (see Chapter 7, Chapter 8 and Chapter 9). Following the CSHE, it is at the scale of the locality that most geographical work on community has taken place. Indeed, 'community' is often used interchangeably with the term 'neighbourhood'. This stress on neighbourhood is in part due to Robert Park's theory of 'natural' areas in cities, in which solidarity is regarded as a function of living in the same place, a product of a given social-ecological system (see section 4.2).

John Cater and Trevor Jones (1989: 169) define **neighbourhood community** 'as a socially interactive space inhabited by a close-knit network of households, most of whom are known to one another and who, to a high degree, participate in common social activities, exchange information, engage in mutual aid and support and are conscious of a common identity, a belonging together' (see Plate 4.1). Four factors are identified which contribute to the emergence of neighbourhood communities. These resonate with the CSHE's claim that communities are 'natural areas':

- *Proximity*: Individuals' face-to-face social networks are assumed to be strongest in their immediate neighbourhood and to decay with distance. In other words, shared activities, relationships and support 'naturally' emerge through the accident of proximity. Research on motherhood, for example, has shown the importance of the neighbourhood social networks which women develop through their children, and the role of these social relationships in shaping their understanding of childcare (Dyck 1990, Holloway 1998). Where neighbourhood social relations do not emerge 'naturally', attempts are often made to foster them artificially – an obvious example being neighbourhood watch schemes (see Chapter 6).

- *Territory*: Humans are territorial, drawing a sense of security, comfort and identity from their immediate locality. As a result, common-sense understandings of shared space and mutual identity often emerge, which are not connected to administrative boundaries. A number of geographical studies have sought to identify and map these territorial definitions of space and to measure community cohesion. Researchers have debated how these solidarities should be gauged. Pacione (1983), for example, has identified six factors: personal attachment to the neighbourhood, friendships, participation in

Plate 4.1 Neighbourhood community epitomized by the village fête
(© Brenda Prince/Format)

neighbourhood organizations, residential commitment, use of neighbourhood facilities, and resident satisfaction. Community solidarity is usually most pronounced in 'working-class' neighbourhoods (Young and Willmott 1962). A sense of shared territory often only becomes evident when the immediate locality appears to be threatened. For example, plans to locate roads or unwanted facilities (secure units, waste disposal sites, and so on) in particular locations often mobilize a sense of community expressed through Not-In-My-Back-Yard (NIMBY) local actions (see section 4.6.3.1).

Box 4.1: 'Working-class' community

Murton, a mining community in north-east England, as described by the journalist John Pilger (1993):

'It was here I first heard the word marra [friend] and crack meaning everything from comradely talk to gossip. They are a remarkable people, preserving and expressing vividly that sense of community that is often spoken of as this society's most abiding strength.'

'It [Murton] was the archetypal pit village with its Democratic Club, Colliery Inn, ribbons of allotments producing champion leeks and pigeons. Everybody knew everybody in an easy freemasonry...Long before governments thought seriously about providing social welfare the miners of Murton were looking after their most vulnerable.'

- *Social homogeneity*: A sense of mutual interest or common experience can be important in the development of community. A number of famous studies (e.g. Hoggart 1957, Young and Willmott 1962) have shown how a shared class identity (including culture, values, language, morality, and so on), and a desire to maintain a particular way of life can underpin a sense of working-class community. These studies implicitly describe working-class community as constructed through a shared sense of male camaraderie which is derived from sharing hard and often dangerous physical work (in the mines or steel mills) and fostered through male-oriented community spaces such as the pub, the working men's club or political parties (in the UK through Labour Party politics). Here, a sense of belonging is fostered by a perceived sense of shared hardship and a pride in the ability of the community to look after its own (Dennis *et al.* 1956). In Box 4.1 the journalist John Pilger describes his impressions of Murton, a working-class mining community in north-east England. Likewise, in a study of an English town given the pseudonym Thamestown, Paul Watt (1998, see also Watt and Stenson 1998) found a strong sense of neighbourhood loyalty and belonging among young Asian men living in an area called Streetville, which he suggested was, in part, a defensive response to the racism they encountered in other neighbourhoods of the town.

- *Time*: Social solidarity or mutual affinity is something which congeals over time. This is most famously illustrated by Young and Willmott's (1962) three-year study of a working-class district in Bethnal Green, London. They found that, despite the fact that housing conditions in this area were often appalling – with no proper bathrooms or heating – there was a strong sense of 'community' identity based on kinship in this East End district. The area was characterized by families of three generations held together by a powerful mother–daughter bond. When daughters got married the newlyweds would live with her parents or nearby, so that the husbands were absorbed into the wives' wider families. The families met informally in the street and would spend weekends together. This close bonding gave them a sense of community in two ways. First, it provided a safety net in that their network of friends and family gave mutual practical support and help to each other (e.g. baby-sitting, lending money, and so on). Second, it gave them a sense of a collective identity.

However, the late 1950s and 1960s were a time of large-scale slum clearance and redevelopment schemes in which people were being moved out of terraced housing in the East End districts of London, such as Bethnal Green, into new estates in the neighbouring county of Essex. Young and Willmott (1962) followed some of the Bethnal Green residents to new homes in a town given the pseudonym Greenleigh. They found that, despite the fact that these people had been rehoused in an immaculate estate with impressive facilities compared to those in their former neighbourhood, they felt that their quality of life had actually deteriorated because they had no sense of community in the new place. In contrast to Bethnal Green, which had been a community that had developed over many generations, the community in Greenleigh had been artificially created and had not had time to jell.

■ 4.3.1 The decline of neighbourhood community?

There is a long history of theorizing the decline of neighbourhood community within the social sciences. Barry Wellman (1979) has characterized debates about its continued existence as community 'lost', 'saved' or 'liberated'.

- *Community lost*: A number of famous writers have argued that social bonds of mutuality have disappeared as result of the transition from feudalism to capitalism, from religion to scientific rationalism and from traditional to legal authority (Cater and Jones 1989). Such arguments are captured in Tonnies' (1967 [1887]) distinction between *gemeinschaft* (community) and *gesellschaft* (mass society) relationships. He defines *gemeinschaft* as a situation where community relationships are tied to social status, public arenas and bounded local territory, based on close contact and emotional ties. This is epitomized by the traditional village community (see Chapter 8). In contrast, *gesellschaft* exists where community relationships are individualistic, impersonal, more rational or contractual, private and based on like-minded individuals. According to Tonnies (1967), social relations were moving (have now moved) from *gemeinschaft* to *gesellschaft* relationships as a result of industrialization, urbanization and mass communication. He lamented what he regarded as a passing of a 'golden age' of community. For Tonnies and other 'community lost' writers such as Wirth (1938: 12), urban residents now have weak, loose-knit social relationships, which are 'impersonal, transitory and segmental', in place of social relations based on solidarity and territorial cohesiveness (see also section 4.5). They see this as contributing to individuals' sense of alienation in the city and to urban social problems (see Chapter 7). Rural communities have therefore been celebrated as the ideal against which urban societies should be compared (see Chapter 8). However, later studies have shown that urban areas may also have positive features which rural areas lack (Frankenberg 1966), and that the presence or absence of 'community' is also contested within rural spaces.

- *Community saved*: In contrast to those who regard community as lost, other commentators have argued that neighbourhood communities have survived and prospered in contemporary industrial societies. The proponents of 'community saved' arguments claim that communities persist because they remain an important source of support and sociability. Examples include Gerald Suttles' (1972) research, which found evidence of strong networks mainly among poor and ethnic minority groups, and Gans's (1962) famous study of an 'urban village'. These writers have praised the persistence of solidarity, communion and territorial cohesiveness.

- *Community liberated*: Barry Wellman (1979) criticized the 'community saved' studies for their narrow focus on the neighbourhood community, arguing that they have only concentrated on communal solidarities in neighbourhoods and through kinship systems. Wider social linkages have either been ignored or have been conceptualized as radiating outwards from the neighbourhood. Those in the 'community liberated' camp reject the local area or neighbourhood as the basis for examining community. Wellman (1979) argues that 'community liberated' starts from the basis that: (a) the separation of home, work and wider kinship networks means that urban dwellers have multiple social networks with weak solidary attachments; (b) high rates of residential mobility weaken community ties and hamper the development of new ones; (c) the development of transport and communications has made it easier for individuals to maintain wider primary ties; (d) the scale, density and diversity of the city and nation state increase the possibilities for individuals to develop multiple loosely bound networks; (e) this heterogeneity of the city and the spatial dispersal mean that it is less likely that individuals with whom an urban resident is linked will themselves be part of close-knit communities. As a result the 'community liberated' argument suggests that primary ties are often spatially dispersed among multiple sparsely knit social networks.

In a study of residents of East New York in Toronto, Canada, Wellman (1979) concludes that, where kin and neighbours have been lost as the basis of strong intimate social ties, they have been replaced by friends and co-workers. These people often live in very different neighbourhoods and are in different social positions, contact being kept both through face-to-face meetings and also by telephone. While some of these friends are only seen socially, others provide a broader range of support and help; some close relations can be counted on in emergency and others cannot; and different contacts are seen over a wide range of time intervals. In other words, he argues that 'East Yorkers tend to organise their intimate relationships as differentiated networks and not as solidarities' (Wellman 1979: 1225). Wellman (1979) claims that, because researchers focus on neighbourhoods, they appear to find densely knit networks that are assumed to be characterized by communal solidarity, yet he suggests that, if researchers step back and also include all those with whom residents are in touch, then the apparent neighbourhood communities may be seen only as clusters of relations in much more complex, sparsely knit and loosely bound networks. He goes on to argue that, as a result of not being an unambiguous member of one solidary community, but rather a member of multiple networks, individuals can experience a disorienting loss of identity: a sense that it is no longer clear to which group they belong. This, he argues, may account for why some studies appear to show that people have lost a sense of community.

■ Summary

- There is a long history of theorizing the decline of neighbourhood community within the social sciences.

- Different authors have argued that social bonds of mutuality have disappeared as a result of transitions from feudalism to capitalism, from religion to scientific rationalism and from traditional to legal authority.

- Other commentators have argued that neighbourhood communities have survived and prospered in contemporary industrial societies.

- Others still reject the neighbourhood as the basis for examining community, conceptualizing it instead in terms of wider socio-spatial linkages.

■ 4.4 A meaningless concept?

The complexity and contradictory nature of debates such as those above (section 4.3.1), about the continued existence of 'neighbourhood community', have provoked other writers to question whether it actually has any meaning as a concept at all. It is argued that communities can exist without a territorial base or that neighbourhoods can have no sense of communal ties or cohesion; that community has no analytical value because it means so many different things to different people and that it is probably only a romanticized concept anyway.

■ 4.4.1 Communities without propinquity

Academics have criticized the notion that neighbourhood necessarily equates with community and the assumption that when community is not found in a neighbourhood it does not exist at all, arguing that 'neighbourhood community' implies a spatial determinism or casual relationship between spatial proximity and social ties or social cohesion (Wellman and Leighton 1979). These writers point out that there are many examples of 'communities' – defined as mutually interactive and supportive groups of people – which have no residential base. Personal mobility and the development of modern forms of communication such as mobile phones, faxes and the Internet mean that people can develop social networks beyond their immediate locality and are more easily able than in the past to maintain these relationships (which are not necessarily face-to-face) over greater distances (Davies and Herbert 1993).

Communities which are not predicated on space have been termed by Webber (1963: 23) **'communities without propinquity'**, while John Silk (1999: 8) labels them 'place-free' or 'stretched-out' communities. These forms of social relations which are based on shared activities, interests or beliefs and are the product of the intentional choices of their members rather than the accident of place, are seen by some feminist critics as less oppressive to women than traditional neighbourhood communities because they enable women to escape the restrictions of family and place which often define and limit women's lives. Such communities may well develop some form of non-residential spatial expression too. For example, in many towns and small cities there is no residential base for lesbian and gay social networks, yet these 'stretched-out communities' often develop some form of territorial base – albeit an 'invisible', transient or temporary one – in the form of bars or social venues (see Chapter 7).

So-called 'virtual communities' are one example of new forms of social relations being created by new technologies that are free of the limitations of place. Cyberenthusiasts argue that information and communication technologies (ICTs) enable users to meet those who share their ideas and interests regardless of geographical barriers of distance and time zones. They are predicted to be an antidote to loneliness and to offer '[n]ew liquid and multiple associations between people . . . new modes and levels of truly interpersonal communication' (Benedikt 1991: 123). Because the social relationships formed in this way are a product of choice rather than luck or spatial proximity, cyberenthusiasts claim that they escape the difficulties of earlier restrictive or repressive forms of community (Willson 1997) and that participants have a stronger sense of social cohesion and commitment than people in face-to-face off-line communities. As Clifford (1988: 13) has commented, 'Difference is encountered in the adjoining neighbourhoods, the familiar turns up at the end of the earth' (Clifford 1988: 13). Indeed, on-line communities of interest can use e-mail campaigns and list servers to mobilize geographically dispersed groups or 'global communities' to pursue common political aims (Riberio 1998) (see Chapter 9). Sherry Turkle (1996) describes them as an attempt to re-tribalize in a new space, while Howard Rheingold (1993) imagines a 'global civil society' with a shared consciousness. In other words, ICT is claimed to be turning community from a local to a global concept (Kitchin 1998).

One of the most prominent advocates of these new forms of 'community' is Howard Rheingold (1993). A member of the WELL (Whole Earth 'Lectronic Link) bulletin board which has several thousand participants (see Box 4.2), he claims that the

---

Box 4.2: Virtual community: some biographies of WELL users

I am a self employed productivity consultant. I live out in the country overlooking the ocean near Bodega bay. The phone, fax and e-mail let me work here, and still be in the business community.

I reside in Seoul, Korea where I practice public relations for the U.S. government.

I am a physician, specialising in women's health, including contraception, abortion, and oestrogen replacement therapy after menopause. I am the medical director of an abortion clinic. I was a member of the Mid-Peninsula Free University in the 70's and organised concerts, including the Dead, Big Brother, Quicksilver, Jefferson Airplane, etc. I'm interested in philosophical/ethical issues surrounding the beginning of life and the end of life and the functional value of rituals and traditions.

I am a 19 year old college student struggling to find myself. I enjoy sitting in a field of dandelions with no socks. I spend too much time playing on my computer. I am an advertising/business major so I will be here for five years or more. I am trying to find the meaning of life . . .

I am a student from Prague, Czechoslovakia, studying in San Francisco's Center for Electronic Art computer graphic and design program.

I am a lawyer, working as a law clerk to 3 state judges in Duluth. I am 31 and single. I graduated from the Naval Academy in 1982, and the University of Minnesota Law School in 1990 . . .

I am a born-again phreak, at age 33. My modem is my life! OK, the weightlifting, the fast car, they are all fun, but the modem is the biggie! As a matter of fact, I met my husband on bbs!

I am a Japanese writer who are [sic] very much interested in ecology and the electronic democracy. I am going to spend two years with studying (joining?) ecological movement and sharing network as a tool for making the new world here in Berkeley.

I work at the only hospital dedicated to the cure and eradication of leprosy in the United States. I also spent 6 months in Romania after the December, 1989, Revolution.

Rheingold 1993: 54–5

emergence of virtual communities like his own can be explained by the erosion of informal public spaces and a sense of 'community' from our off-line worlds (see Chapter 6). For Rheingold (1993: 3), on-line social relations offer the opportunity to enjoy many of facets of 'real' social relations which are disappearing from off-line public spaces:

> People in virtual communities use words on screens to exchange pleasantries and argue, engage in intellectual discourse, conduct commerce, exchange knowledge, share emotional support, make plans, brainstorm, gossip, feud, fall in love, find friends and lose them, play games, flirt, create a little high art and a lot of idle talk. People in virtual communities do just about everything people do in real life, but we leave our bodies behind. You can't kiss anybody and no-body can punch you in the nose, but a lot can happen within those boundaries. To the millions who have been drawn into it, the richness and vitality of computer-linked cultures is attractive, even addictive.

Perhaps somewhat ironically, however, those members of WELL who live in driving distance of the San Francisco Bay area have consummated their on-line community by holding face-to-face meetings in the form of picnics and parties.

While Rheingold's (1993) study focused on a small Californian Bulletin Board service, other writers have portrayed different forms of cyberspace as communities including Usenet groups (Baym 1995), Geocities (Bassett and Wilbert 1999, see also Chapter 7) and immersive textual environments such as MUDS and MOOS (Reid 1995, Turkle 1996). As Sherry Turkle (1996: 244) notes: 'Women and men tell me that the rooms and mazes on MUDS are safer than city streets, virtual sex is safer than sex anywhere, MUD friendships are more intense than real ones, and when things don't work out you can always leave.'

In contrast to these cyberenthusiasts, cybercritics dismiss the notion of 'virtual community', warning against the dangers of assuming that communication equals community. They point out that on-line participants are very transitory and that, because communication is disembodied, participants cannot know 'who' they are talking to. Indeed, on-line users may 'cycle through' different identities (see Chapter 2). Text-based communications are also condemned as superficial and as at best 'pseudo-communities' (McLaughlin *et al.* 1995) and their participants as potentially apathetic and disengaged from the politics of off-line life (Willson 1997). Only face-to-face meetings are regarded as being able to produce truly intimate friendships. According to cybercritics, there is no sense of social responsibility or accountability on-line – if people do not like what is being said or who is participating it is easy to log on and then off. Flaming and dissing (insults and abuse) on-line are also common. These criticisms are summed up by Sardar (1995: 787–8 cited in Kitchin 1998: 88), who writes:

> A cyberspace community is self-selecting, exactly what a real community is not; it is contingent and transient. In essence a real community is where . . . you have to worry about other people because they will always be there. In cyberspace you can shut people out at a click of a button and go elsewhere. One therefore has no responsibility of any kind . . . Cyberspace is to community what Rubber Rita is to woman.

Box 4.3: Propinquity without community

'Our sales pitch was that Bradley Stoke was going to be a new town, a new concept on community living. It was going to be like a town you see in the soap operas, and lots of people were taken in by this. But there's no community spirit here, people don't want to know their neighbours. If I had known that it was going to end up like this, then I wouldn't have bothered moving here … I tried to organise a Neighbourhood Watch scheme for my road but nobody could be bothered. That reflects the attitude around here.'

'Everything is new here and everyone is from somewhere else … There are no family networks and there are no bonds between people. You can't leave the kids with the grandparents around the corner because the grandparents are probably living miles away. People are lonely and isolated in Bradley Stoke. Rather than do anything about it they have become apathetic and inward looking … There's very little to be positive about apart from the fact that it's close to the motorway and you can get in and out of the area quite easily.'

Chaudhary 1995, *The Guardian*, 10 June

■   **4.4.2 Propinquity without community**

The notion that neighbourhood equals community has also been criticized by authors who argue that *propinquity without community* is common in contemporary Western societies. Although 'neighbourhood offers a potential territorial base it is neither sufficient nor necessary to ensure common goals, common action and a common identity' (Cater and Jones 1989: 168). Indeed, people who work together spend more time in close proximity to each other than those who live in the same street (Etzioni 1995). Box 4.3 highlights the example of Bradley Stoke, a new town near Bristol in the UK. Here, despite the fact that the houses have been designed and laid out in neat cul-de-sacs to foster a sense of togetherness, residents complain about the lack of community. This, in turn, can contribute to the residents perceiving a lack of informal control within the neighbourhood, which can heighten their sense of vulnerability and fear of both personal and property crime (Valentine 1989).

The very developments in communication and transport which are credited with creating the possibilities for the emergence of communities without propinquity have also been criticized for the prevalence of propinquity without community. For example, some commentators have suggested that ICTs contribute to the loss of community, rather than being the solution to the problem. By encouraging people to become more home-centred, using ICTs to communicate with friends, go shopping, and so on, new technologies are keeping people off the streets. This, in turn, leads to the abandonment and demise of 'public' space and the creation of a more polarized society (see Chapter 3 and Chapter 6). As McCellan (1994: 10) cited in Kitchin (1998: 90) puts it:

> rather than providing a replacement for the crumbling public realm, virtual communities are actually contributing to its decline. They're another thing keeping people indoors and off the streets. Just as TV produces couch potatoes, so on-line culture creates mouse potatoes, people who hide from real life and spend their whole life goofing off in cyberspace.

■ 4.4.3 Analytical value?

Other critics have claimed that 'community' is a *meaningless word* because it describes so many different things: a sense of belonging that may or may not be attached to territory, prisons, schools and other total institutions (see Chapter 5), non-workplace relationships, face-to-face relationships, and so on (Stacey 1969). Hillery (1955) found 94 different definitions in the academic literature. As Buttimer (1971) points out, 'there are few concepts in the sociological literature which have been so variably and ambiguously defined'. As a result, Cater and Jones (1989: 170) suggest that 'it is a verbal ragbag which can mean anything to anyone and therefore has very little descriptive, still less analytical value'.

■ 4.4.4 Romanticized vision

It is difficult to develop theoretical generalizations about community from particular studies of individual neighbourhoods because they are non-comparable and they often represent romanticized portraits of social relationships (Bell and Newby 1971, Rose 1990). The emotive power and warmth of community relations is commonly held up as a social ideal and a resource (e.g. Hoggart 1957), especially in relation to working-class communities (see section 4.3). Yet, as Cater and Jones (1989) point out, these neighbourhoods were not necessarily stable. They write, 'Historically, the stereotype is one of a lost Golden Age of settled working-class communities undisturbed for generations until the post-war onslaught of urban renewal, suburbanisation and mass culture. While . . . it is unarguable that community was a highly valuable working-class resource, it is also true that the nineteenth and early twentieth centuries were times of great upheaval for much of the working class' (Cater and Jones 1989: 174). At this time commercial and industrial expansion into traditional working-class neighbourhoods in the UK meant that the poor were commonly evicted, while other households who were unable to pay the rent would 'shoot the moon' (move furniture from apartments under the cover of darkness before the landlord collected the rent). This instability in the local population must have had a disruptive impact on community formation (Stedman Jones 1983).

There is also some evidence that, contrary to the sentimentalized imagery of working-class mutual aid, these social relationships were not always fraternal and social but could rather be characterized by various forms of oppression and structural inequalities in terms of class, patriarchy, etc, which constrained individuals' rights to choose their own way of life. Indeed, studies of working-class communities have tended to romanticize male camaraderie (section 4.3) and to obscure the experiences and contribution of women (a notable exception being the work of Young and Willmott 1962). Meg Luxton's (1980) research on women in a Canadian aluminium smelting town and Kathie Gibson's (1991) work on gender relations in an Australian mining community go some way to redressing this wrong, however.

Cornwall (1984) argues that so-called classic community studies, such as that by Young and Willmott (1962), represent communities in a harmonious way because the respondents interviewed in the research had a vested interest in describing their community in glowing terms. She points out that these people were experiencing a period of rapid social change, with all the uncertainty and anxiety which that entails, therefore, she argues, it is not surprising that they should hanker after their lost way of life and recall it with nostalgia in terms of stability, security and harmony (Cornwall 1984). In contrast, in Cornwall's (1984) own research in the East End of London, respondents recalled a past of economic struggle and of brutal social relations (especially men's treatment of women). Their accounts bear little resemblance to the images of warmth and sympathy that are represented in Young and Willmott's (1962) study.

There is also little indication in the so-called classic studies of the extent to which 'community' might have had different meanings for different members and the extent to which the meaning of community may have been negotiated, contested and reworked by particular groups in different times and places. Rose (1990), for example, observes that even though the 1920s working-class community of Poplar, London, was characterized by neighbourliness and mutual aid, the elderly were often isolated and conflicts were still common between different streets and between children/youth and adults.

In the face of these criticisms and those cited in sections 4.3.1, 4.4.1, 4.4.2 and 4.4.3 above, there would appear to be some evidence to suggest that 'The local urban community is a *romantic vision* which, if it ever existed at all, is now well down the path towards oblivion' (Cater and Jones 1989: 170).

■ Summary

- Debates about the continued existence of 'neighbourhood community' (in 4.3.1) have provoked the question of whether it has any meaning as a concept at all.

- Critics of 'neighbourhood community' point out that communities can exist without a territorial base and that neighbourhoods can have no sense of communal ties.

- Others argue that 'community' has no analytical value because it means so many different things and that it is over-romanticized.

■   **4.5 Imagined community**

In the late 1980s and 1990s the concept of 'community' began to be retheorized. In his book *Imagined Communities*, Benedict Anderson (1983) argued that communities are not based on fact or territory but are mental constructs. Focusing on the national, rather than a local scale, Anderson proposed that all nations are **'imagined communities'** (see Chapter 9), '[i]magined because the members of even the smallest nation will never know their fellow-members, meet them, or even hear of them, yet in the minds of each they carry the image of their communion' (Anderson 1983: 15) and imagined because people within a nation often have a deep sense of comradeship or identity with others even though, in reality, there may be exploitation and inequality between fellow citizens. Although Anderson (1983) conceptualizes communities in terms of structures of meaning, he is quick to point out that these have very real and material consequences, noting, for example, that the emotional pull of 'the nation' is so strong that people will fight and die for these 'imagined communities'.

Conceptualizing 'community' not as a social structure, but as a structure of meaning, which can be problematized in terms of who defines it, how and why, is, Gillian Rose (1990) argues, also a useful way of thinking about social relations at scales other than the nation. She points out that, although 'communities' may be imagined, at the same time it is important to remember that they are not idealist because imaginings are grounded in specific social, economic and political circumstances. She therefore defines imagined communities as 'a group of people bound together by some kind of belief stemming from particular historical and geographical circumstances in their own solidarity' (Rose 1990: 426). This argument is best illustrated through Gillian Rose's (1990) own study of Poplar in the East End of London in the 1920s and Claire Dwyer's (1999a) work which demonstrates how ethnocultural diasporas represent another form of imagined community made up of interlinked place-based communities which cross-cut national boundaries.

■   4.5.1 Imagining working-class community in Poplar, London, UK

In interviews recalling the 1920s, residents represented Poplar in East London to Gillian Rose (1990) as resembling a large social and convivial village, recalling that everyone knew everyone else and that people helped each other out and shared resources. There were strong family networks amongst women and a well-developed informal economy. This neighbourliness had a moral basis too – church attendance was high. However, interviewees also recalled social distinctions between different streets, the isolation of the elderly and that inter-neighbourhood conflicts, particularly between children/youth, were common.

The Poplar Labour Party at the time placed great emphasis on its local roots, ensuring that its candidates were all local residents, while its policies reflected the sense of neighbourliness outlined above. In 1921 all but six of its councillors were jailed for spending the rates (a form of tax) on local poor relief. Most of the residents of Poplar strongly supported their action, which furthered a local sense of communion, pride and shared identity. Rose (1990) therefore argues that Poplar in the 1920s was an imagined community. She writes that 'its collective imagining was based upon specific, materially grounded cultural and political discourses: inter war working-class mutual aid, the localism of neighbourliness, a certain form of political organisation, and the democratic socialism of the Independent Labour Party . . . These came together in a political act the very essence of which was its civic loyalty, and thus was neighbourhood loyalty overlaid with a borough-based identity and Poplar borough imagined as a community' (Rose 1990: 433). It is important to recognize, however, that this was a process that was not inevitable and, indeed, neither was it permanent. The death or removal of many of the politicians in the late 1920s and the increasing control of the state over local government saw a waning of this sense of unity.

■ 4.5.2 Imagining the British Asian Muslim community

Claire Dwyer (1999a) suggests that 'community' is at the heart of the politics of multi-culturalism (i.e. minority communities are seen as having shared customs, place of origin, and so on, which are defined in opposition to a hegemonic national community which is often seen as homogeneous). It is also central to anti-racism (i.e. it is used to define collectivities on the basis of shared experiences of racism). Nevertheless, Dwyer points out that 'ethnic communities cannot be imagined as existing in an organic wholeness with self-evident boundaries' (Dwyer 1999a: 54). Rather, she explains, they are better understood as imagined communities whose very boundaries and norms are in an ongoing process of negotiation and contestation.

Drawing on the experiences of young Muslim women from a small town in southern England, Dwyer explores how discourses of 'community' are negotiated by young British Muslim women in the construction of their identities. All of the girls in her study described themselves as part of a local 'Asian community' – which they imagined in opposition to British society, on the basis of its different religious practices and cultural values, although they also considered themselves British (see Dwyer 1997, and Box 4.4). This gave them a sense of security and freedom from racism in contrast to their imaginings (and indeed experiences) of other places as white, which they regard as hostile and where they feel 'out of place'. Yet this imagined community also had drawbacks in that the girls felt under constant surveillance to ensure that they behaved in an appropriately gendered way within its spatialized boundaries. Paradoxically, then, this community is imagined as a place of both safety and oppression.

The girls also sought to connect themselves to a wider globalized Muslim identity (often in opposition to narrow constraints of the local Pakistani community) by

---

Box 4.4: Discourses of community

Habiba: 'Well within the Asian community, Hertfield does cater for it. And the different religions such as Hinduism, Sikhism and all things like that, there are things like halal meat shops. You know like normal everyday foods that we need, that we eat, that shall I say British society doesn't . . . you know it doesn't really cater for us. But within the community, with the shops around it, we're really helping ourselves. I think in a way, by producing things that help us and encourage us.'

Nazreen: 'But when we got out of Hertfield, right, all these people, right, they look' [laughs].
Zhora: 'They stare at you really badly.'
Husbana: 'They just generally see you in Asian clothes or a scarf on your head, or you know, your colour, I suppose. And you feel out of place, well I feel out of place.'

Alia: 'If we go out you know someone always sees you and "Oh God look at her she's out there on the street, let's go and tell her parents." [agreement] And you get home, and before you get home, the gossip is around the whole town you know . . . I mean even if you're not doing anything wrong . . . People are just looking for an excuse to wag their tongues about.'

Rozina: 'Yeah if you just walk down the street and you've got trousers on, and one lady said, "I saw you", and she said, "I saw S's daughter and she was wearing trousers", and that's all they do, they gossip.'

Dwyer 1999a: 58–9

Anita [Sri Lankan Buddhist and Trinidadian Muslim parents, doesn't fit in with 'Asian' or 'White' oppositional communities at school even though she is a Muslim]: 'I don't belong anywhere. I'm just by myself. In our class there are white people, and there are Asian people and then there is me.'

Dwyer 1999a: 61

---

using their knowledge from reading and studying Islam to challenge some of their parents' restrictions on their behaviour. Yet within these imaginings of community there were also considerable differences between the girls. While they all identified as Muslims, the religion varied in its meaning and significance to them. Some were devout and practised their faith, others did not. Two of the girls, one from a Sri Lankan Buddhist and Trinidadian Muslim background, and another, whose father was Pakistani and mother an Afro-Caribbean convert to Islam, felt that they did not fit into this imagining of community because it was predicated on an assumed opposition between Asian and white, Muslim and non-Muslim.

These two different examples (4.5.1 and 4.5.2) both demonstrate that communities are imagined through debate and dialogue. Although these structures of meaning are fluid and contested, they are important to their 'members' and have wider political meaning.

■ **Summary**

- The concept of 'community' has been retheorized as a mental construct or imagining.

- Communities are imagined because members carry an image of communion even though they will never know all their fellow members, and because they have a sense of comradeship despite unequal and exploitative relationships between them.

- These structures of meaning are fluid and contested but are still important to their 'members' and have wider political meanings.

■ **4.6 Community politics**

The many positive connotations associated with the term 'community' – such as comradeship, solidarity, mutual support, and so on – mean that it is a concept which has been mobilized in different forms by political parties, social movements and individual groups of citizens, to achieve different ends at various moments in time.

■ 4.6.1 From mutuality to community 'service' and the state

Yeo and Yeo (1988) have traced the use of the term 'community' within the UK. They identify three key distinctions in the way that the concept has been mobilized: *community as mutuality*; *community as service* and *community as state*.

- *Community as mutuality*: Socialists in the UK in the mid-nineteenth century used community to represent a vision of supportive mutual social relations. 'In the socialist community, the family principle was to be extended beyond blood ties' (Yeo and Yeo 1988: 233). This vision of a different social order involved enlarging the home to create single communities with shared interests rather than individual family units with separate or competing interests (Yeo and Yeo 1988).

- *Community as service*: Also in the mid- to late nineteenth century, the British middle classes began to develop a different version of community – as service.

Yeo and Yeo (1988) argue that community service was particularly important to professionals such as doctors and lawyers, while middle-class women used public work or 'community service' as a way of escaping the home (see also Chapter 3). This notion of community as a service provided by middle-class philanthropists to the poor began to displace the dominant understanding of community as mutuality. It was a view of community imposed from above, that was based on unequal social structures, as opposed to the community as mutuality created within a framework of more ethical relationships. A different example of women's community work, what it means to those involved, and the possibilities it offers for retheorizing public/private, home/work dualisms is found in Moore Milroy and Wisner's (1994) study of Kitchener-Waterloo, Ontario, Canada between 1900 and 1980.

- *Community as state* in part emerged to displace the dominant understanding of community as service. Yeo and Yeo (1988: 246) explain that 'In an age of majority democracy (from 1867 but particularly after 1918 when all men over the age of 21 and some women over 30 had the vote) the community could mean much more than simply the state and its inhabitants. Now the community could be seen as the citizenry, possessing, determining and licensing the legitimate interests of the nation.' The term 'community', which resonates with warmth and mutuality, has often been appropriated and attached to state projects which are compulsory and punitive like the community tax, community schools, community care, community development projects, community policing, and so on (see also Chapter 6). A more positive example is found in the work of Jeremy Seabrook (1984), however. He traced an imaginative initiative taken by the local authority in Walsall, West Midlands, UK, to tackle its housing problems and local alienation by decentralizing its neighbourhood policy in order to give tenants living in local authority housing estates control over the places in which they lived. The boundaries of the neighbourhood communities within which these Neighbourhood Offices were located were defined by the people who lived there, rather than being established according to existing administrative divisions. In this way, the local state attempted to mobilize an ideal of community for social and political ends.

4.6.2 Communitarianism

Communitarianism is a social movement which sees community as a resource that can be used to create a more balanced and democratic society. In the eyes of communitarians, 'community' has been undermined and weakened in contemporary Western societies by liberalism, which is understood to have promoted individualism and self-determination at the expense of social obligations and the general 'good of society'. In other words, people are characterized as selfishly pursuing their own

interests and being obsessed with their own rights rather than recognizing their responsibilities and obligations to others. As a result, communitarians argue that the moral
fabric of communities in contemporary Western societies is coming apart at the seams.
The breakdown of traditional nuclear families, a rise in criminality, drug-taking, welfare dependency, cultural conflicts, economic insecurity, social disorder, and political apathy are all attributed to the fact that institutions and political processes are
being corrupted by the rise of individualism (Etzioni 1993, Bowring 1997).

Communitarians regard community as the solution to a range of 'real' and 'perceived' problems in contemporary Western societies (Low 1999). For example,
communities are imagined to draw on social ties to encourage members to abide by
shared values such as 'not littering the street with trash' and to 'police' themselves
by monitoring each other's actions and chastising those who violate these shared
codes of behaviour (Etzioni 1993). In these ways, they are assumed to produce their
own moral order (see also Chapter 6), and to have a moral voice (i.e. individuals
will feel guilty and ashamed if told off by other members of the community), so that
they only need to resort to the state, in the form of the police and judicial system,
when all else fails (Smith 1999). Etzioni (1993: 248, cited in Low 1999: 90)
explains that communities are 'the webs that bind individuals, who would otherwise be on their own, into groups of people who care for one another and who help
maintain a civic, social and moral order'. In other words, they offer a warm secure
alternative to the anonymity of modern society (Sayer and Storper 1997).

Communitarianism is therefore a social movement which aims, firstly, to shore
up communities in the face of their perceived decline, and secondly, to change political representation to enable the voices of communities to be heard and to let them
determine policy (Low 1999). Community in these terms is identified at a range of
scales from families, neighbourhoods and regions up to the level of the nation state.
It is a vision of a 'community of communities' (Etzioni 1995: 31). Its proponents,
such as Etzioni (1993) believe that decision making and managing problems should
be done in communities at the smallest possible scale, and that the state should only
intervene when communities cannot sort out their own problems. This has some re-
sonance across the political spectrum. Both Bill Clinton, when he was president of
the USA, and the British prime minister, Tony Blair, have talked about the importance of communities rather than individuals or class, while communitarian ideals
also appeal to right-wing politicians in the USA, like Ross Perot and Pat Buchanan,
because of the focus on 'taking back Government' (Low 1999).

Yet, Etzioni's (1995) idea of a 'community of communities' envisages not only
strengthening small-scale communities, but also ensuring that they feel part of the
supracommunity at the national scale. In this respect, he believes that the over-
arching community must have a set of shared values such as democracy, individual
rights and mutual respect. In other words, there must be mutual acknowledgement,
equality and a recognition of differences within and between communities at the
national scale. This conceptualization partially reflects Doreen Massey's (1991)
vision of a less defensive, more outward-looking 'global sense of place', in which

she shows that place-based communities are not static and fixed but rather are produced and remade through globalized webs of connection, which link the local and global (see Chapter 1 and Chapter 9).

One way forward communitarians suggest is to identify communities which exist and to try to strengthen them, for example through voluntary work. Etzioni (1991) imagines this in some form of national domestic service akin to the American Peace Corps or the British Voluntary Service Overseas (Silk 1999). In addition, communitarians promote a range of policies such as giving more family leave, providing grants for voluntary groups and even decentralizing government policy to community organizations.

Etzioni has, however, been criticized for holding a rather romantic, nostalgic and idealized view of communities (see section 4.4.4) which naively contains an implicit assumption that they are all homogeneous and have a moral voice. Indeed, his conception of community is very slippery. Murray Low (1999) observes that families are referred to as communities when he is discussing divorce, neighbourhoods as communities in relation to drugs, companies as communities when labour issues arise, and the nation as a community in terms of free speech. While Etzioni wants to form communities of communities at larger scales, composed of countless face-to-face discussions, he offers no clear mechanisms for how his vision should be put into practice (Low 1999). Etzioni's vision is very anti-statist (the state always being seen as a last resort), yet it is not clear why he thinks that it is such an unsatisfactory form of regulating social life (Low 1999). Low (1999) points out, for example, that central political issues such as resource distribution and family or social policy cannot really be handled on a small scale by poorly defined communities, and that, given that it is hard to get consensus at local level, it is difficult to imagine what mechanisms could be used to get at larger-scale core values.

4.6.3 Community-based activism

Community-based activism takes many different forms, including that which is neighbourhood-based and that which is based around the shared interests of 'stretched-out communities' (Silk 1999: 8). The urban sociologist Manuel Castells (1983) has used the term **urban social movements** to describe the range of city-based social and political community groups that are involved in struggles to achieve resources, service and social change. While urban social movements have traditionally been regarded as progressive, they can also be defensive, reactionary and exclusionary (see also section 4.7).

4.6.3.1 NIMBYISM

Urban and rural spaces are full of competing and colliding interests and activities. As a result, conflicts often arise between them over the use, occupation and control of

Plate 4.2   Not-in-my-back-yard protest (© Maggie Murray/Format)

space by different groups or organizations (see Chapter 7 and Chapter 8). When neighbourhoods are faced with actions or decisions over which local people feel they have no control, or which they believe will impact negatively on their homes and lives, community activist groups often emerge to fight the perceived external threat (see Plate 4.2). These threats may be '*physical*', such as new roads, or industrial complexes, which may affect property prices, personal health and safety or the quality of the environment more generally (e.g. Allison 1986); or '*social*', in which the threat is perceived to come from a feared or 'othered' group (see Chapter 2). For example, plans to site mosques, mental health facilities, sex offender units, sites for travellers, or red-light districts within residential areas are common catalysts for activist groups to form (e.g. Hubbard 1999). Hostility towards such proposals can produce covert or overt prejudice, with 'communities' seeking to exclude undesirable 'others' and to maintain the boundaries or purity of their space (see Chapter 6). These issues are no less pressing in rural than in urban areas (Dalby and McKenzie 1997). Indeed, siting disputes in built-up environments often result in developments being displaced to rural or remote areas – where they are just as vociferously opposed (see Chapter 8).

Both physical and social threats become symbols of a common purpose and therefore of commonality or community identity. In the process, powerful discourses of the place or neighbourhood as 'home' are mobilized (Routledge 1996). For this reason, neighbourhood community activism which is aimed at resisting an external threat is often dubbed NIMBYISM (Not-In-My-Back-Yard). Sometimes these protests are based on progressive ideals, at other times they are fought to protect narrow or dominant class interests (Watson 1999). While these actions are often credited with reviving 'community spirit', the outward expression of commonality can also conceal or silence other differences between community 'members'. These points can be illustrated through the research of Dalby and McKenzie (1997) and Peter Murphy and Sophie Watson (1997).

In 1989 a local corporation proposed to quarry granite from the Kluskap mountain near St Ann's Bay in Cape Breton, Nova Scotia, Canada, for use as gravel in road construction projects in the USA. In doing so, it gave a focus to local concerns about the future of the 'community' and how traditional cultural practices could be preserved. Both the Eskasoni First Nation – Mi'kmaq (Sacred Mountain Society) – and the non-First Nation settlements – Save Kelly's Mountain Society – (SKMS) mounted opposition to the plans. The Mi'kmaq campaigned around the spiritual and cultural significance of the Kluskap as the sacred abode of the Mi'kmaq god. The SKMS challenged the sustainability of the proposals, which they regarded as a threat to the environment and wildlife. Their joint efforts were successful, and in 1992 the corporation abandoned its plans. The Mi'kmaq have since followed this victory by fighting (with the First Nations Environmental Network) to have the Kluskap designated as a Protected Area by the Department of National Resources (Dalby and McKenzie 1997).

This example illustrates how a shared concern over the potential destruction of the local environment by global economic processes mobilized a sense of commonality and shared identity among a community under threat, and encouraged them to search for alternative economically and culturally sustainable forms of local economic development. However, this apparent unity also concealed axes of difference, both between Mi'kmaq and the white residents, and within each of these two respective groups. The Mi'kmaq were split by several fissures of difference between those who held allegiance to the Grand Council, and others with an allegiance to a system of band chiefs and councils, between traditionalists and Catholics, and between different traditionalists. The white community was likewise divided, in this case between those who supported the quarry plans in an area of high unemployment and those who opposed them (Dalby and Mackenzie 1997).

Whereas the Mi'kmaq and the SKMS mobilized to fight a physical threat to their environment, in Sydney, Australia, communities have been galvanized by efforts to resist plans to construct mosques and temples in the suburbs. Community opposition to the site selection process was mobilized around both perceived physical and perceived social threats. Residents argued that the visual intrusion of distinctive temple and mosque buildings would sit awkwardly in a space where so-called 'Christian' architecture was the 'norm'. A lack of car parking, and the congestion, crowds and noise which would potentially be generated by these social meeting places were used as further justifications for NIMBY objections to the siting plans. In discussing their social fears, some community activists made openly racist comments about 'white' neighbourhoods being 'contaminated' by Vietnamese people using their facilities. In some cases, the residents were actually able to legitimate their 'racism' by invoking particular planning norms and regulations to enforce their objections to the proposed buildings (Murphy and Watson 1997).

■ 4.6.3.2 Shared interests of 'stretched-out communities'

Community activism which is not neighbourhood-based can also develop as a result of the marginalization or exclusion of the interests of particular social groups. 'Stretched-out' communities often emerge to fight discrimination, to campaign for civil, social or political rights, or to protect or gain access to public services. In other words, they are often mobilized by a desire for **social justice**.

This form of community activism is usually autonomous in that the activists are not associated with specific political parties (although they may sometimes form strategic alliances, especially to gain funding), and they operate across a range of spatial scales, some even mobilizing on a global scale via the Internet (see, for example, Chapter 9, Box 9.7). Many – though not all – have a collective style of leadership, emphasizing mutual co-operation and the sharing of responsibilities for organizing and decision making.

One example is the British Women's Aid Movement, which emerged in the 1970s in response to a growing awareness of the problem of domestic violence (see Chapter 3). At this time it was less common for women to be in full-time paid employment and therefore they were both more economically and more psychologically dependent on male partners than perhaps they are today. This problem was compounded by the way local authorities allocated housing to homeless people. At this time the authorities were generally not very sympathetic to homeless women seeking accommodation, regarding families with a male and a female partner as more deserving of public housing than other household forms. This made it very difficult for women experiencing domestic violence to leave abusive relationships. In response to this social problem, a group of women in Chiswick, London, established a refuge as a place of safety for women who had experienced domestic violence (Rose 1978).

One woman in particular, Erin Pizzey, took the initiative to mobilize other groups of women round the country to follow suit by writing a book and appearing in the media. Public demonstrations and graffiti campaigns were also used to draw the attention of the public to male violence (see Plate 4.3). Eventually, a nationwide federation of refuges was established, including specialist refuges for women from ethnic and cultural minority communities, and what became known as the

Plate 4.3   March against male violence

Women's Aid Movement developed (Rose 1978, Hague 1998). In other words, the Chiswick activists mobilized a distinct social group, women, to challenge the provision of a public service, namely housing for those experiencing domestic violence, by taking direct action to achieve change. This organization operated independently of the formal political party system and was non-hierarchical in structure, establishing a strong base at grassroots level, with each refuge having a high degree of autonomy.

While the enthusiasm of key actors is a strength of this form of community activism, the reliance on the commitment of particular individuals means that if they suffer burnout or 'campaign fatigue', the movement may falter. Likewise, the issues around which these 'stretched-out' communities are mobilized are often triggered by a particular event which sparks the momentum of activists. For example, when the British Conservative government announced its intention to include a section in the *Local Government Act* of 1988 banning the promotion of homosexuality from schools, the lesbian and gay 'community' mobilized nationally (and indeed even internationally, drawing on the support of activists around the world) to fight its imposition by organizing protest marches, petitions, direct action, and so on. After the Act was passed and the issue faded from public attention, this solidarity and momentum dissipated. These forms of community activism often wax and wane as different issues emerge as important at different times.

■ **Summary**

- The positive connotations of the term 'community' mean that it has been mobilized in different forms, to achieve different ends at various moments in time by political parties, social movements and individual groups of citizens.

- Communitarianism is a contemporary social movement which sees community as a resource that can be used to create a more balanced and democratic society.

- Communitarian visions have been criticized for idealizing communities, naively assuming they are homogeneous and have a moral voice, and for not specifying the mechanisms necessary to put their visions into practice.

- Community-based activism takes different forms, for example NIMBYISM, and actions based on the shared interests of 'stretched-out communities'.

■ 4.7 Community: a desirable ideal?

A radical challenge to the notion that community is a desirable ideal has come from the feminist writer Iris Marion Young (1990c). She criticizes the concept of community for three main reasons: it *privileges unity over difference*, it *generates exclusions* and it is an *unrealistic vision*.

- *Privileges the ideal unity over difference*: While it is understandably an admirable goal to want social relationships to be based on mutual identification, closeness and comfort, Young argues that the result of privileging unity is often to encourage people to suppress other differences. In other words, 'communities' are often predicated on one identity, e.g. class, gender, sexuality or race, which becomes a single rallying point. Yet, 'we have multiple and sometimes contradictory subject positions and are sometimes torn between identifications, often moving between identifications in different situations and places' (Pratt 1998: 26). Individuals can thus paradoxically feel both 'inside' and 'outside' a community (see Rose 1993), sometimes regarding it as a positive experience, at other times as stifling or oppressive (see the discussion of paradoxical space in Chapter 1 and Chapter 8).

- *Generates exclusions*: Ironically, the effect of attempting to secure unity and regulate differences in order to consolidate a sense of sameness or oneness is often to produce the counter-effect of fragmentation and dispersion. Community politics necessarily involves drawing boundaries between insiders

– those who are part of the community – and outsiders – those who are not. These understandings of sameness and difference are often constructed around bodily differences (see Chapter 2). Indeed, a desire to be with people like ourselves or with whom we identify, and a fear or dislike of those who are different from ourselves are often the basis of political sectarianism, bigotry, hatred and discrimination. As Cornwall (1984: 53) observes: 'There is a strong sense of community in Bethnal Green [UK], but it should be noted that where there is belonging, there is also not belonging, and where there is in-clusion, there is also ex-clusion. In East London, the dark side of community is apparent in a dislike of what is different, which finds its clearest (but no means its sole) outlet in racial prejudice.' Intolerance and parochialism are an oppressive feature of some communities. As Etzioni (1995: 146, cited in Smith 1999: 27) suggests, 'one of the gravest dangers in rebuilding communities is that they will tend to become insular and indifferent to the fate of outsiders'. An example of how 'community-building' can generate exclusions is evident in the attempt of lesbians in Sydney, Australia, to purchase a space in the inner city which could become the focal point for their 'community'. The desire of lesbian-identified transsexuals to participate in the 'Lesbian Space Project' provoked some women to argue that transsexuals should be excluded on the grounds that they were not 'real' lesbians. Rather than securing the boundaries of the lesbian community, these attempts at exclusion triggered hostilities between community members, political re-alignments and ultimately its fragmentation (Taylor 1998).

• *Unrealistic vision*: Visions of organizing society based on small face-to-face decentralized community groups are unrealistic and will never transform contemporary politics (see section 4.6.2, Chapter 3 and Chapter 8) – not least because there is never one universal shared concept of 'community' anyway. Most individuals' identities cut across several different communities (see first bullet point above). Different individual members of so-called communities therefore will always perceive and experience the community differently. Their conceptions of community may overlap but they may also contradict or sit oddly with each other (see Box 4.5).

Having criticized the concept of community, Young (1990c) then goes on to argue that instead of trying to celebrate 'sameness' we should be celebrating difference. She points out that city life is all about being together with strangers who are dif-ferent from ourselves. An unoppressive city should therefore be defined by its open-ness to what she describes as 'unassimilated otherness'. In other words, people should be able to move between neighbourhoods without being aware of where boundaries lie. Young (1990c) ends by calling for a politics of difference which celebrates the distinctive cultures and characteristics of different groups, rather than a politics of community which artificially attempts to establish unity and homogeneity between group identities.

Box 4.5: Different experiences of a lesbian 'community'

'There's a very strong sense of community I suppose because there's only a few of us we stick together whether we like each other or not sometimes. I mean I often smile 'cos like we're less choosy who we mix with because we're a small community. So I find I invite people to parties...simply because they're there and I haven't got the heart to leave them out.'

'I have found it quite difficult because part of the circle I mix with are child-free women who have made that decision for themselves and are quite definite that that's what they've always wanted and just occasionally I've felt their sort of intolerance of children and their lack of understanding of what it means to be a mother.'

'There were very different groups, there was a group of young women, all or some of whom had been in care or in trouble with the law. There was a political group and all these little factions that knew each other. I didn't realise at first that you were supposed to show allegiance to one particular group. And I got tarred with lots of different brushes, because I just spoke to everybody. Because I had a leather jacket and spiky hair I was labelled a rough dyke but at different times I was called a radical feminist. I mean all these labels, it was hopeless because some of them just wouldn't talk to me, especially some of the politicos who'd written me off.'

'I mean we used to go to gay venues and things and I did used to get looks – "Oh dear, what's she come as, she's come in drag." And I had one woman say to me once when I was all dressed up. She was being very sarcastic and it did quite get to me at the time. And I thought shall I change my image? Shall I have my hair cut? Should I do this, should I do that to fit in with them?'

Valentine 1995:104–5

■   **Summary**

- Community is being challenged as an undesirable ideal because it privileges unity over difference, generates exclusions and is an unrealistic vision.

- Rather than celebrating artificial attempts to establish unity and homogeneity between group identities ('sameness') we should be celebrating the distinctive cultures and characteristics of different groups ('difference').

■   Further Reading

- The work of the Chicago School of Human Ecology and its critics is summarized in most social geography textbooks but it is worth reading the original publications (most of which are still available through libraries) and its early critics, as well as contemporary ones: Park, R., Burgess, E. and McKenzie, R. (1967) [1925] *The City*, 4th edn. University of Chicago Press, Chicago; Cortese, A. (1995) 'The rise, hegemony and decline of the Chicago School of Sociology in the 1920s and 30s', *Sociological Review*, **44**, 474–94; Firey, W. (1945) 'Sentiment and symbolism as ecological variables', *American Sociological Review*, **10**, 140–8; Jackson, P. (1984) 'Social disorganisation and moral order in the city', *Transactions of the Institute of British Geographers*, **9**, 168–80, and Sjoberg, G. (1960) *The Pre-Industrial City, Past and Present*, The Free Press, New York.

- The most classic study of neighbourhood community is Young, M. and Willmott, P. (1962) [1957] *Family and Kinship in East London*, Penguin, Harmondsworth. Davies and Herbert, and Wellman provide good overviews of many of the debates about the decline of neighbourhood community: Davies, W.K.D. and Herbert, D. (1993) *Communities Within Cities: An Urban Social Geography*, Belhaven Press, London; Wellman, B. (1987) *The Community Question Re-evaluated*, University of Toronto, Toronto and Wellman, B. (1979) 'The community question: the intimate networks of East Yorkers', *American Journal of Sociology*, **84**, 1201–31. Rheingold, H. (1993) *The Virtual Community: Homesteading on the Electronic Frontier*, Addison Wesley, Reading, MA is an important advocate of virtual communities.

- Anderson, B. (1983) *Imagined Communities: Reflections on the Origin and Spread of Nationalism*, Verso, London is the book that has been most influential in the retheorization of community. Recent key papers on imagined community and communitarianism are included in a special issue of *Environment and Planning A*, 1999, vol. **31**.

- Finally it is important to be aware of Iris Marion Young's critique of the concept of community as a desirable ideal: Young, I.M. (1990c) 'The ideal of community and the politics of difference', in Nicholson L.J. (ed.) *Feminism/Postmodernism*, Routledge, London.

■   **Exercises**

1. Think about your own sense of 'community'. Do you feel part of a neighbourhood community, a student community or 'stretched-out' communities? Why? How would you define those who are included in your imaginings of community? Who

is excluded? How are you positioned within different communities? What tensions exist within and between them?

2. Go through some back copies of local or national newspapers. What examples can you find of political parties, social movements or individuals mobilizing the concept of 'community' for different ends? How do they each represent their vision of community? What voice do they have and to what extent are they heard? To what extent might these be regarded as romantic or idealized views of 'community'?

■ Essay Titles

1. Critically assess the significance of the Chicago School of Human Ecology for geographers' understanding of 'community'.

2. Critically evaluate Gillian Rose's claim (1990) that: 'The concept of community as a way of describing a particular social reality has fallen into general disfavour and geographers appear to have dropped the term almost entirely.'

3. Critically examine the use of the term 'public' by geographers in relation to community.

4. Using examples, critically evaluate Iris Marion Young's (1990c) claim that 'community is an exclusionary concept and an undesirable ideal'.

5 Institutions

■ 5.1 Institutions

Institutions, which to a certain extent both stand in for, but also stand apart from the home, have received relatively little explicit attention from geographers, despite the fact that they 'constitute an intriguing type of social arena' (Mennell, Murcott and van Otterloo 1992: 112). With the exception of Robin Flowerdew's (1982) *Institutions and Geographical Patterns* and the recent special issue of *Geoforum* (2000), institutions have been, to borrow a phrase from Chris Shilling (1993: 9), an 'absent presence' within the discipline. Present in that geographers have studied spaces such as the workplace or the asylum, but absent in the sense that they have rarely identified or reflected on what Chris Philo and Hester Parr (2000) term the 'institutional dimension' of these environments.

Traditionally, institutions have been thought of in terms of bricks and mortar, as material built environments, such as prisons, asylums or workhouses, which are designed to control and improve bodies and minds (Philo and Parr forthcoming). Philo and Parr divide this sort of geographical work into two different strands. First is the *geography of institutions*, where the focus has been on understanding and exploring what are effectively geographical patterns of social and spatial separation. This research has focused on why and how institutions such as prisons and asylums have been located away from mainstream society. Second, work in *geography in institutions*, work has examined the internal layout of these spaces and has highlighted the ways in which they are designed with the intention of achieving social control over the occupants or particular therapeutic outcomes.

However, Philo and Parr (2000) go on to argue that these two ways of looking at institutions are now being rethought. They suggest that 'institution' is a very slippery term, which is difficult to define. Flowerdew's (1982) edited collection on institutions tended to identify them as largish collectivities, that have internal structures, rules and procedures in which there are tensions between the structures or social systems of the institutions and the agency of participants within them. While the institutions in Flowerdew's (1982) collection were largely taken as pre-given, 'their own coherence, roles and rules [were] in effect taken as largely formulated in advance' (Philo and Parr 2000), some of the papers in the *Geoforum* (2000) special issue complicate this understanding by thinking about how institutions come into being, are sustained and transformed. In this sense, institutions are regarded not as stable or fixed entities but as things which emerge in practice and which, instead of being thought of as impacting on people, places or situations in set ways, are understood to both transform and *be transformed* by them. In these terms, institutions are being redefined, not as fixed structures – indeed, they may not even be located in material buildings, but may instead take the form of dispersed networks of resources, knowledge and power – but rather as dynamic, fluid and precarious achievements (Philo and Parr 2000).

Although geographers are beginning to study a wide range of institutions, such as the BBC, building societies, hospitals, or local authorities, this chapter focuses on just four examples of institutions: school, workplace, prison and asylum. The school and workplace are partial institutions, where people spend only part of their day (with the exception of boarding schools). The asylum and the prison have traditionally been defined as total institutions or carceral landscapes where undesirable others are exiled (Gleeson 1998), although deinstitutionalization means that asylums at least, are now being reconfigured through a range of different agencies and policy frameworks. All four institutions seek to 'place' the body geographically and temporally (Parr 1999: 196) and to discipline it. Michel Foucault's (1977) work on discipline and surveillance (see Chapter 2), in which he uses Bentham's model of an ideal prison, the 'Panopticon', to illustrate his ideas, is a recurring theme in each of the four sections which make up the rest of this chapter (see Driver 1985). Although the institutions discussed here represent spaces or power structures which are designed to achieve particular ends, the four sections each demonstrate that the school, workplace, prison and asylum are not pre-given or stable entities. Rather, each section emphasizes the dynamic, negotiated and fragile nature of these institutions.

■ 5.2 Schools

Two worlds make up the school. First, there is the world of the institution. This is the adult-controlled formal school world of official structures: of timetables, and lessons organized on a principle of spatial segregation by age. Then there is the informal world of the children themselves: of social networks and peer group cultures.

■ 5.2.1 The adult world

Schools are places where children are not only cared for but also contained. The principle of universal education emerged in the UK during the nineteenth century. Industrial capitalism at this time was characterized by the brutal exploitation of child labour in factories. This began to concern middle-class reformers and philanthropists. They regarded children as a natural resource to be nurtured and protected, and were worried that their brutalization was both dehumanizing and would lead to moral and social instability (Valentine 1996c). At the time, working-class children were likened to packs of 'ownerless dogs' (May 1973: 7) roaming the streets, stealing, behaving immorally and making a nuisance of themselves (see also contemporary debates in Chapter 6). Education was perceived by the philanthropists as a way of instilling discipline, respect for order, and punctuality in working-class children before they assimilated the deviant ways of their parents. Schools were therefore conceived of as 'moral hospitals' (May 1973: 12). Not only were they intended to impose middle-class values on the population as a whole, but they also had the added benefit of helping middle-class parents to control their own children.

During this same period the first statutory distinctions were made between adults and children in the form of legislation recognizing juvenile delinquency, and reformatory schools were introduced as a way of instilling moral values in delinquent youths (Ploszajska 1994). These reformatory institutions were developed on the principle that the social and physical environment could influence behaviour. Boys' reformatories were located in the countryside away from the corrupting environment of the city; whereas, because girls were expected to aspire to domesticity, institutions for them were established in the suburbs. The design and layout of these schools were also intended to facilitate the discipline and surveillance of the children (Ploszajska 1994).

Similar concerns about the need to educate working-class children in the normative regimes of polite society in order to protect the middle classes and their offspring were also evident in the USA in the nineteenth and early twentieth centuries. Alongside the development of education and reformatory schools, the Playground Association of America was established, with the aim of keeping children off the streets and of transforming 'street urchins' into respectable adults-to-be (Gagen 2000).

The compartmentalization of children into the compulsory institutional setting of the school, where their use of both time and space is controlled by teachers, has contributed to the development of a contemporary understanding of children as in a process of 'becoming', as physically vulnerable and passive dependants in need of care by adults and protection from the adult world, and as differentiated from adults in a deferential and hierarchical way (Valentine 1996c, Smith and Barker 2000). This process has also been exacerbated by the fact that, as a result of the increase in the number of women in the paid labour market and the corresponding expansion of out-of-school childcare services, today's children are spending increasing amounts of time in institutionalized settings (which can be both liberating and oppressive for them) (Smith and Barker 2000).

Compulsory schooling is, or at different times has been, the basis for the delivery of welfare services such as inoculation against common illnesses or the provision of nutritionally balanced meals; and is a place where individual children can be monitored for signs of neglect or abuse (Valentine 2000). Beyond this social role and the academic goal of achieving exam passes, schools are also spaces where children are acculturated into adult norms and expectations about what it means to be a 'good' citizen. Children are expected to learn to conform to authority and, in doing so, to become compliant and productive workers of the future (Rivlin and Wolfe 1985, Aitken 1994). This is a process through which gender, class and racialized roles and identities are also (re)produced (e.g. see Willis 1977, Krenichyn 1999, McDowell 2000a, 2000b) but it is also one through which schools can challenge young people's racism and explore the meaning of whiteness in young people's lives through anti-racist education (see Nayak 1999a, 1999b).

In *Learning to Labour*, a classic ethnographic study of British working-class 'lads' from an industrial area, Willis (1977) explores young people's opposition to and subversion of the school system ('dossing', 'blagging', 'wagging', 'having a laff' and so on). In cataloguing their counter-school culture, Willis argues that the 'lads' limit their own opportunities for social mobility, and so reproduce their own class position. In this way, the 'lads' serve the interests of capital by perpetuating an unskilled labour force. Willis (1977: 3) writes: 'it is their own culture which most effectively prepares some working class lads for the manual giving of their labour power; we may say there is an element of self damnation in the taking on of subordinate roles in Western capitalism' (see also McDowell 2000a, 2000b).

Education is a process which involves not only shaping children's minds but also their bodies. The body is embedded in a whole range of practices through which schools ensure children's integration into the dominant culture. For example, teachers 'civilize' children, promoting particular forms of bodily control and comportment (to dress properly, to sit still, to be quiet, to have table manners, to be polite, to be punctual, to respect traditions such as saluting the flag, etc) to enable them eventually to be admitted into adult society (Elias 1978, see Chapter 2). Educational institutions are therefore a hotbed of **moral geographies** – of moral codes about how and where children ought to learn and behave (Fielding 2000: 231) (see also the section on home rules in Chapter 3 and Chapter 6).

Fielding (2000) argues that teaching, learning and management in UK schools are constructed 'through the moral beliefs and practices of the governors, headteachers, teachers, learning support assistants, the Local Education Authority (LEA), the Office for Standards in Education (Ofsted) and central government'. Through this moral framework messages are sent out to school staff about what it means to be a 'good' teacher and to children about what behaviour is expected from them. However, these messages may be interpreted differently: by particular schools, which each evolve their own specific educational ethos, by individual teachers through the exercise of their professional autonomy, and through the individual agency of children in responding to them. In particular, children are not passively moulded by unidirectional processes of socialization but actively contest and rework adult frameworks (although there are limits to this, hence the number of children who are suspended and expelled from schools).

Drawing on empirical work in primary schools, Shaun Fielding (2000) demonstrates how members of staff with contrasting teaching styles manage the space of their classrooms differently to create different geographies (in terms of the design of the classroom, the spacing of desks and the spatial freedoms allowed to the children) and through the response of the children to them, different moral orders (see also Rivlin and Wolfe 1985 and Valentine 2000). In this way, although schools as institutions represent spaces or power structures which are designed to achieve particular ends, they are not fixed, uniform or stable entities. Rather, through the adoption of different teaching practices and styles, teachers can be active agents in forming pupils' relationships with the school. Schools are thus, as Philo and Parr (2000) argue, best understood as dynamic and precarious achievements.

■ 5.2.2 Children's cultures

It is within the context of peer group culture – or structures of meaning – that young people learn how to mark themselves out as same or different from others and to manage tensions between conformity and individuality (James 1993). To be socially competent is to be acknowledged as 'one of the crowd', rather than being the anonymous one among the crowd, yet also to not express inappropriate individuality and therefore be excluded as an outsider (James 1993). In this way, young people's identities are embedded in complex networks of relations in which the power to permit or withdraw friendship – to include or to exclude – is central to children's school cultures.

Allison James (1993) argues that a ruthlessly patterned hierarchy characterizes children's worlds. While there is no necessary relation between physical difference and marginality or outsiderhood, different bodily forms are given significance in terms of social identity by other children. She picks out *height, shape, appearance, gender* and *performance* (see Chapter 2) as marking out the boundaries of normality. Children who deviate from the 'norm' are immediately labelled 'different' by their peers, though the consequences for their individual identities of the social meanings that are ascribed to this 'difference' vary (see Figure 5.1).

While physical development or size/age is the most institutionalized principle for grouping children at school, amongst the children themselves gender is perhaps the most important basis for constituting social groups. Within schools there is often a strong sense of gender opposition (boys *v* girls) in which children mark out and ritualize gender boundaries (Thorne 1994). This is evident in the way pupils take up and contest different spaces within school grounds at break and lunch times (Krenichyn 1999, Valentine 2000).

Kris Krenichyn's (1999) study of a New York high school, for example, found that the boys dominate the basketball court, and that girls are made to feel unwelcome

Figure 5.1 A ruthlessly patterned hierarchy characterizes children's worlds

there. Instead, most of the girls take up spaces on the stairwells, fearful that if they gain access to alternative privileged spaces such as the 'games room', the boys will overrun them. Other studies of British schools have suggested that different girls' friendship groups cluster in different 'desired places' (such as the toilets or behind outbuildings) and that these relationships and attachments are critical to girls' identity formation and sense of self-worth (Gilligan 1982).

Contemporary adult understandings of childhood as a time of 'innocence' and 'vulnerability' (see Chapter 2) mean that schools are often imagined to be desexualized institutions (Epstein and Johnston 1994). Yet sex and sexuality are important in a whole repertoire of child–child and even pupil–teacher interactions, including name calling, flirting, harassment, homophobic abuse, playground conversation, graffiti, dress codes, and so on (Haywood and Mac an Ghaill 1995). Adult heterosexual cultures are refracted through children's peer cultures. In particular, a number of studies have demonstrated that girls are under pressure to construct their material bodies into particular models of femininity, that they are judged more harshly on their bodies than boys, and that this, in turn, has an effect on their self-esteem, self-confidence and self-identities (Holland *et al.* 1998).

Gordon *et al.* (1998) highlight the desire of girls to be seen as 'typical' or 'normal'. They argue that girls must manage their bodies to stay on the right side of the slippery boundary between being acceptably attractive (not fat, no spots, no

surplus hair, fashionable, etc) without being overly sexualized (for example, by being labelled a 'slag' or a 'tart' – terms that imply sexual promiscuity). In this way, 'normality' is understood through its opposite 'the place not to be, or more accurately . . . the girl not to be' (Hey 1997: 135).

While for girls it is bodily appearance and shape that are most important to their ability to produce the successful heterosexual feminine identity, for boys it is bodily performance that is crucial to their ability to maintain a hegemonic masculine identity (Mac an Ghaill 1996, see Box 5.1). Coakley and White (1992) argue that boys associate sporting prowess with being a 'successful' male, and gain kudos from participating in competitive and aggressive leisure activities. In particular, boys are expected to be tough, and be able to 'handle' themselves physically or to be able to occupy and take up space through a verbal or intellectual performance (see also Chapter 2).

Compulsory heterosexuality, misogyny and homophobia play an important role in policing and legitimizing such hegemonic male heterosexual identities. Like girls, boys suffer pressure to conform to what is a narrow and constraining conception of sexuality. Haywood and Mac an Ghaill (1995) argue that it is dis-identification with other male students – to be 'a poof' or 'a Paki', for example, is also to be a non-proper boy – that enables heterosexual males to produce their own identity. Through such peer group relations masculine identities are therefore differentiated to produce a hegemonic masculinity in relation to subordinated and racialized masculinities as well as femininities.

Box 5.1: School days

Breaktime
Girl 2: Like the year 9s they sort of have their own territories, like Y9 girls have the toilets – we can't go in [giggling] and Y9 boys have like the corners in the playground or down in the shrubberies.
Girl 1: We normally go to the same place now. I can remember when I first came we always used to stay up here underneath the bridge because we were scared to go down into the football [pitch], or into the basketball courts – we were always scared to go down there, so we always used to stay near the car park [i.e. near teachers].

(Lower school pupils)

Girl 1: At break times it's just really horrible.
Girl 2: Yeah. You get squashed, you can't get out.
Girl 1: Especially when you're the smallest people in the school. It's really awful.
Interviewer: What happens then?
Girl 1: All the bigger ones push in. They usually get banned from coming to the tuckshop if they push in, so it stopped that problem, but I don't usually go, I usually pay someone else to go for me.

(Year 7 girls)

Embodied femininities and masculinities
Girl: Lots of girls, they just say 'Oh I can't eat any more chocolates. They give me spots', or something. 'I can't have fizzy drinks', or something. 'Oh I've got a pot belly', or something. They go on and on sometimes – they want to eat it and then they'll say, 'No I can't, I can't eat it.' And then half of them aren't even fat or anything or overweight who'll complain about their stomach.

(Year 10 girl)

Boy 1: You have to show that you can take it.
Boy 2: It's like you've got to sink a few tinnies [cans of alcoholic drink].
Boy 1: We go up the offy [off-licence] at break for 'em.
Interviewer: What about the girls – do they join in too?
Boy 1: It's only the lads that have got to drink.
Boy 2: It's only us that do it, girls are too busy worrying about their figure 'n that.

(Year 10 boys)

Valentine 2000

Plate 5.1 Masculinities and femininities in the classroom are played out through normative models of heterosexuality. (© Jacky Fleming)

The highly gendered character of school cultures described above can have important consequences for both girls' and boys' educational outcomes. Melissa Hyams (2000) argues that for young Latina women in Los Angeles, completing high school – with all that this achievement represents in terms of future employment opportunities – is dependent on the way that the girls negotiate their gender identities and sexual morality. In exploring the way they talk about their experiences of be(com)ing high school students, she suggests that for the young Latina women 'There is an integral relationship between their gender and sexual identities conceived in terms of "victimisation" and "loss of control" and their historically low academic achievement and attainment.' Likewise, Holloway *et al.* (2000) suggest that the heterosexual economy of the classroom may shape young peoples' technological competence and have repercussions for their future employment prospects. This study of British children's use of Information and Communication Technologies (ICT) shows that the different attitudes of male and female pupils towards computers and patterns of ICT use (which generally favour males over females) are negotiated through competing masculinities and femininities which are played out within the school through normative understandings of heterosexuality (see Plate 5.1).

■ **Summary**

- Schools are places where children are cared for but also contained.

- Universal education was established in order to keep children off the streets and to transform working-class 'street urchins' into respectable adults-to-be.

- The compulsory institutional setting of the school has contributed to differentiating children from adults in a deferential and hierarchical way.

- At school children are acculturated into adult expectations of what it means to be a 'good' citizen. Schools are thus a hotbed of moral geographies.

- Although schools are designed to achieve particular ends, they are not fixed, uniform or stable entities. Rather, through different teaching practices, staff can be active agents in forming pupils' relationships with the school.

- Within peer groups children learn to mark themselves out as the same as or different from others and to manage tensions between conformity and individuality. Thus, their identities are embedded in complex networks of relations.

- Schools are often imagined to be desexualized institutions, yet compulsory heterosexuality and homophobia play an important role in policing and legitimizing hegemonic gender identities.

■ 5.3 The workplace

Like schools, organizations are institutions which attempt to shape the bodies and identities of those who work in them, and which develop specific gendered and sexualized cultures.

In the 1980s feminists argued that 'Our whole world is gendered, from shampoo and tissues and watches to environments as local as the "ladies toilet" and as large as a North Sea oil rig. Things are gendered materially (sized or coloured differently) and also ideologically' (Cockburn 1985). This process, it was argued, was extended not only to work tools (e.g. the typewriter is feminine and the crowbar masculine) but also to occupations themselves. Such is the association made between gender and sexuality, some writers have argued, that the heterosexuality of a person perceived to be in the 'wrong' gendered job would automatically be questioned. For example, nursing is gendered as a female job and, as a consequence, male nurses are commonly assumed to be effeminate and therefore gay. This carry-over of gender/sex-based expectations in the workplace is what Nieva and Gutek (1981: 59) termed 'sex role spillover'. In other words, the gendered characteristics of workers and occupations are mutually constituted.

In a classic study of the printing industry Cynthia Cockburn (1985) shows how both *essentialist* (see Chapter 2) and *moral* explanations were used to justify men's claims that women were unsuited to the skilled and relatively well-paid work of print composition, although the men showed no concern about women doing less remunerative jobs that involved similar skills but to which men had no claim. These essentialist justifications included assertions such as that women had weaker spines than men and so would not be able to stand up for long periods of time; that women were too soft and afraid of getting hurt to do what is a potentially dangerous job; that women were too irrational for an occupation that required logical and problem-solving abilities; that women had an innate aversion to machinery. The social and moral justifications rested on assumptions that it was logical and proper for the male head of the family to be the breadwinner and so well-paid jobs should be the preserve of men; and that women would be coarsened by working alongside men because they would be subject to swearing and the general sexist abuse and so would lose their femininity.

Louise Johnson (1989) found evidence of similar arguments in her study of the Australian textile industry. Here skill was differentiated by sex and connected to sexed bodies. In the mending room speed and nimbleness, combined with dexterity and a domestic skill, produced a highly specialized and supposedly feminine task. Such a skill, although developed on the job, was never seen as such, but was seen as an innate part of being a woman. In contrast, men's skills were understood as acquired through conscious effort and training which enhanced their innate capacities such as technical competence and mechanical affinity. Johnson (1989: 137) concluded that men's and women's bodies were characterized in different ways to the benefit of men over women and thus that 'an awareness of the sexed body in space, especially as it is lived and historically constituted is . . . a vital addition to any understanding of the workplace'.

In another study of the textile industry, this time in Bradford, UK, Peter Jackson (1992) highlighted how the specific division of labour here represented constructions not only of gender but also of race. In these workplaces employers alleged that the manual dexterity (smaller hands and nimble fingers) of Pakistani immigrants made them particularly suited to textile work. In most cases employers also used blatantly racist stereotypes to claim that 'immigrants' [sic] were more easily able to endure the tedium of certain kinds of work than white employees and to justify their segregation on the night shift. Jackson (1992) also pointed out, however, that ethnic brokers, who recruited workers through formal migrant associations in Pakistan and acted as interpreters in Britain, played a role in perpetuating these stereotypes because if the management wanted 'hard-working', 'docile', 'nimble-fingered' workers then it was in their interests to provide them – even to the point of internalizing and exaggerating the racist stereotypes of their employers.

Figure 5.2 The gendering of work can be resisted over time. (© Jacky Fleming)

Both studies of gender segregation and of the racialization of the labour force within workplaces share a recognition that these processes are not stable or consistent. Rather, there have been/are very different constructions of gender or racisms, each historically specific, and articulated in different ways within the societies in which they appear (see Figure 5.2). As Halford and Savage (1998) observe, social practices within organizations can also reassert or challenge the gendering/racialization of work over time.

The postindustrial service economy which has emerged in the late twentieth and early twenty-first centuries has brought new ways of working. The recognition in the studies of traditional industries (above) that gendered and racialized characteristics of workers and occupations are mutually constituted 'has been extended to the very structure of employees' bodies' (McDowell 1997: 120). Within contemporary Western economies aesthetic and emotional components of labour now increasingly have more value than their technological capabilities (Lash and Urry 1994). Indeed, in relation to interactive service work Robyn Leidner (1991: 155–6) argues that: 'Workers' identities are not incidental to the work but are an integral part of it. Interactive jobs make use of their workers' looks, personalities, and emotions, as well as their physical and intellectual capacities, sometimes forcing them to manipulate their identities more self consciously than do workers in other jobs.'

Box 5.2: Panoptic management

'You come in some days thinking, "Oh God I wish I wasn't here, I just feel so awful" and everyone around you seems so confident and they're looking really good and you just think to yourself, "What am I doing here? How can I do this?" . . . It can really make a difference to how you feel about yourself and your work.' [female flight attendant]

'The pilots aren't weighed . . . It's just us . . . Of course it's because we're women . . . the male crew are weighed but if they're a pound or two over, they don't take it as seriously as they do with us . . . they weigh us more frequently and take it more seriously if we're over.' [female flight attendant]

'This happens all of the time. Many of the girls are very attractive, but very young and so we have to make sure that they can handle it. On night flights . . . some business passengers can tend to drink a lot and you have to know how to handle them. You should always humour them. After all they are the passengers, and they are paying our wages'. [female recruitment interviewer/flight attendant]

Tyler and Abbott 1998: 442, 443, 446

As a result, it is increasingly difficult to make a distinction between subjects (workers) and objects (commodities and services).

Having control over workers' corporeal capacities, and intervening in their lives to develop aspects of their identities as an occupational resource, have therefore become part and parcel of many organizations' strategies (e.g. Tyler and Abbott 1998). Bodily 'norms' and standards – in which being slim, attractive and able-bodied are seen as aesthetic ideals for both women and men and as the physical embodiment of productivity and success – are an aspect of many organizations' recruitment criteria (McDowell 1995). In this culture it is not surprising that numerous studies reveal that discriminatory practices against disabled people are commonplace (Morrell 1990, Hall 1999).

Through employee dress codes, appraisals, performance reviews, counselling and stress management employees are also being encouraged, or forced, to be self-**reflexive** (Lash and Urry 1994). The assumption is that workers must manipulate their bodily performances to 'embody' the organizations in order to become and remain employees of particular companies (McDowell 1997). In a study of women with multiple sclerosis Dyck (1999) found that some went to great lengths to manage their bodies at work in order to conceal their deteriorating physical capacity.

This is epitomized by Tyler and Abbott's (1998) study of the recruitment, training and work of flight attendants at two airlines (see Box 5.2). They found that the employees' bodies were considered by these organizations to be a material signifier of the personality of the airlines themselves. Flight attendants were expected to maintain the airline look (to have poise, be elegant, and so on). At recruitment sessions Tyler and Abbott (1998) observed applicants being rejected for being too old, having messy hair, bad skin, poor posture, bitten finger nails, prominent teeth and so on. Bodily norms were also evident in training programmes, uniforms and grooming regulations that covered weight to height ratio, clothes, shoes, hair and make-up. These were maintained through what they term 'panoptic management' (see Chapter 2) which they define as forms of 'managerial control through which employees internalise beliefs which generate behavioural conformity' (Tyler and Abbott 1998: 440). Such forms of control included regular weight checks, on the basis of which attendants could be disciplined and even sacked if they persistently failed to conform to the airlines' height/weight regulations. These were also subtly reinforced by informal peer pressure.

In a similar study, this time of merchant banks in the City of London, Linda McDowell (1995 and McDowell and Court 1994) found that both women and younger male employees spend time and money on 'body work' and that consequently they were remarkably uniform in their appearance. However, the effort and resources which are devoted to maintaining a particular state of embodiment are not usually recognized as part of the labour process and so are not remunerated as 'waged labour' (Shilling 1993).

The purpose of bodily discipline and a controlled presentation of the self in interactive service work is to create a positive interaction with clients in order to bring transactions to a successful conclusion. Abbott and Tyler (1998) observe how female flight attendants are expected to flirt with business class passengers, while maintaining what in the industry is defined as an 'intimate distance'. In this way their sexuality is commodified and exchanged as part of the service the airline sells to its customers. It is not only heterosexuality but also homosexuality that can be commodified in this way. In a wonderful ethnographic account of working in a restaurant, Phil Crang (1994: 691) remarks upon how waiting staff are recruited on the basis of their performative personalities (i.e. for being extroverts, chatty, witty, etc), noting that suitably camp gay men are often employed to provide what he terms 'a bit of homosexual frisson'.

In some institutions nothing is left to chance and interactions are carefully scripted for employees in order to routinize and standardize their contacts with clients. These routines, such as the Combined Insurance script in Box 5.3, teach workers to stand, move or use eye contact in particular ways, etc. In doing so they bring into play workers' emotions, attitudes and sexuality (Leidner 1993). An extreme example of this form of routinization is the US company Amway Corporation. Leidner (1993) claims that this institution tries to mould its employees' lives by shaping their family life, political convictions, religious beliefs and even friends in a process which it terms 'duplicating' (see also Chapter 3). The (hetero)sexuality of most organizations means that such institutional practices can exclude or alienate lesbian and gay employees (Hearn *et al.* 1989, Valentine 1993).

Box 5.3: Routinization

The Combined Insurance Company script for employees

After ringing the doorbell wait for them to answer with your side to the front of the door. Do a half turn when they open it. It's almost as though they catch you by surprise – non confrontational – be casual.

Lean back a little when they open the door. Give them space. Attitude – be a loose goose.

To get inside use the Combined Shuffle:
1. Say 'Hi, I'm John Doe with Combined Insurance Company. May I come in?' Handshake is optional. Break eye contact when you say 'May I come in?'
2. Wipe your feet. This makes you seem considerate and also gives the impression that you don't doubt that you'll be coming in.
3. START WALKING. Don't wait for them to say yes. Walk right in – BUT – be very sensitive to someone who doesn't seem to want you to come in. In general act like a friend.

Leidner 1993

In these ways institutions seek 'to gain a pervasive influence in every area of the employee's life' (Casey 1995: 197). Surveillance – the open-plan design of work-spaces, electronically mediated panopticism such as the use of closed-circuit television or electronic tills, and bodily screening through drug and alcohol testing – is used to enforce or encourage compliance. In this sense the body has become an important regulatory issue in the contemporary workplace (Casey 1995).

Employees' bodies are not, therefore, merely reflections of wider social relations but are a product of organizational dynamics and the ability of these institutions to wield power and construct meanings. However, it is worth remembering, as Susan Halford and Mike Savage (1998) point out, that employees also take direct and covert forms of action to resist their employers.

■ 5.3.1 Workplace culture

In conjunction with the contemporary focus on employees' embodied perform-ativity there has also been a rise in corporate identification and sociality: being part of the 'team' or 'family' (Casey 1995). Paul Du Gay (1996) points out that an awareness of the potential value of workplace culture dates back to the 1930s when it was argued that managerial elites should attempt to foster a strong sense of work group identification in order to fulfil their employees' desire for a sense of belonging and to counter the alienation of modern industrial life. Other writers have claimed that informal activities in the workplace forge a work culture by drawing on outside interests and relations in which work, home and leisure are elided (Roy 1973). According to Du Gay (1996: 41), in the contemporary workplace, culture is important because it is 'seen to structure the way people think, feel and act in organisations'.

Workplace meals and social events (e.g. Christmas meals, annual dinner-dances, work picnics or sports days) are examples of the sorts of activity that are used by employers to align workers' identities and bodies with the organization's goals and to foster a sense of belonging. Michael Rosen (1988) argues that the intention behind such events is to establish a model of the workplace as a moral caring place where employees not only work but are 'loved'.

While work meals and parties are 'social' events they are also organizational events or activities, so that, although workers are invited 'to party and have fun', these experiences are at the same time institutionalized. The choice of an expensive hotel as the venue is often used to signal the success of the business and the 'identity' to which employees should aspire (Rosen 1985). The events themselves have both explicit and implicit functions in (re)producing the shared meaning systems of an organization's culture (Rosen 1988). Speeches and award ceremonies are often an important part of these ritual meals. The public rewarding of staff (for example, for productivity, sales figures or loyalty) signifies the employer's values, acknowledges that individual employees have internalized the organizational culture and so have been successful, while also offering the prospect of similar rewards to the other staff if they too embrace the goals of the institution. These sorts of events do not necessarily work by directly motivating employees to rush back to the office to work harder but have a more subtle effect by shaping the way the workforce think and feel about the organization (Rosen 1988).

Seating arrangements at such events can be used to serve institutional ends. Social divisions may be reproduced and therefore reinforced by setting the tables according to workplace hierarchies or these can be arranged to level social relations in order to relax or diffuse tensions between staff and managers. Likewise, alcohol, by relaxing employees and making conversation flow, can be used to unite colleagues and create a temporary intimacy or sense of belonging. It allows workplace tensions to be diluted, permitting everyday antagonisms and uncertainties to be recognized and acknowledged without challenging the status quo, while also serving as a convenient excuse if any 'social norms' are transgressed (Rohlen 1974).

Shared social activities and events can also be important in producing wider employee–employee relationships as well as employee–employer relationships. Work and non-work (home life, leisure activities, etc) are intimately related (see also Chapter 3). Lunch together, a drink in the pub at the end of the day, an office sweepstake, informal outings to clubs and sports events, even workplace banter, jokes and larking around can define the identities of employees and the relationships between them. More insidiously, by blurring the distinction between social identities/

friendships and work identities/relationships, these informal workplace cultures can define particular jobs as 'masculine', 'feminine' or heterosexual. As a consequence, those whose social identities do not fit in within the dominant workplace culture are marked out as not belonging within the institution or occupation (Valentine 2000). In her study of merchant banks in the City of London, Linda McDowell (1995) demonstrates how a workplace culture predicated on the use of highly sexualized language, practical jokes, and even actual sexual harassment produces these institutions as heterosexual masculine environments in which women are positioned as an inferior and embodied sexualized 'other'. In particular, the heterosexuality of many workplaces means that lesbians and gay men often chose to conceal their sexual identity in these environments (Valentine 1993).

Stress management courses, for example, aim to dissipate workers' frustrations, improve their performances and produce employees who are committed to work and are less likely to express grievances (Newton 1995). Employers often look favourably on those who participate in stress management-style courses, indeed promotion may be dependent on a willingness to participate in these sorts of programme. In this way, 'cultures of excellence' are transferred onto bodies of subjects (Du Gay 1996).

However, just as some employees resist the ability of institutions to construct their embodied identities so too everyday mundane acts of resistance are used to challenge the (re)production of institutions' and employees' workplace cultures.

■ Summary

- The specific division of labour within organizations can represent constructions of gender and race.

- The gender segregation or racialization of labour forces are not stable or consistent processes. Rather, each is historically specific and articulated in different ways. Social practices reassert or challenge the gendering/ racialization of work over time.

- Within contemporary Western economies aesthetic and emotional components of labour now have more value than their technological capabilities.

- Organizational strategies can involve control over workers' corporeal capacities, and interventions in their lives to develop aspects of their identities as a resource. However, these practices are also contested and resisted.

- Social activities are important in producing relationships between organizations and their employees, and between employees.

The prison is a highly regimented environment. Inmates, especially those on remand, spend a large amount of their time literally locked up in a cell. All inmates receive a basic daily monetary allowance paid through a credit system (no cash changes hands). There are also opportunities for work in a range of different occupations for which they are paid and/or receive other privileges. The kitchen is one of most valued places to work because it brings privileges in terms of food, as well as money. The income inmates receive or earn enables them to purchase 'luxury' items, such as toiletries, confectionery, some foodstuffs and fruit, from a prison 'shop'.

Loss of autonomy and the ability to make choices about things like what and when to eat serve as a constant reminder of the inmates' distance from 'home'. The commodities available in the prison shop therefore become important connections with the outside world – a way of imagining or remembering home – and of linking inmates with other people across space and time. They also provide opportunities for them to resist the institutional regime by enjoying food outside the confines of the institutions' set meal times, which are commonly much earlier than most people would eat on the 'outside'. A gap of up to sixteen hours or more between the evening meal and breakfast is not uncommon (Valentine and Longstaff 1998).

As a total institution the prison can literally be inscribed upon inmates' bodies (see Chapter 2). A lack of exercise, being indoors, an institutional diet, depression, can all have bodily manifestations such as weight gain, weight loss, changes in pallor or skin condition, diarrhoea or constipation. Downes (1988) coined the term 'depth of imprisonment' to describe the extent to which inmates experience a sense of invasion by the institution as a result of their material conditions. However, these bodily inflections of the prison are also resisted through the purchase of skincare products and foodstuffs from the shop, and use of the prison gym during 'association time' when inmates are free to exercise, make phone calls or to hang out on the landings.

The limited access inmates have to money or material artefacts means that within this institution other goods become meaningful or useful commodities that have exchange value. A number of studies have shown that goods are not necessarily used up in one act of consumption but can have long and complicated biographies in which their meanings are transient and products have different meanings for different owners (Miller 1987). This is also true of certain items within the prison. For example, fruit can be turned illicitly into alcohol and the silver foil on some confectionery is used for taking drugs. Other items have more obvious uses and value. Phone cards, protein-rich foods and dietary supplements, tobacco and drugs are all precious commodities within the black economy of prisons. Both phone cards and tobacco can be bought from 'the shop', although the amount that each individual is allowed to purchase each month is restricted. As a result, some inmates will buy fruit and chocolate and trade them for phone cards to build up a collection. Then, at the end of the month, when other inmates have used up their allowances for these goods, those who have stockpiles of what are now scarce commodities are able to double or treble their value and trade them back for other more valuable goods, such as drugs. In this way, what are known as card barons, baccy barons and drug barons are established on different prison wings (see Box 5.4).

These trading networks are based on a restricted sphere of exchange because the barons can set up monopolistic barriers by refusing to accept payment for transactions except in particular forms; they have a steady clientele who have no way of transacting with others outside this system and no other means of investing their wealth. This black economy is protected and reproduced by the barons. For example, they may deliberately lure inmates working in the kitchens into debt and then use this to blackmail them to steal food from the stores (Valentine and Longstaff 1998).

Violence and bullying between inmates is endemic. Figures for the UK suggest that a third of inmates have had possessions stolen or damaged; one in eight has been physically assaulted and one in fifteen sexually assaulted (King and McDermott 1995). These conflicts can be a product of relationships inside the prison but they can also be a product of trouble – such as debts – started outside, or of racial or 'gang' tensions on the street. Racism is rife both on the part of prison staff (who in the UK are invariably white), in terms of verbal abuse, the allocation of work, etc and on the part of some white inmates.

Association time, particularly meal breaks, provides potential flashpoints within prisons because a large number of inmates come together at the same time and because conflict can occur as a result of the meanings and value of some food items within the black economy of the prison. The ability of prison officers to exercise their authority at meal times is fundamental to their ability to maintain order in the prison as a whole (Valentine and Longstaff 1998).

Officers and inmates are often assumed to be located in a fixed relationship in which power is regarded as the property of the officers, who rule the prison in a continuous and permanent way. Yet, this limited conception of power as an institutional phenomenon does not explain the range of power relations which permeate institutions such as prisons. It is Foucault (see also Chapter 2) whose work has contributed most to developing contemporary understandings of power. His work questions a focus on institutionalized centres of power in which power is seen as a force which operates in a unidirectional way. Rather, he conceptualizes power as a diffuse force which permeates all levels of society and which is (re)produced in an indirect and often erratic fashion via multiple mediatory networks (McNay 1994). Foucault (1980: 98) describes it thus:

> Power must be analysed as something which circulates, or rather as something which functions in the form of a chain. It is never localised here or there, never in anybody's hands, never appropriated as a commodity or piece of wealth. Power is employed and exercised through a net-like organisation. And not only do individuals circulate between its threads; they are always in the position of simultaneously undergoing and exercising this power.

Box 5.4: Prison life

Trading networks
Grant: 'There's a big thing on thieving down the kitchen – what a lot of people do, they get like, say, a lump of cheese and then get like a phone card – or half an ounce of tobacco or whatever.'

Colin: 'You can sell a KitKat [a chocolate bar] wrapper on some wings for £5. If they've got heroin and nothing to chase it on ... What I used to do is to get a KitKat, you know, on exercise on C wing, near a window, open a KitKat right careful and I could see them all watching me, all drug addicts, they'd all be watching me and I'd throw it, you know, the wrapper and the foil and they used to go like that – the bloke was after the paper before you'd got it out.'

The black economy
Prison Officer: 'Well you'll always have baccy barons, now it's drugs barons – they'll get all the drugs they'll have it back in and people go and put their interest rates up extortionately, like 100 per cent interest rates and people get so far in debt they cannot pay it back, so it comes down to paying in other ways, and one of the ways if they work in the kitchens is: "Well you get me this, or you get me that" – whether it be meat, chickens, milk powder so then they get into trouble – if we catch them they'll either be nicked or sacked out the kitchen. But then they're in the unfortunate position they go back to the wing where the person will readily get at them.'

Flashpoints
Prison Officer: 'The biggest flash point in any prison is down to the food. If the food is total crap you have got trouble. You can usually tell, most prisons that go, it's down to the food. If you can serve decent quality with a substantial portion, the lid'll stay on.'

Eating In
Don: 'I couldn't do with eating in a dining room, it's tension and it's horrible. Oh no.'
Interviewer: 'Why?'
Don: 'I've done it both ways, you know, where you eat with a group of people – not like in a restaurant, that's different – but with prisoners. I wouldn't like to eat me meal with a load of different prisoners ... I like to be in me cell, I'm on me own, I can just pick and eat what I want. People talking to you and looking at other people and little things wind you up. Little things like that wind me up. People get set on.'

Valentine and Longstaff 1998: 141, 143, 145, 147

The black economy creates continuous and multiple networks of relations between inmates and between inmates and officers.

When the inmates are allowed out of their cells for meal breaks and association time the officers use a range of spatial strategies (such as the ordering and segregation of different groups or individuals when food is collected or eaten) to maintain surveillance over the inmates. Likewise, the credit transactions operated in the prison shop allow officers to observe who is buying what and therefore to anticipate which inmates may be dealing in particular commodities and which may be in debt to the barons (Valentine and Longstaff 1998).

Foucault's work suggests a rather depressing picture of inmates as helpless in the face of the covert and unwelcome gaze of the staff (Sibley 1995a). Yet, much of the officers' surveillance is overt – with officers patrolling the landings and looking into cells – rather than covert. It is also not necessarily unwelcome. Inmate violence means some individuals will deliberately seek out the protection of the watchful eye of the officers.

However, officers do not always intervene, despite the fact that they are aware of illegal trading activities or bullying. Some may condone or collude in trading networks in return for kickbacks or because they receive overtime payments if they have to carry out extra duties that result from illegal activities. These are just some of the many connections that form webs of power which link officers and inmates in a series of complicated alliances and collusions.

The official view is that prison gangs should be stamped out, yet officers often take a more pragmatic view, arguing that they may have little negative effect on prisons as a whole but may have a positive effect because the barons also keep a watchful eye on each other (Hunt et al. 1993). If one tries to advance their sphere of influence in the wings, others will try to check them. These complex webs of power between individual inmates and groups of inmates can therefore benefit prison officers by enhancing their ability to keep order in the prison.

Indeed, studies of prisons suggest that social relations within them have to be oriented towards compromise and accommodation because officers have to secure the day-to-day co-operation of inmates in the routine functioning of the prison, yet have few resources to reward or punish those who are already behind bars (Sparks and Bottoms 1995). Sykes (1958) argues that officers have to practise what he calls 'penological realpolitik' or jail-craft to win the co-operation of inmates, and although other writers disagree with this analysis of the prison system they do recognize that inmates are bound together in intimate alliances which involve daily compromise and negotiation (Irwin 1980). For example, officers are sometimes wary of cracking down on everyday 'illegal' activities because repressive measures may prove counterproductive, causing inmates to be resentful and unco-operative, and threatening the delicately balanced and precarious social system within the institution. Individual officers can also be intimidated by the barons and gangs whose power inside the institution may stretch beyond its porous walls, placing officers' families at risk.

In this myriad of ways, inmates and staff are therefore locked in intimate alliances, conflicts and collusions based on compromise and negotiation. In particular, the food system of the prison appears to represent an endless play of dominations in which the distribution of power between officers and inmates and between individuals and groups of inmates is dynamic, relational and constantly open to modification.

■ **Summary**

- The limited access inmates have to money means that within prisons other goods become meaningful commodities with exchange value.

- Trading networks create a black economy within prisons which are dominated by powerful inmates: 'the barons'.

- Officers and inmates are often assumed to be located in a fixed relationship in which power is regarded as the property of the officers. Yet, this conception of power does not explain the range of relations which permeate prisons.

- Rather, the black economy creates continuous and multiple networks of relations between inmates and between inmates and officers, which lock them in intimate alliances and collusions based on compromise and negotiation.

- Rather than being a stable and fixed social system, the prison is perhaps better understood as a delicately balanced and fragile institution.

■ **5.5 The asylum**

In the nineteenth and early twentieth centuries those with mental ill-health were confined and contained in remote, isolated asylums. Indeed, higher-class private asylums were often situated in the countryside. Though ostensibly located in a therapeutic landscape, these rural retreats effectively hid their residents' existence from the world both by spatially separating them from urban centres of population and by concealing their presence behind the facade of the idyllic rural (Parr and Philo 1995). This meant they were also a convenient place for Victorian husbands and fathers to wrongfully confine troublesome wives and daughters.

These big old hospitals have come to symbolize a geography of stigmatization. 'It is in the segregation of individuals with MIH [mental ill-health] within these "sites of insanity", that it is possible to see the laying of a foundation for not only their spatial, but also their *social* exclusion from mainstream society' (Milligan 1999: 222).

A sense of abandonment or entombment is notably evident in some historical accounts of asylums (Parr and Philo 1995).

Indeed, asylums are places where, as Hester Parr and Chris Philo (1995) argue, patients are taught to accept medical definitions of their experiences, in other words to 'learn' to be ill. They quote Chamberlain's (1988: 120, cited in Parr and Philo 1995: 203) account of his treatment in which he explains:

> For six months I was in and out of hospital (several times involuntarily), was given large doses of 'tranquillising' drugs, and was generally made into a mental patient. I was told, and I believed, that my feelings of unhappiness were indications of mental illness. At one point, a hospital psychiatrist told me that I would never be able to live outside a mental institution . . . When I was defined as 'ill', I felt 'ill', and I remained 'ill' for years, convinced of my own helplessness.

These institutions were not only designed to contain madness but also to cure it. The medical and psychiatric professions have tended to adopt essentialist explanations (see Chapter 2) for the origins of mental ill-health, regarding it as the product of the physical or chemical malfunctioning of the body/mind. Treatments have ranged from blood-letting and herbs in the nineteenth century to powerful drug therapy in the twentieth and twenty-first centuries. Asylums, like other institutions, seek to discipline their patients to produce 'docile bodies'. This is done in two ways. 'The mind and body in ward settings is manipulated and regimented, not only with bio-chemical treatments which alter the very chemical constitution of the body, but also by a ward routine which continually seeks to "place" the body, geographically and temporally' (Parr 1999: 196).

Through the control of patients' use of space and their activities (such as set ward meal times) their agency is diminished. However, hospitals are made up not only of public spaces of surveillance where individuals' identities are moulded to conform with institutional models, but also of free spaces where patients can retain some sense of individuality by transgressing rules about behaviour, pastimes, etc (Goffman 1961). Hester Parr (1999) illustrates the fact that patients are not helpless subjects in the face of medical control and surveillance by pointing to examples where individuals demonstrate agency by demanding food outside meal times or refusing to wash. She argues that, in these small ways, individuals can resist the disciplining of minds/bodies within asylums and challenge medical categorizations and the practicalities of the ward routines. These gestures also point to the fact that patients are capable of exercising advocacy for others and taking wider political action (Parr 1999).

Asylums were the main sites of care for those with mental ill-health from the nineteenth century through to the mid-twentieth century. However, the development of psychotropic drugs in the mid-1950s and criticisms of the asylum as a repressive institution in the 1960s (e.g. Goffman 1961) resulted in a policy of **deinstitutionalization**. As a result, the spatial and social segregation of the institution has been replaced by a model of community care (see also Chapter 3), in which support for those with mental ill-health is provided through a range of smaller facilities such as

day care centres, drop-in clinics, night shelters and soup kitchens, perhaps punctuated by short stays in hospitals.

Chris Philo and Hester Parr (2000) argue that 'the dispersal of the asylum has
simply resulted in the opening of smaller but equally isolating places to inhabit for
the mental patient (the drop-in, the shelter, the clinic); a fractured web of deinstitu-
tionalised but in many respects still "institutional" geographies'. Here, the institution is reconfigured through a range of different medical, social and legal agencies
and policy frameworks.

In these new institutional spaces of care '[t]he politics of appearance and the locat-
ing of services are large concerns of both service providers and users' (Parr and Philo
1995: 214). Buildings are chosen to be accessible but also anonymous in terms of
their physical appearance. The aim is to integrate patients into the 'community' in
order to lessen the stigmatization of 'madness' (Philo and Parr 2000). Here, the term
'mad' is reappropriated by critical geographers (e.g. Parr and Philo 1995, Parr 1999)
– in the same way that 'queer' has been reappropriated by the lesbian and gay community – as an alternative to 'mentally ill', which is too closely associated with the
medical–psychiatric pathologization of alternative mental states. The use of 'mad'
also acknowledges the fact that people with different mind/body experiences may
not conceptualize themselves as ill (Parr 1999).

The 'consumer'-oriented nature of the contemporary health service is enabling
the users of mental health care services to have a greater say in their planning and
delivery (Parr 1997, 1999). In this way, while these institutions are still places where
an individual 'is assumed to have a clear identity as a patient, a user, an ex-user
or staff – and they are specific places of treatment that comprise for some people
who access them a difficult and challenging arena in which to find and to express
individual identities . . . [they] also serve as sites for collective reworking of the
status and identity of the users of psychiatric services' (Parr and Philo 1995: 216).
Paradoxically, however, by entering into dialogue with powerful agencies in order
to promote their rights, demonstrate the heterogeneity of users' experiences and make
services more 'user-friendly', consumer-citizens (see Chapter 9) are also absorbed
into the medical–psychiatric institutional establishment (Parr and Philo 1995).

It is also worth highlighting the fact that although these new institutional spaces
of care are intended to promote the inclusion of the 'mad' within the wider community, new forms of spatial exclusion still emerge. For example, neighbourhood
NIMBY groups often resist the siting of mental health care facilities in residential
areas (see Chapter 4). Those with mental ill-health are also frequently subject to violence from the public and are arrested or detained because of their behaviour on the
streets (see Chapter 6). In the drive to close large-scale, oppressive institutions, governments have overlooked the lack of understanding and poor levels of support which
are available within 'the community'. This lack of care is felt most keenly by those
people who had internalized their identities as patients and for whom the asylum
represented 'home' (Parr and Philo 1995). Not surprisingly, some 'mad' people would
prefer to live in institutional environments. This is not a call for a return to the

isolated and exclusionary asylums of the past, but rather a recognition that some experienced the asylum as a safe, peaceful place, a place of refuge and protection where they were distanced from the stresses of everyday life (Milligan 1999). Christine Milligan (1999) suggests that the provision of such institutions, for those who want them, could take the form of smaller therapeutic retreats, rather than large-scale, spatially segregated total institutions.

■ **Summary**

- In the nineteenth and early twentieth centuries those with mental ill-health were confined to isolated asylums which were designed to contain and cure 'madness'.

- Asylums seek to discipline patients to produce 'docile bodies' but individuals can resist medical control and categorization and take wider political action.

- Following a policy of deinstitutionalization, a model of community care has replaced the spatial and social segregation of the asylum.

- The institution is now reconfigured through a range of different medical, social and legal agencies and policy frameworks.

- These new institutional spaces of care are intended to promote the inclusion of the 'mad' within the community but new forms of exclusion have emerged.

■ Further Reading

- Two of the most important geographical works on institutions are R. Flowerdew's (1982) *Institutions and Geographical Patterns* (Croom Helm, London) and the recent special issue – especially the introduction by Philo and Parr – of *Geoforum* (2000). More broadly, Foucault's work on discipline and surveillance underpins much of the theorization of institutions. See Foucault, M. (1977) *Discipline and Punish: The Birth of the Prison*, Allen Lane, London. His work is assessed in Driver, F. (1985) 'Power, space and the body: a critical assessment of Foucault's *Discipline and Punish*', *Environment and Planning D, Society and Space*, 3, 425–46.

- Good starting points to read about schools are: Thorne, B. (1994) *Gender Play*, Rutgers University Press, New Brunswick, NJ; James, A. (1993) *Childhood Identities: Self and Social Relationships in the Experience of the*

Child, Edinburgh University Press, Edinburgh; Holloway, S.L. and Valentine, G. (eds) (2000) *Children's Geographies: Living, Playing, Learning*, Routledge, London (especially the chapters by Gagen, Fielding and Smith and Barker). A classic study of how working-class young men reproduce their class position through the schooling system is Willis, P. (1977) *Learning to Labour: How Working Class Kids Get Working Class Jobs*, Saxon House, Westmead. Gender and heterosexuality in relation to children and young people's cultures are the focus of Holland, J., Ramazanoglu, C., Sharpe, S. and Thomson, R. (1998) *The Male in the Head: Young People, Heterosexuality and Power*, Tufnell Press, London, and Haywood, C. and Mac An Ghaill, M. (1995) 'The sexual politics of the curriculum: contesting values', *International Studies in Sociology of Education*, 5, 221–36; and they are also featured in papers in a special issue about children and institutions in volume 32 of *Environment and Planning A* (2000).

- Cynthia Cockburn's work on the print industry remains a classic study of the gendering of an occupation: Cockburn, C. (1985) *Machinery of Dominance*, Pluto Press, London. In terms of interactive service work, key sources include Leidner, R. (1993) *Fast Food, Fast Talk: The Routinisation of Everyday Life*, University of California Press, Berkeley, CA, and McDowell, L. (1997) *Capital Culture: Gender at Work in the City*, Blackwell, Oxford. Both Du Gay, P. (1996) *Consumption and Identity at Work*, Sage, London, and Halford, S. and Savage, M. (1998) 'Rethinking restructuring: embodiment, agency and identity in organisational change', in Lee, R. and Wills, J. (eds) *Geographies of Economies*, Arnold, London, offer perspectives on organizational cultures and change. A good, more general source is Lee, R. and Wills, J. (eds) (1998) *Geographies of Economies*, Arnold, London, which includes chapters on the issues discussed here as part of a broader take on economic geography. Among the journals where research on the social relations and culture of the workplace is published are *Economic Geography, Environment and Planning A*, and *Environment and Planning D: Society and Space*.

- The prison and the asylum have received less attention within geography than the school or the workplace. Useful case studies include: Valentine, G. and Longstaff, B. (1998) 'Doing porridge: food and social relations in a male prison', *Journal of Material Culture*, 3, 131–52; Parr, H. and Philo, C. (1995) 'Mapping "mad" identities', in Pile, S. and Thrift, N. (eds) *Mapping the Subject: Geographies of Cultural Transformation*, Routledge, London, and Parr, H. (1999) 'Bodies and psychiatric medicine: interpreting different geographies of mental health', in Butler, R. and Parr, H. (eds) *Mind and Body Spaces: Geographies of Illness, Impairment and Disability*, Routledge, London. The classic work on asylums, however, is a sociological study: Goffman, E. (1961) *Asylums: Essays on the Social Situation of Mental Patients and Other Inmates*, Penguin, Harmondsworth.

■ Exercises

1. Think about your experiences of school as a child. Write a short account of a particular memory of your schooldays which stands out in your mind. Exchange your account with some friends. Discuss what you have written. What does it tell you about the nature of the institution, its moral geographies and your peer group cultures?

2. Watch the first 20 minutes of the film *Disclosure*. As you do so, consider whether the jobs in the firm were gendered, and, if so, how? Think about the different ways in which the characters were dressed – what did their appearances say about their different roles in the workplace? How many examples of sexuality at work did you notice? How did these shape the production of different spaces? How are 'public' and 'private' spaces produced within this workplace?

■ Essay Titles

1. To what extent do you agree that institutions have been an 'absent presence' within the discipline of geography?

2. Critically assess Fielding's (2000) claim that schools are a hotbed of moral geographies.

3. Critically evaluate the extent to which all work might be described as 'sex work'.

4. Focusing on the example of *either* the prison or the asylum, discuss the extent to which it might be described as a total institution.

The street

6.1 The street

The street 'symbolise[s] public life, with all its human contact, conflict and tolerance' (Boddy 1992: 123). As a result, it has been romanticized and celebrated as a site of political action, an environment for unmediated encounters with strangers, and a place of inclusiveness. Indeed, it has been argued that, by claiming space in public, marginalized social groups themselves become part of the public and that in this sense spaces such as the street are essential to democratic politics. However, within contemporary Western societies the white middle classes are increasingly apprehensive about the proximity of demonized 'others' (such as youth, the homeless, those with mental ill-health) in public space and the dangers of encountering them on the streets. Their desire to be insulated from 'otherness' is, some writers claim, leading to their withdrawal from public life and this, in turn, is contributing to the degeneration of public space.

In response, cities are embarking on a range of measures which are aimed at halting this perceived decay and which are designed to create safe and ordered 'public' environments. These include the widespread introduction of private security forces and electronic surveillance in everyday places and urban renewal schemes which involve privatizing and commodifying 'public' space.

Pessimists complain that these measures are killing the democratic mix and vitality of the streets and that these transformations signal the end of truly 'public' space. They oppose the privileging of white middle-class lifestyles in defining the nature of 'public' space. For these writers, genuinely open-minded space should be gritty and disturbing, not safe and ordered, because for a space such as the street to be 'public' it must, by definition, involve encounters between strangers that will produce conflict and dissonance.

Other writers, however, have sounded a note of caution in the face of this debate between what appear to be two oppositional and irreconcilable visions of public space. They observe that geographers are in danger of thinking of spaces such as the street either as spaces of political expression and celebration, or as sites of repression and control, and that to do so is to overemphasize the extent to which public space is being 'controlled' and to underemphasize its contestation.

All these points are explored in this chapter, which begins by examining the history of the street as a democratic space. It then goes on to consider citizens' fear of crime and disorder. The following sections examine both the construction of a moral order on suburban streets and the way that particular social groups are defined as 'dangerous others'. Attention is then turned to the policing of the street by the police force and private security industry, and through electronic surveillance and community initiatives. The final section of the chapter evaluates the conflicting visions of public space evident in contemporary debates about fear of crime and policing.

■ 6.2 The democratic street?

The notion of public space can be traced back to the Greek agora. This was 'the place of citizenship, an open space where public affairs and legal disputes were conducted . . . it was also a marketplace, a place of pleasurable jostling, where citizens' bodies, words, actions, and produce were all literally on mutual display, and where judgements, decisions and bargains were made (Hartley 1992: 29–30, cited in Mitchell 1995: 116). The agora was also a place of exclusion, however. Only those of power, standing and respectability were free to participate in this public sphere. Slaves, women and foreigners were all denied citizenship.

The history of public space in North America and Britain is also one of meetings, display and politics, and one of exclusion and struggle (Marston 1990). Historically, women, non-white men and sexual dissidents have been denied access to public space and have had to fight their way (e.g. through the campaigns of the suffragettes, civil rights movement, lesbian, gay and bisexual activists) into the public sphere and to win the right to representation as part of the political public (Mitchell 1995).

In the process these excluded groups have taken to the streets in order to press their claims for their rights. The street has therefore become an important organizing ground and space where different causes can be seen and heard by others (Mitchell 1996). Mitchell (1996) argues that this public visibility was particularly important in the nineteenth century for organizations such as Industrial Workers of the World (a radical union and social movement) because the US press did not

report their strikes or demands, either ignoring their actions altogether or reporting them from the perspective of employers. The street was therefore an important
forum in which these sorts of organizations could communicate their arguments to
the wider public. The success of some of these struggles in turn showed other marginal groups the importance of taking up the space of the street in order to stake a
claim for their rights. Doing so transformed the street into a space *for* representation (Mitchell 1996).

Over time these sorts of struggles have expanded understandings of the 'public'
to embrace a wider social spectrum of the population. Public space is now commonly
understood as a popular space to which we all have access, a space where we meet
strangers (Sennett 1993), a space where we encounter 'difference' (Young 1990b),
a space of unmediated interactions where people can just go and 'be' (Mitchell 1996).
Sharon Zukin (1995: 262) summarizes 'the defining characteristics of urban public
space [as] proximity, diversity and accessibility', claiming that 'The question of who
can occupy public space, and so define an image of the city, is open-ended' (Zukin
1995: 11).

Public spaces like the street are often romanticized and celebrated as a site of authentic political action, as a place of inclusiveness where people can come together and
therefore as inherently democratic spaces (Sorkin 1992). Don Mitchell (1995: 115)
argues: 'By claiming space in public, by creating public spaces, social groups themselves become public . . . Insofar as homeless people or other marginalised groups
remain invisible to society, they fail to be counted as legitimate members of the polity.
And in this sense, public spaces are absolutely essential to the functioning of democratic politics.' Indeed, because public spaces such as the streets are constantly being
used, invested in or claimed by different groups, the street is constantly changing
(Goheen 1998). In this sense, the dynamism and diversity of the streets and public
culture of the city are often contrasted with the sterility and homogeneity of the suburbs. However, in celebrating the democratic nature of the street it is also important to remember that the courts, the law, the actions of the police, the claims of
other social groups and even vigilantes can all restrict the rights of individuals or
groups to occupy the streets.

Within geography there are numerous examples of different groups claiming
and contesting the public space of the street. Just two examples are the Notting Hill
Carnival in London (Jackson 1988), and the attempts of lesbian and gay activists
to march in the St Patrick's Day Parade, in South Boston, USA (Davis 1995) (see
Plate 6.1).

The Notting Hill Carnival was initiated in the 1950s with the aim of trying
to reverse the decline of the neighbourhood. This had become associated with bad
housing, prostitution, and, in the late 1950s, with racial conflict. In the 1960s the
Carnival became a way for the local working-class residents to protest and organize
against their poor housing and social conditions, and was also used to mobilize
residents against the construction of a motorway (M40) flyover through the area
(Jackson 1988). In the 1960s and 1970s a large number of Afro-Caribbean people

Plate 6.1   Boston, USA, St Patrick's Day parade (© Peter Erland)

settled in the neighbourhood and the Carnival began to develop a clear Trinidadian identity (Jackson 1988).

This was a time of rising unemployment and of social polarization and tensions between blacks and whites. In the late 1970s many second-generation, British-born Afro-Caribbean young people began to feel extremely alienated and disillusioned with British society. In 1976 the Carnival erupted into violent riots and was to become associated with some of the key events in the politicization of 'race' within the UK. While prior to this trouble the policing of black communities had not been regarded as a problem and the Carnival had been regarded as a harmonious event, Paul Gilroy (1987) argues that the violence – dubbed a 'race riot' by the press – marked a turning point, with black people as a whole being criminalized and **institutionalized racism** exposed. Subsequent reporting of the event, for example, was explicitly racist, focusing on the lawlessness and violence of black youths, the instability of West Indian families and immigration (Jackson 1988).

In the following year the police argued for the Carnival to be taken off the streets and held in a stadium where they claimed it could be 'contained' and 'controlled'. They were unsuccessful, however, and the Carnival continues to be a neighbourhood event. Following the events of 1976, there was an ideological battle for control of the Carnival between two rival organizations. One group regarded it as a political event which could be used to press for social and economic reforms. The other regarded it as a creative event that would be destroyed by blatant attempts to use it in these ways, arguing that by remaining overtly 'non-political' it could still be politically and culturally effective (Jackson 1988). The Carnival retains at its heart a paradox. It is both socially integrative, bringing hundreds of thousands of black and white people onto the streets together, while also serving as a vehicle for the expression of protest and resistance in which the freedom and mobility of the street is central to the use of music and masquerade to suspend the dominant cultural order and to create a symbolic space 'beyond the reach of racism' (Gilroy 1987: 73).

In the early 1990s Irish lesbian, gay and bisexual residents of South Boston (GLIB) sought to participate in the annual St Patrick's Day Parade. Ethnic street parades are an important means for social groups to spatially articulate their identity and to symbolically define the social and cultural character of particular neighbourhoods. For GLIB, the street parade was an important space in which to assert the visibility of sexual dissidents within the Irish community. This was a time when **queer politics** was emerging in North America and Europe, advocating the use of performative acts such as kiss-ins, mock weddings and queer shopping outings to subvert and challenge heterosexist assumptions about the straightness of the streets (see also Bell and Valentine 1995a, Valentine 1996a).

The desire of GLIB to be included in the street parade was read by South Boston residents as both a threat to local understandings of what it meant to be Irish and to the neighbourhood itself. GLIB were eventually allowed to join the march – although they encountered a barrage of hostility and abuse – because the courts ruled that the parade was a public event. By very visibly participating in the march, gay men, lesbians and bisexuals inserted themselves into a heterosexist space in order to claim the right to be seen, heard and included within the Irish community. In doing so, they redefined the space, 'not just for the moment, but in general, by bringing new meaning and debates to the discourse of Irish and Irish-American identity' (Davis 1995: 297).

Examples such as this highlight the use of the street as a democratic space for articulating difference and for legitimating the rights of particular social groups to be part of the 'public'. Yet, at the same time, within contemporary Western societies there are increasing concerns about the (in)civility of encounters between strangers and the proximity of demonized 'others' on the streets. These are throwing up questions about who has the right to use and occupy public space and how the streets should be policed and controlled. The next sections explore these debates.

■   **Summary**

- The history of public space is one of exclusion and struggle.

- Historically, groups such as women, non-white men and sexual dissidents have had to fight their way into the public sphere and win the right to be part of the political 'public'.

- In the process, the street has become an important organizing ground for excluded groups and a space where their arguments can be communicated to a wider public.

- These struggles have expanded understandings of the 'public' to embrace a wider social spectrum of the population.

- The defining characteristics of public space are now understood to be proximity, diversity and accessibility. As such it is celebrated as inclusive and inherently democratic.

■   **6.3 Streets of fear**

Street violence is regarded as an increasingly common problem in most North American and European cities. Although statistically it is young men who experience most interpersonal violence (URL 6), it is women who are regarded as the group most at risk. The crimes women fear most are sexual violence or assault by strangers. Statistically, such incidents are relatively rare (especially in comparison with figures for domestic violence: see Chapter 3), although at some point in their lives most women encounter more minor forms of sexual harassment in public space, such as verbal abuse, wolf-whistling or flashing (see Chapter 2) (Wise and Stanley 1987). These 'everyday' incidents are linked on a continuum with extreme forms of violence and therefore are often regarded as a potential precursor to more serious forms of abuse. In other words, an 'everyday' awareness of the possibilities of harassment can contribute to some women's perceptions of insecurity on the street (Valentine 1989). These anxieties are compounded for those who believe themselves to be unable to defend themselves against a male assailant. Some individuals consider themselves to be unable to do so, perhaps because they are elderly, ill or have a physical impairment which restricts their vision, mobility or strength. Others imagine that all women would be unable to defend themselves against a man because they regard all men as larger and stronger than all women (even though this is not the case and also overlooks the fact that self-defence techniques do not necessarily require bodily strength).

The media play a part in exaggerating the extent of violent crimes such as rape and murder (Smith 1984). Crime stories are easy to obtain, the human interest angle sells newspapers and they are also a useful editing device. The selective reporting of violent incidents also generates images about *where* and *when* women are at risk (Valentine 1992). By disproportionately publicizing attacks which occur in public space, the media help to create a perception that the street is a dangerous place, despite the fact that statistically women are more at risk from domestic violence (see Chapter 3). This maps onto the historical public–private dualism in which women have been associated with the private space of the home and men with the public world of the street (see Chapter 1 and Chapter 3). This dualism has been used to draw a distinction between 'respectable' women and 'less deserving' women and between 'sensible' women and 'reckless' women. For example, in cases where women have been attacked in public space at night the police and the media have sometimes implied that they are, to a certain degree, responsible for their own fate, and have warned other women to avoid putting themselves in similar situations of vulnerability (Valentine 1989).

Women's perceptions of fear on the street are closely associated with their perceptions of who occupies and controls the space (see Box 6.1). Fear is closely associated with disorder. Graffiti (see Cresswell 1992 and Ley and Cybriwsky 1974), litter, groups of young people or the homeless on the street are often read as a sign that the space is not looked after or controlled, either formally by the police or private security forces, or informally by local residents, passers-by, store keepers, and so on. This, in turn, erodes women's confidence that anyone would notice or intervene to help if they were attacked (Valentine 1989). Some commentators argue that neighbourhood community is in decline (see Chapter 4) and that this is also contributing to people's perceptions of vulnerability in public space. Consequently, initiatives such as Neighbourhood Watch (section 6.6.4) are being launched to try to generate this sort of informal social control artificially.

Fear of crime increases at night because the use of the street changes so that that this space is produced in a different way. During the daytime the street is usually occupied by people from all walks of life going to school, work, shopping, and so on. However, at night, not only does darkness reduce visibility and increase the feeling that potential attackers might be able to strike unobserved, but there are also fewer people on the streets and they are usually dominated by the group women most fear: unknown men.

As a result, many women adopt precautionary strategies to keep themselves safe at night. These might be *time-space avoidance strategies* to distance themselves in space and time from perceived danger (for example, by avoiding going out at night after a particular time, or by using a car or taxis to avoid walking in public space after dark), or *environmental response strategies* adopted by women when they are in public space (such as walking confidently, carrying a rape alarm or knife, etc). While most women might use a combination of strategies to keep themselves safe, not all women have the same ability to adopt particular precautions. For example, those who work shifts may be unable to avoid public space at night, those on a low income do not necessarily have the option of using private transport or taxis. As a result, some women's fear structures their use of space more than others. Indeed,

---

Box 6.1: Fear of violence

'Occasionally I go into town in the day and I park in the Renault multi-storey. When I'm going up there I think "ooh", and if there's anybody waiting on the steps, I mean it's not dark, there are lights but it's lonely. Occasionally I run up, I think any-body could jump out on you, but I just push it out of my mind and hurry up.'

Reading, UK, young woman

'When there's no houses or people or anything. Where I know I'm so vulnerable and that I haven't got a chance, that really scares me. Subways for example. I just don't walk under them not even in the day. The amount of times I've gone to walk under them and there have been people down there smoking or whatever, and that's awful because once you've committed yourself to walking down there then you've got to carry on.'

Reading, UK, young woman

'I never go into town at night. if I go anywhere, or do anything it's always in the morning, or early afternoon, if it's anything else we do it as a family.'

Reading, UK, married woman

'Darkness doesn't bother me – it's just that most normal people are in by that time and there's little activity around apart from gangs and men going to the pub.'

Reading, UK, single woman

'I mean I like jogging. As soon as it starts getting dark you can't do that on your own. You can't jog in a park for fear of being jeered at, or that one particular bloke will do something. It's wrong, we should be able to go out at night, we're human beings as well.'

Reading, UK, young woman

Cited in Valentine 1990: 290, 299, 300, 301

---

although women may be frightened by an awareness of potential violence on the street, many regard it as important to control these emotions so as not to let these fears restrict their use of public space, and to resist discourses which construct women's place as in the private sphere (Mehta and Bondi 1999).

Recent research suggests that it is not only women who are fearful of violence on the street. There are also growing concerns in North America and Europe about children's safety in public space (Valentine 1996b) and about the impact of fear of crime on the elderly (Pain 1999). Well-publicized cases of murders and violent assaults which have been motivated by racism (Plate 6.2) and homophobia have also drawn attention to an escalation in 'hate crimes' (Herek and Berrill 1992, Valentine 1993). In 1997 the Federal Bureau of Investigation (FBI) recorded 9861 hate crime offences in the USA (URL 7).

Plate 6.2 London's racist heartland

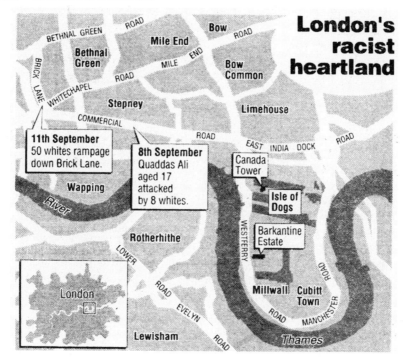

In a study of young people's leisure activities in a British town, given the pseudonym Thamestown, Paul Watt (1998, see also Watt and Stenson 1998) found a strong degree of localism amongst Asian young men who spent much of their time in Streetville, the neighbourhood where they lived. In contrast, the white middle-class youth (see also Nayak 1999b's discussion of skinhead style, whiteness and masculinity) who participated in the research hung out in the town centre and out-of-town-venues. Watt (1998) attributed this difference not just to the fact that the Asian young people had a strong sense of loyalty and belonging to where they lived (see Chapter 4), but also to the fact that this was a place where they felt safe from the sort of racist aggression and harassment which they encountered in other parts of the town. Indeed, the research also found some evidence that the Thamestown white youths actively used violence to exclude other ethnic groups from their neighbourhood streets (Watt and Stenson 1998), although it did not uncover quite the same degree of territorialism – termed by Webster (1996: 26) 'ethnic apartheid' – that has been evident in studies of other English towns and cities (e.g. Webster 1996, Keith 1995).

In a similar vein, a fear of queer-bashing deters many lesbians and gay men from articulating their sexuality in everyday public spaces and so contributes to produ-cing the street as a heterosexual space (Valentine 1993, Valentine 1996a).

Indeed, surveys suggest that fear of crime is becoming a more general anxiety across all sectors of the population. Describing people's experiences of the streets of New York, USA, Siegel (1993: 3) writes: 'They view the streets, sidewalks, parks, plazas and subways of New York as an arena of combat rather than civility, where the goal is survival rather than enjoyment. Many said the distress of their daily rounds could be measured by the number of times they were hassled between home and work.'

One consequence of such fears, Lyn Lofland (1973) argues, is that urban residents increasingly try to avoid the public realm by creating a private symbolic shield around themselves in public space. They minimize expressivity or body contact with other people, choose seats in public places or position themselves to signal that they have no intention of interacting with anyone else, they minimize eye contact with strangers, and at the slightest sign of trouble they flee. As a result, she argues, there has been a deterioration in the quality of encounters on the streets; in turn, people are choosing to lead more home-centred lives (see Chapter 3) or are avoiding public space, for example preferring to use private rather than public transport or to shop in malls policed by closed-circuit television and security guards rather than on the street. This is leading to a change in the way public space is produced. As a result, the streets are seen as less controlled and so more dangerous and so a spiral of avoidance, decline and abandonment is setting in.

This decline in public space is also being seen as threatening private space because our ability to enjoy and feel safe at home depends to some extent on our ability to keep the dangers imagined in public space at bay (Goheen 1998). The concept of 'defensible space' (Newman 1973, Coleman 1990) – which is predicated on a belief that the design and layout of buildings and neighbourhoods can foster surveillance and informal control – has been developed by architects in an attempt to reduce residential crime (although it is heavily criticized for being **environmentally determinist**). The next section looks at how fears about the dangers on the street are played out in suburban neighbourhoods.

■ **Summary**

• Women are considered to be the group most fearful of crime but surveys suggest that it is becoming a more general anxiety across the population.

• Fear of crime is closely associated with perceptions of who occupies and controls the space of the street, and with disorderly environments.

• Fear of crime is leading people to avoid public space. This is changing its balance of use and a spiral of avoidance and abandonment is setting in.

• Public space-based fears undermine perceptions of safety in private space too because this depends upon keeping 'public' dangers at bay.

■ 6.4 The moral order of the suburban streets

Beyond the centre of cities lie numerous smaller communities collectively known as suburbia. More people live in these environments in North America and the UK than anywhere else and those who do tend to be disproportionately drawn from the ranks of the wealthy and the powerful (Baumgartner 1988). The 'suburbs' have been the chosen residence of the respectable middle classes in the UK since the Victorian era, although the shift in the US population from central cities and rural districts to the suburbs did not occur *en masse* until the 1970s and 1980s (Baumgartner 1988).

The main role of the suburbs from the Victorian era onwards has been one of social segregation. In nineteenth-century Britain the suburbs provided a refuge for the middle classes from the dirt and the disorder of industrial cities and their working-class inhabitants. Likewise, contemporary suburban dwellers are also looking to escape urban problems such as crime, congestion, and undesirable 'others'. Siegel (1993) suggests that the white middle classes are increasingly fearful of 'difference'. He writes, 'The modern world is full of people who are terrified of other people, socially, sexually, or politically different from themselves.' While the city is a space characterized by heterogeneity and disorder (see Chapter 7), homogeneity, stability and peacefulness are usually regarded as the defining qualities of the suburbs (although it is important to recognize that, in practice, the suburbs, like other spaces, are complexly differentiated: see Chapter 7). In many respects they draw on 'imagined' or idealized notions of community which are associated with English 'village' life (see Chapter 4 and Chapter 8).

In a study of 'Hampton', a pseudonym for a quiet residential suburb of 16 000 people situated between New York City and an unidentified smaller city on the eastern seaboard of the USA, M.P. Baumgartner (1988) argues that the suburbs have a particular moral order based on an overwhelmingly powerful and widely understood pattern of restraint and non-confrontation. She argues that this moral order is possible in Hampton because it is socially homogeneous. Its white middle-class residents have established 'norms' or appropriate ways of behaving towards each other and have little contact with 'other' groups who are regarded as unpredictable and threatening.

Compared with cities, suburbs such as Hampton have few public places which draw people onto the streets where they might encounter strangers (for example, it has only one cinema, one bowling alley, a handful of restaurants and shops). Residents are very home-centred and prefer to use cars rather than to walk or use public transport. As a result, the streets are usually quiet (Baumgartner 1988). Residents expect those in public places to behave in unobtrusive and restrained ways, while they also have high standards of cleanliness and aesthetics in relation to the built environment. Despite very low levels of predatory behaviour by strangers, Baumgartner (1988: 104) observes that the citizens of Hampton are 'greatly concerned about crime and wary of people who seem out of place. They see strangers as potentially dangerous and anyone who is too conspicuous as alarming'.

Although there is minimal contact between residents, who tend to keep themselves to themselves, most know each other by sight, or from their embodied identities are clearly recognizable as 'belonging' in Hampton. Strangers are understood to be easily visible because they do not fit in with the homogeneity of the suburb.

The value placed on non-confrontation and restraint within Hampton means that residents frequently draw on the police for help rather than personally exercising informal social control over the streets. On average, they make about the same number of phone calls to the police for 'assistance' per head of the population as do the residents of the city of Chicago (just under two calls per day per 1000 people). Most of these calls, however, are not to report violence or crime but are for what is termed non-legal assistance in that they are reporting the presence of strangers in the neighbourhood and requesting police surveillance (Baumgartner 1988). Such tactics represent what David Sibley (1988) has termed in another context 'the purification of space', which he defines as 'the rejection of difference and the securing of boundaries to maintain homogeneity'. Through using the police in this way, residents are able to exclude undesirable others who are regarded as polluting their environment and so maintain the moral order of the suburbs.

■ Summary

- The main role of the suburbs is to provide a refuge for the middle classes from urban problems such as crime and undesirable 'others'.

- While the city is characterized by heterogeneity and disorder, homogeneity, stability and peacefulness are regarded as the defining qualities of the suburbs.

- The suburbs have a particular moral order based on a pattern of restraint and non-confrontation.

- Strangers are understood to be easily visible because they do not fit in with the homogeneity of the suburbs.

- Suburban residents commonly use the police to exclude undesirable others and so maintain the homogeneity and boundaries of their 'community'.

■ 6.5 Dangerous 'others'

Sections 6.3 and 6.4 have both hinted at the fact that, at different times and in different spaces, particular groups have become demonized as 'dangerous' and as a threat to other citizens and to the moral order of the street. These dangerous 'others' include children and youth, black people, those with mental ill-health and so on. This section explores some of the processes through which some of these groups become defined as a 'problem'.

■ 6.5.1 Children and youth

The notion of a '**moral panic**' – 'invented' by the sociologist Stanley Cohen to explain the public outcry caused by the clashes between 'mods' and 'rockers' in England in the mid-1960s – is a useful concept for thinking about how various youth cultures and young people's behaviour in general is often viewed by adultist society as 'criminal' or 'deviant'. Cohen (1972: 9) defines a moral panic thus:

> A condition, episode, person or group of persons emerges to become defined as a threat to societal values and interests; its nature presented in a stylized and stereotypical fashion by the mass media, the moral barricades are manned by editors, bishops, politicians and other right-thinking people; socially accredited experts pronounce their diagnoses and solutions; ways of coping are evolved or (more often) resorted to; the condition then disappears, submerges or deteriorates and becomes more visible.

The media play a pivotal role in moral panics by representing a deviant group or event and their effects in an exaggerated way. They begin with warnings of an approaching social catastrophe. When an appropriate event happens which symbolizes that this catastrophe has occurred, the media paint what is often a sensational and distorted picture of what has happened in which certain details are given symbolic meanings. In turn the media then provide a forum for the reaction to, and interpretation of, what has taken place. The public then become more sensitive to the issue raised, which means that similar 'deviations', which might otherwise have passed unnoticed, also receive a lot of publicity. This spiral of anxiety can eventually lead to punitive action being taken against the 'deviant' group or event by relevant authorities. Angela McRobbie (1994) suggests that, in this sense, moral panics are about instilling fear into people – either fear to encourage them to turn away from complex social problems or, more commonly, fear in order to orchestrate consent for 'something to be done' by the dominant social order.

Moral panics are related to conflicts of interest and discourses of power and are often associated with particular 'symbolic locations', such as the street. These panics are frequently mobilized in relation to particular groups of young people, such as mods and rockers, when they appear to be taking over the streets or threatening the moral order of the suburbs. In this respect, McRobbie (1994) argues, moral panics are increasingly less about social control and more about a fear of being out of control and an attempt to discipline the young. This process often involves a nostalgia for a mythical 'golden age' where social stability and strong moral discipline were a deterrent to disorder and delinquency. Indeed, Ulf Boethius (1995) even goes so far as to suggest that, by criticizing the young and their lifestyle, adults

Figure 6.1 Youth violence is becoming increasingly politicized

By permission of Bob Gorrell and Creators Syndicate, Inc.

defend their own more disciplined way of living and try to convince themselves that they have not lost anything by becoming adults. Such moral panics have been evident in both the USA and the UK at the end of the twentieth and the beginning of the twenty-first centuries.

Across the USA, young people are currently the subject of popular suspicion and anxiety. One of the key triggers of this contemporary concern is the perceived omnipresence of youth gangs on the street, in which 'gang' has become a codeword for 'race' and a symbol too of drugs, guns, graffiti, gangsta rap and violence (Lucas 1998). The media have played a key role here in exaggerating the number of gangs and in distorting their activities through racialized constructions of youth violence. While violence is presented as a black problem 'one of the terrifying visions of white suburbanites is that of the migration of drive-by-shooting gang bangers . . . to the more prosperous (white) peripheral neighbourhoods' (Dumm 1994: 185) (see Figure 6.1). In this sense, moral panics about youth gangs are indicative of more general processes of social polarization. The issue has become increasingly politicized, resulting in the inclusion of measures to prosecute gang members and treat juveniles as adults in the *Violent Crime Control and Law Enforcement Act* of 1994 (Lucas 1998).

Over 1000 US cities and smaller communities have introduced curfews on teenagers which require young people aged under 17 years old to be off the streets by 10.30–11 pm. In the first two months following the imposition of a curfew in New Orleans, 1600 teenagers were picked up and detained until they could be collected by their parents. The Mayor of New Orleans justified this action in a radio interview, saying, 'It keeps teenagers off the streets. They need it, there's too many teenagers hanging around the streets.' It is not a view shared by the American Civil Liberties Union, who condemned the measure for targeting a powerless group of people who do not vote in order to enable politicians to claim that they are doing something about crime. They challenged the legality of curfews but lost when the Supreme Court ruled that the first amendment does not give teenagers a generalized right of association that permits them to be out after hours and that official discrimination on the basis of age is possible.

Mike Davis (1990) suggests that in southern California residential curfews are deployed selectively against black and Chicano youth. He writes: 'As a result of the war on drugs every non-Anglo teenager . . . is now a prisoner of gang paranoia and associated demonology. Vast stretches of the region's sumptuous playgrounds, beaches and entertainment centers have become virtual no-go areas for young Blacks or Chicanos' (Davis 1990: 284). He cites several examples to make his point. In the first, an off-duty black police officer, Don Jackson, took some 'ghetto kids' into an exclusive white area. Here, despite carefully obeying the law, the group were stopped and frisked, and Jackson was arrested for disturbing the peace. In the second example, a group of well-dressed black members of Youth for Christ were on a visit to the Magic Mountains amusement park when, with no justification, they were surrounded by security guards and searched for 'drugs and weapons'.

Whereas in the USA this moral panic about young people is focused on gangs and is explicitly racialized, in the UK there has been a more general anxiety about what a British television documentary dubbed 'the end of childhood'. It began in the mid-1990s with the murder of a two-year-old, Jamie Bulger, by ten-year-old boys. Although extremely unusual, this murder was not completely unprecedented and quickly became a reference point for other cases of violence committed by children. Other evidence, from statistics on bullying, joy-riding and reports of the mugging of the actress Elizabeth Hurley by a group of teenage girls, was mobilized by the media to fuel popular anxieties about the unruliness of young people (Valentine 1996c). Commenting on this phenomenon, newspaper columnists and writers made claims such as:

> The child has never been seen as such a menacing enemy as today. Never before have children been saturated with all the power of projected monstrousness to excite repulsion and even terror (Warner 1994: 43).

> There is a growing uncertainty about the parameters of childhood and a mounting terror of the anarchy and uncontrollability of unfettered youth (Pilkington 1994: 18).

While children have been labelled as the problem in this UK moral panic, the blame for their behaviour has been laid at the door of parents, schooling and the state. All three stand accused of having made children ungovernable by eroding the hierarchical relationship between adults and children. It is argued that parents have traditionally had 'natural' authority over their offspring as a result of their superior size, strength, age and command of material resources. This authority was sustained by the law and religion but also by everyday 'norms' about the appropriate behaviour of adults and children (Jamieson and Toynbee 1989). However, towards the end of the twentieth century understandings about what it means to be a parent are alleged to have changed, with adults voluntarily giving up some of their 'natural' authority in favour of closer and more equal relationships with their offspring. This has been accompanied by a general shift in both legal and popular attitudes to young people, away from an 'adults know best' approach towards an emphasis on the personhood of the child, and children's rights. In addition, the more liberal line adopted by the state towards the physical punishment of children has also removed this tool, which adults could previously use to enforce their authority.

This moral panic about both child violence and the ungovernability of children is being used to justify adults' perceptions that contemporary young people – particularly teenagers – are a threat to their hegemony on the street. Teenagers value the street as the only autonomous space they are able to carve out for themselves away from the surveillant gaze of parents and teachers (Valentine 1996b). Hanging around on street corners, underage drinking, petty vandalism, graffiti (Cresswell 1992) and larking about are some of the ways that young people can deliberately or unintentionally resist adult power and disrupt the adult order of the streets and the suburbs. These uses of space are read by adults as a threat to their property and the safety of children and the elderly, but at the same time they believe that they no longer have the authority to regulate or control teenagers' behaviour. As a result, this moral panic is being used to mobilize a consensus which (as in the USA) is being used to justify various strategies to restrict young people's access to, and freedoms in, 'public' space (Valentine 1996b, 1996c), although these measures have not yet extended as far as the US-style curfews described above.

While these moral panics in the USA and the UK appear to be a contemporary phenomenon, it is worth noting that adults' fears of, and hostility towards the younger generation have been extremely widespread over the centuries. Pearson (1983) observes that there have been complaints about young people's moral degeneracy and criminality for over 150 years. He cites examples of scares about 'cosh boys' and 'Blitz kids' during the Second World War as well as a string of moral panics about young people's misuse of leisure time in the inter-war years.

■ 6.5.2 The black body

In 1991 a black American, Rodney King, was brutally beaten in the street by white police officers in Los Angeles, USA. The event was captured on video and was used in evidence in the trial of the policemen the following year. Despite this visual record of their vicious and violent actions, the officers were acquitted by a white jury, which sparked widespread rioting in Los Angeles.

In trying to understand why the video was read against visible racism Judith Butler (1993b) argues that King was regarded by the jurors as threatening and dangerous to the police, despite the actual video evidence of what happened, because the black body is circumscribed as dangerous prior to any gesture such as the raising of a hand or gun (see also Chapter 2). The police were therefore understood to be structurally placed to protect whiteness against violence, where violence was understood to be the imminent action of the black male body. In these terms the actions of the police officers could be read by the white jurors as *not* violence because, in their eyes, the black body was a site and source of danger and so the police were right to subdue this body even before it had done anything. In other words, the white jurors' belief in the inherent danger of blackness – even though black people have far more to fear from white people than the other way round – allowed them to rearrange the circumstances to fit their own conclusions.

The Rodney King case is a notorious example of the criminalization and demon-ization of the black body, but similar racist constructions of the allegedly inherent criminality of ethnic minority groups structure everyday encounters on the street between the police and minority communities in cities across North America and Western Europe. Peter Jackson (1994) cites the example of Toronto, Canada, where the city's reputation for tolerance and safety was damaged in the early 1990s by a series of police shootings of black suspects which led to a rapid deterioration in relations between the force and members of the city's black (West Indian) communities.

A task force was set up to investigate police–community relations across Ontario; this invited submissions from groups and individuals. These, and newspapers and media commentaries of the time, exposed the extent of popular constructions of the link between 'race' and crime and the associations made connecting 'black crime' to particular neighbourhoods and streets within the city.

These constructions were, in turn, challenged by sections of the black com-munity. For example, an editorial in the black newspaper *Contrast* (17 August 1988) argued:

> This murder [the police shooting of Lester Donaldson] is a continuing story of policemen who are not trained to be sympathetic or to negotiate with brains and not brawn. It's the story of policemen who carry the racist myths around that blacks are savages, incapable of reasoning and therefore must be dealt with in a savage manner (cited in Jackson 1994: 227).

Likewise, the Jamaican-Canadian Association made the point in *In Focus* (1990) that:

> We resent the implication that because we are black we are inherently more criminal, therefore we must be policed differently – with guns drawn and the emergency task force at the ready . . . Black men have died under very questionable circumstances – shot while eating supper, half crippled in a very small room, shot for speeding, shot for driving a stolen car (cited in Jackson 1994: 225).

The Task Force concluded by making a number of recommendations which included increasing police training, the recruitment of more officers from ethnic minority groups and setting up an independent civilian-run police complaints procedure. While there was some grudging acknowledgement that individual officers may have been racist, the police were unwilling to acknowledge the existence of **institutional racism** within the police force, or to recognize that constructions of black criminality needed to be understood within the framework of wider social relations that are characterized by real inequalities of power between black and white people (Jackson 1994).

■   ### 6.5.3 People with mental ill-health

From the Victorian era to the late twentieth century people with mental ill-health were commonly segregated in asylums and hospitals (see Chapter 5). These were intended to be places for the treatment or 'normalization' of 'patients' while at the same time protecting the public from the 'mad' by socially and spatially containing and isolating them from the rest of society (Parr and Philo 1996).

From the 1960s onwards policies of de-institutionalization were adopted throughout Western societies. As a result, the spatial and social segregation of the asylum has been replaced by a model of community care. On the one hand, this can be seen as a positive way of improving care for the those with mental ill-health. On the other hand, the failure to provide adequate support in the community means that ex-psychiatric patients sometimes struggle to cope when leaving institutions. This has created new forms of marginalization in terms of poverty, homelessness (see Chapter 3), which has meant that many people with mental ill-health end up isolated and living in hostels or on the street. Here they face stigmatization and exclusion.

Understandings of what is and what is not acceptable behaviour in 'public' space are socially constructed. Powerful agencies such as the police, private security forces, and even gangs govern behavioural codings of the streets. For the homeless, socially legitimated private space does not exist. 'For those who are *always* in the public, private activities must necessarily be carried out in public . . . public parks and streets begin to take on aspects of the home; they become places to go to the bathroom, sleep, drink, or make love – all socially legitimate activities when done in private, but seemingly illegitimate when carried out in public' (Mitchell 1995: 118).

The social exclusion experienced by those who are homeless is compounded for those who also suffer from mental ill-health. Discourses of impurity, difference and disorder mean that those with mental ill-health are not permitted to occupy public space unconditionally (Parr 1997). Their experiential worlds are neither understood nor tolerated. The different ways in which those with mental ill-health may dress, behave and use space are usually regarded as 'inappropriate', and because their behaviour appears to be unpredictable or disorderly it is often read as threatening to so-called 'normal' citizens. The fact that these usages of the street may constitute for people with mental-ill health necessary and legitimate uses of public space is ignored

---

Box 6.2: Transgressing normative codes of behaviour

'The thing is in crisis, it's just that I feel really good, I feel high, and to be quite honest all I have said to you I would say it in the street! Now that is offensive... of course you have laws against that... You just can't go in town and make a speech, y' know, they say this is a free country, but you are not able to do it! You can be arrested, but then I don't care at that time, I lose all inhibition, but then as I am saying my thinking is a bit warped... as soon as I say something I go back to a mental hospital, so I try to stay away from that environment.'

'One mentally ill, all mentally ill... no one sees the individual anymore... you have no individualistic things about you... is mad... is mad [laughs]. If I am enjoying myself, then I am enjoying myself. It might not suit somebody else: '"He's funny y' know, he's been in hospital y' know?" Fact is, I'm enjoying myself, if I'm freaking out... if I'm having a good time at a club – I'm having a good time! That is why I am there, to have a good dance y' know? But because it's freaking out... "He's crazy, he's been in a hospital"... What can you say? What can you do? Do you govern your life around that?'

Parr 1997: 441–2

---

or denied. As a consequence, individuals can be arrested, restrained or even institutionalized for transgressing 'normal' codes of behaviour (Parr 1997).

In this way, 'mad' people can be denied the freedom enjoyed by other citizens to use and occupy everyday public spaces such as the street. In Box 6.2 Darren (one of Parr's respondents) describes how his instinctive behaviour on the street – for example, he shouts abstract sentences at passers-by – is regarded as unacceptable and leads to him being taken into custodial care. Hester Parr (1997: 451) explains that 'individuals with mental health problems may have a contradictory relationship with public space in the city as they try to reclaim territory for unmodified behaviour, while also experiencing the sane-itising regulations that emerge from common sociocultural codings and understandings of how the self should be presented in everyday life'.

■ Summary

- At different times and in different spaces particular groups are demonized as a threat to other citizens and to the moral order of the street.

- Young people are the subject of popular anxiety in the USA and the UK. In the USA there is a moral panic about youth gangs which is explicitly racialized. In the UK fears are about child violence and the ungovernability of children.

- Black bodies are demonized as inherently criminal and violent.
- The different ways in which those with mental ill-health may behave and use space are often regarded as 'inappropriate' and as threatening.

■ 6.6 The policing of the street

This section examines the role of the police, private security forces, electronic surveillance and community initiatives in the production of the street as a 'public' or 'privatized' space.

■ 6.6.1 The police

The police control the street in three ways. First, the physical presence of uniforms on the street provides symbolic reassurance of order and control (see Box 6.3). Second, the police use active surveillance to monitor behaviour of those in public. Third, they intervene to establish order. Most of this 'peace keeping' is done without invoking the power of the law through arrest but through 'rough informality' such as controlling people's behaviour in space or moving people or groups between spaces (although this form of policing relies on individuals and groups complying with officers' requests). By using these three techniques, the police create order and predictability on the streets and are perceived both to act as a deterrent to crime and to reduce citizens' fear of crime. Rubinstein (1973: 166, cited in Herbert 1998: 233) goes so far as to claim 'For the patrolman [*sic*] the street is everything; if he loses that, he has surrendered his reason for being what he is.'

During the late twentieth and early twenty-first centuries, in response to increasing crime rates, a growth in the size of communities and developments in technology, police forces in North America and Europe have become increasingly professionalized, efficient, and sophisticated organizations (Smith 1986, Herbert 1998). Advances in radios, computers, cars and helicopters have all allowed the police to shift from a strategy of patrolling the streets on foot to what is known as unit beat policing. Although ideally this involves using a combination of car and foot patrols, in reality it has produced car-based police forces which react to rather than deter crime in what has been dubbed a 'fire brigade model of policing'. In association with this change, a powerful occupational culture has emerged within police forces, which has contributed to devaluing the role of the 'beat bobby' (Smith 1986). Consequently, although the police can now 'construct a mobile and expansive net

---

Box 6.3: Policing

I've always liked to see policemen on the streets, I've always found it reassuring. I mean in Woodley you sometimes see them strolling through the shopping parade. I mean only last week I saw two of them walking through together. I mean they only had to walk through there once and it was reassuring.

Reading, UK, middle-class woman

I think you would be helped to feel more confident if you knew there were police-men [*sic*] on the beat. I mean if it returned to the bobby on the beat days, and you'd pass him [*sic*] on the way to the shops and he'd say 'Hello Mrs Webb ...' and you'd know somebody was actually around then you'd feel much safer.

Reading UK, middle-class woman

I think it would be nice to have police on the beat. I think if you knew there were patrols going round it would help a bit, put people [criminals] off. Not that it would necessarily protect you more, but if someone [trouble makers] was aware that there were police around, which is why I think the trouble has gone out of the town centre because they are actually increasing police patrols.

Reading UK, young woman

Valentine 1989: 185

---

of state power', at the same time they have become removed both organizationally and operationally from communities (Herbert 1998: 225).

Susan Smith (1986) argues that this divide between police and citizens is widest in inner cities and amongst racial and cultural minorities. She argues that black people are disproportionately likely to be stopped by the police and that the police are often reluctant to act energetically where the victims of crime are from racial and cultural minority groups. As a result of these apparently unjust law enforcement practices these communities can become increasingly closed to the police because people are unwilling or afraid to co-operate with enquiries. This, in turn, means the police tend to fall back on stereotypes and on 'military'-style policing of these areas. These tactics are then met with resistance from local people, which can result in the creation of what the police term 'no-go areas'.

Steve Herbert (1998) cites the example of Smiley and Hauser in Los Angeles, USA. The residents of this neighbourhood are mostly from low-income minority groups and there is a long history of police–community antagonism. The police regard it as an area of illegal activity and hostility and make periodic attempts to reclaim their authority through increasing their presence on the streets and through more aggressive geopolitical tactics (Herbert 1998). In turn, gangs use their local geographical knowledge to outwit or ambush officers, removing street signs to make it more difficult for police to find their way round, and chalking their insignias on the streets to be read by police helicopters.

In the face of criticisms that they have become organizationally and operationally divorced from the public, the police have begun to revive the notion of 'community' policing as a part of a strategy to reduce citizens' fear of crime rather than actual crime. Pilot schemes in the USA and the UK have involved recontacting victims, newsletters, citizen patrols, and neighbourhood watch schemes (see section 6.6.4). 'Community policing' is also being used in an attempt to improve relations between police forces and citizens in the neighbourhoods which the police regard as 'no-go' areas. These initiatives aim to defuse local tensions and to encourage members of the community to provide more information to the police about local crime. In Smiley and Hauser in Los Angeles, USA, the police have held meetings to try to find out what the concerns of the local residents are and to seek ways of working with them (Herbert 1998).

Community policing itself has also been criticized, however:

- It is viewed as a nostalgia exercise, which harks back to an imagined 'golden age' of discipline and order.

- It is an impractical initiative because communities are not 'real' and 'definable' but rather are imagined and contested (see Chapter 4).

- These schemes are often subverted by a lack of resources and internal resistance from the police, who are reluctant to patrol the streets on foot when other occupational roles provide better career prospects.

- In areas which have a long history of police–community tensions, residents are often so mistrustful of the police that they are reluctant to participate in community programmes, or fear retaliation from gangs if they do co-operate (Herbert 1998).

■ 6.6.2 Private security

Private security includes personnel (in the form of guards, store detectives, investigators, escorts, couriers, and so on), hardware (such as alarms) and closed circuit television cameras (which are discussed in section 6.6.3).

The definition of private security personnel includes all those who are privately employed in jobs with security functions that can be distinguished from both the police and other members of the public who perform 'policing' roles as part of their job, such as ticket collectors or park wardens (Stenning and Shearing 1980). Their numbers have expanded rapidly in North America, Europe and Australia over the last few decades. In the USA private security officers are estimated to outnumber police officers three to one (URL 8). Private security guards are everywhere, from the city streets and shopping malls to offices and suburban neighbourhoods. The

National Association of Security Companies estimates that the USA as a nation spends 73 per cent more on private security than it does on public law enforcement (URL 8). In Australia recent surveys suggest that there are approximately 100 000 people employed as security personnel, which means that they outnumber the state and federal police by more than two to one (URL 9).

The pervasiveness of private security leads Stenning and Shearing (1980) to argue that 'there is a strong probability that if a member of the public is going to be subject to police-like powers by anyone in society that it will be from a private security person rather than a public police officer. In some areas private security is becoming a serious competitor with the public police and the single most important instrument of social control and law enforcement.'

The driving force behind the rise of the private security industry has been the perceived failure of the police to deal adequately with people's fear of crime and a desire to have the sort of 24-hour presence of an officer on the street to act as a deterrent which police officers do not have the time or resources to provide. Writing in the UK newspaper *The Guardian*, the journalist Joanna Coles (1993: 18) argues that: 'The police's monopoly on law and order is under threat as never before. The growth of private firms patrolling public space is inevitable. If the police won't respond to public fears the market will.'

In the USA Mike Davis (1990) argues that private security has become a **positional good** defined by income. Through local homeowners' associations most affluent neighbourhoods in Los Angeles have contracts with private police forces. Signs such as 'Guarded Community' and 'Armed Response' are an everyday sight on suburban lawns, with some neighbourhoods becoming 'gated communities' which are walled off from the public, with access restricted to guarded entry points (Davis 1990). Davis (1990) claims that this growth in the security industry has less to do with personal safety than with the desire of the white middle classes to be insulated from those whom they define as 'unsavoury' or dangerous 'others' (see section 6.5). Indeed, he points out that private security may only serve to fuel citizens' fears.

Private security forces mimic police strategies for controlling space, such as wearing uniforms to symbolize authority, and using 'rough informality' to regulate behaviour on the streets, which produces some confusion about the interface between the two sectors. Private security guards do not have the same powers as the police. Rather, they have the same powers of citizens' arrest as the public, although where they are invested with the authority of a property owner to control private land, they do have the right under civil law to evict trespassers (Kelling 1987).

The growth of privately owned commercial spaces in the city, such as shopping malls and leisure centres, is effectively privatizing activities which once took place in the public space of the street. While the police force is not necessarily barred from these privatized landscapes, they do not form part of the regular beat of police patrols. In other words, large parts of our cities are becoming 'mass private spaces' under the effective control not of the police exercising public law enforcement powers derived through the law and criminal justice system, but of private security forces working for corporate employers whose powers are derived from the rights of private property ownership.

Although under the law in Canada a public place is defined as 'any place to which the public have access as of right or by invitation, express or implied', recent court cases have ruled that owners of shopping malls 'do have the right to withdraw permission to be in such a place from any member of the public at any time, and that anyone who refuses to leave under these circumstances, whatever may be the reason for being asked to leave, commits trespass' (Stenning and Shearing 1980). As a result, when you move out of a publicly owned place into a privately owned so-called 'public' space your rights, freedom and privacy are reduced.

The powers of private security guards to manage everyday 'public' spaces are a problem for two reasons. First, while the police force has specific entry requirements, rigorous selection procedures and training, there are no such controls in place to regulate the employment of private security personnel. A survey of Britain's 8000 private security firms (who employ around 162 000 people), carried out by the Chief Constable of Northumbria, found that their employees were responsible for committing approximately 2600 offences per year. In one firm alone 11 out of the 26 employees had a total of 74 convictions for offences including rape, possession of firearms, housebreaking and assault (Travis and Campbell 1994). US research has produced similar evidence of inadequate background checks on employees, and information that the training of guards often amounts to little more than watching introductory videos (URL 8).

Second, because the law only defines criminal acts as offences, there is little community or societal consensus as to what constitutes 'unacceptable' or 'disorderly' behaviour in public. In many cases there is a clash between collective community values and individual rights around the issue of 'order' and 'peace'. Giving private security services carte blanche to regulate 'public' venues such as neighbourhood streets or shopping malls allows a few legally unauthorized individuals to determine for themselves what behaviour, and from whom, is permissible. Not surprisingly, it is those groups who are demonized as a threat to other citizens and to the moral order of the street – for example, those with mental ill-health, the homeless, teenagers and ethnic minorities, etc. – who receive the rough end of this private policing, being disproportionately stopped and moved on in 'public' places (see section 6.5).

Describing the contemporary downtown 'renaissance' in Los Angeles, USA, Mike Davis (1990: 231) observes that 'the occasional appearance of a destitute street nomad in Broadway Plaza or in front of the Museum of Contemporary Art sets off a quiet panic; video cameras turn on their mounts and security guards adjust their belts . . . It [the downtown renaissance] is intended not just to "kill the street" but to "kill the crowd", to eliminate that democratic mixture on the pavements and in the parks.'

Even police forces are concerned about the extent to which private security forces pose a threat to civil liberties. Speaking at the British Police Federation conference, its chairman, Alan Eastwood, told his audience that 'the expansion of the

Plate 6.3 CCTV is an increasingly pervasive feature of the landscape (© Becky Kennison)

private security industry into the realm of public policing is something to be deplored, to be resisted, to be stopped'.

6.6.3 Electronic surveillance

In 1980 the Miami Beach Police Department, USA, implemented what was then termed a 'video patrol' programme of its main retail areas by putting cameras on top of traffic light poles and monitoring their transmissions. This doubled the surveillance coverage they could provide by deploying officers on the street. Despite technical problems, the 'video patrol' programme appeared to produce a significant reduction in crime and an increased perception of safety amongst local citizens (Surette 1985).

From these modest beginnings, what is now known as closed-circuit television (CCTV) or video surveillance (Plate 6.3) has, like private security patrols, become an increasingly pervasive feature of the urban landscape throughout most affluent Western societies (Squires 1994). By 1996 all Britain's major cities (with the exception of Leeds) had installed CCTV, while a further 200 police and local authority schemes were evident in smaller towns. Commenting on this phenomenon, the journalist Duncan Campbell (1993) writes:

If, in the last 24 hours, you have gone shopping, travelled to work, visited a post office, taken a train, watched a football match, put petrol in your car, visited the off-licence, or walked through a city centre, you will have played at least a small part in one section of Britain's film industry that is experiencing astonishing growth: video surveillance.

The agenda behind the widespread uptake of CCTV has often been an economic rather than a social one (Fyfe and Bannister 1998). In the 1980s many city centres were in decline as a result of the twin forces of de-industrialization and the development of out-of-town retail and business parks. In a bid to resist this decay and attract inward investment, many cities embarked on regeneration projects. As security was one of the major advantages attributed to out-of-town developments, many included CCTV in a whole package of measures designed to attract people and businesses back into the city (Fyfe and Bannister 1996, 1998).

Nick Fyfe and Jon Bannister (1996, 1998) cite the example of Glasgow, Scotland, where a survey of businesses revealed that fear of crime and actual damage or theft to property was putting people off visiting the city and causing businesses to locate elsewhere. The Glasgow Development Agency (a government quango which promotes economic development in the city) tackled the problem by developing a public/private partnership to finance a CCTV system they called Citywatch, using the slogan 'It doesn't just make sense – it makes business sense.' The GDA projected that this initiative would bring up to a quarter of a million more visitors to the city a year, create 1500 jobs and generate an extra £40 million for city-centre businesses (Fyfe and Bannister 1996, 1998).

Although video surveillance systems are assumed to work by acting as a deterrent and by increasing citizens' perceptions of safety, there have been few evaluations of how successful they actually are. There is some evidence that they may improve crime detection rates (Short and Ditton 1996), which was highlighted when this media technology was used to trace two boys who abducted a toddler from a UK shopping centre and murdered him. However, it is difficult to gauge whether CCTV cuts crime rates or just displaces crime elsewhere.

More often than not CCTV is used to track and even harass drunks, the homeless, 'suspicious' youth and others whose appearance, posture or conduct do not conform to 'normal' codes of behaviour (see section 6.5). Norris and Armstrong's (1997 cited in Graham 1998) work reveals that CCTV control rooms are sites of racism and sexism, where certain types of young people are labelled 'yobs', 'toerags', 'homeless low-life' or 'drug-dealing scrotes'. In this way, the operators impose a 'normative space-time ecology' on the city by stipulating who 'belongs' where and when, and treating everything else as a suspicious 'other' to be disciplined, scrutinized, controlled (Graham 1998: 491). In other words, video surveillance represents another subtle form (see also section 6.6.2) through which the management of so-called 'public' space is being privatized and the democratic functions of the street undermined (Reeve 1996). Boddy (1992: 150) comments that 'the potential impact of CCTV is the imposition of a middle-class tyranny on the last significant urban realm of refuge for other modes of life'.

This trend is causing the UK civil rights group Liberty concern. Liberty points out that there is no legislation to provide statutory controls over how video cameras may be used (for example, there are fears that cameras on the street could be used to track individuals' movements or monitor their homes), how long films may be kept and who may have access to them. Indeed, in September 1994 the UK Department of the Environment announced plans to change the law to allow the installation of CCTV without the need for planning permission. In response, Derek Sawyer, a member of the Association of London Authorities Police Committee said, 'A lot of people, particularly in London, want to lead private lives, the thought of security cameras checking who visits them in their homes, for instance, is an unsettling one' (*The Guardian* 1993). The criticism was shrugged off by the then prime minister, John Major, who dismissed the civil libertarians' complaints about surveillance, saying that he was more concerned with the freedom of people to walk the streets without fear and that only those with something to hide need worry.

Surveillant-simulation technologies already mean that human subjects can 'effectively be reduced to their time-space electronic trails or representations, as their movements and behaviour are logged, tracked, and, increasingly, mapped via systems linking CCTV, computerised tracking systems, GISs [Geographical Information Systems], and mobile and fixed phone networks' (Graham 1998: 490). This trend is set to continue as technological developments in tagging, time-space tracking systems, advanced face recognition software and biometric scanning offer even more radical possibilities for new forms of socio-spatial control and repressive socio-spatial practices (Lyon 1994, Graham 1998). Davis (1990: 253), for example, claims that 'Having brought policing up to the levels of the Vietnam War and early NASA, it is almost inevitable that the LAPD [the Los Angeles Police Department] and other advanced police forces will try to acquire the technology of the electronic battlefield and even Star Wars. We are at the threshold of the universal electronic tagging of property and people – both criminal and non criminal (small children, for example) monitored by both cellular and centralised surveillances.'

In this sense, Bentham's Panopticon (1995) (see Chapter 2 and Chapter 5) – a prison design involving a central tower whose complex structure of screens, blinds and lighting prevents its inhabitants from telling when they are under the gaze of the guards – which has become the symbol or image of 'all-seeing power' and of self-surveillance, could be brought too close to reality for comfort.

6.6.4 Community initiatives

The term 'Neighbourhood Watch' covers a wide range of community-based activities supported by the police, which are aimed at preventing neighbourhood street crime (such as street robberies and car theft) and protecting homes from property crime. Neighbourhood Watch schemes encourage residents to take responsibility for the security of their own property and that of the local street. This is done on the one hand by the use of Neighbourhood Watch street signs, stickers in windows,

Plate 6.4   Neighbourhood Watch encourages residents to take responsibility for the street (© Becky Kennison)

newsletters, police talks, marking property and home security surveys; and on the other hand by active surveillance and reporting suspicious activity to the police (Plate 6.4). The intention is to reduce the opportunity for crime to occur and to increase the probability of detection if it does happen.

One of the first Neighbourhood Watch schemes was the Community Crime Prevention Program which was implemented in Seattle, Washington, USA, in 1975. This went on to become the model for subsequent programmes in the USA and later in the UK too (Bennett 1992).

Neighbourhood Watch schemes were first introduced in the UK in 1984. By the early 1990s 115 000 schemes had been established, covering five million households – approximately 28 per cent of the population. This take-up compares favourably with Canada, where 25 per cent of the population are members of Neighbourhood Watch schemes, and the USA, where the figure is only 19 per cent (the figures are reported in Bennett 1992). Only a quarter of all schemes are initiated by the police; the majority are driven by neighbourhood communities (see Chapter 4).

The North American programmes differ from the early British schemes because they usually involve citizen patrols of the street – 'block watches'. But in September 1994 the British government also initiated a campaign called Street Watch. Launching it, the Home Secretary urged British citizens to volunteer to patrol the streets and to be the 'ears and eyes of the police' by reporting anything suspicious.

However, he also made it clear that the Street Watch patrols would have no special powers and that members of the public should not risk 'having a go'.

Neighbourhood Watch and Street Watch schemes have been criticized because:

- Most schemes are initiated by residents in white, middle-class areas which are low in crime, rather than by those who live in low-income, high-crime neighbourhoods.

- The co-ordinators are not selected by the police but are volunteers – usually older, white, middle-class, owner occupiers, who are married with children (Skogan and Maxfield 1981, Bennett 1989) – who receive very little training and, in many cases, devote very little time to Neighbourhood Watch.

- The success of schemes depends on the motivation and enthusiasm of the local co-ordinators and the neighbourhood community. Initial eagerness is often not converted into active participation (Husain 1988). There is no evidence that Neighbourhood Watch members are more vigilant than anyone else. Indeed, the evaluation of schemes in London found disappointingly low levels of street surveillance and participation in home security measures (Bennett 1992).

- Police support in terms of money, the provision of materials (residents often have to pay for street signs) and the time officers can spend with co-ordinators is often inadequate.

- While initially successful, many schemes fail in the long term. Evaluation of US schemes in Seattle, Washington, and in Hartford, Connecticut, showed that, although victimization rates were reduced at the outset, this effect declined within 18 months (Bennett 1992). More surprisingly, a study of five Neighbourhood Watch schemes in Chicago, USA, found that crime declined in only one area and that in three out of the five areas it actually rose (Rosenbaum 1987).

The Street Watch scheme has been criticized by the police as 'vigilantism by the back door'. Vigilantism, defined as 'organised, extralegal movements, the members of which take the law into their own hands' (Brown 1975: 95–6), is already on the increase in North America and the UK. Citizens' disillusionment with the failure of the police to patrol the streets and to deal with crime has resulted in local groups taking action to dispense summary justice. For example, in Bierley, Bradford, UK, local men dealt with car thieves by breaking their fingers and stripped a joyrider and dumped him naked on a local moor at night (Pendry 1993). As such, vigilantes typically subscribe to the view that, on occasions, it is necessary to break the law in order to uphold it. Senior police officers have warned that well-intentioned Street Watch volunteers could face dangerous situations and become targets for retaliation.

Community schemes as a whole have also been criticized because they can potentially produce social exclusions and because of the way they might be appropriated by the state for anti-democratic purposes. First, they focus on the spectre of trouble from 'outside', privileging the ideal of sameness over difference. Rosa Ainley (1998: 94) observes, for example, that: 'Neighbourhood Watch schemes are run on [the] principle [that] . . . members of the scheme are pure, while anyone outside is potentially contagious and therefore worthy of suspicion.' The effect of attempting to secure unity or the boundaries of community by regulating difference can therefore produce exclusions and trigger violence and hatred (see also Chapter 4).

Second, some groups have greeted the advent of 'community policing' with suspicion rather than enthusiasm because they interpret it as a potential threat to civil liberties. Bridges (1983: 32) argues that chief police officers have 'set about reorganising their forces the better to penetrate and spy on the community and to suppress social and political unrest'. These critics fear that community policing might become a way of undermining political pressure groups and that police involvement in community activities could become a form of surveillance, perhaps leading to preemptive action against potential (but not actual) criminals. Some have even claimed that extending police powers into 'communities' could be interpreted as one element in a drift towards a police state.

■ Summary

- In the twentieth and twenty-first centuries police forces have become increasingly professionalized, efficient and sophisticated organizations.

- Advances in technology have allowed the police to shift from foot patrols to unit beat policing, but this has removed them from communities.

- The perceived failure of the police to deal adequately with people's fear of crime has led to the rapid expansion of the private security industry.

- The private security industry imposes 'a 'normative space-time ecology' on the city by defining who 'belongs' where and what is 'appropriate' behaviour. In this way, it can undermine the democratic functions of public space.

- Community initiatives include activities supported by the police which are aimed at preventing neighbourhood street crime and protecting homes.

- These have been criticized for producing social exclusions and because they may be appropriated by the state for anti-democratic purposes.

■ 6.7 The contested street: the end of 'public' space?

Section 6.2 outlined how the public space of the street has been celebrated as a democratic and inclusive space whose defining characteristics are 'proximity, diversity and accessibility' (Zukin 1995: 262). The subsequent sections, however, have shown how the public realm, rather than being a social order of civility, sociability and toler-ance, has increasingly become one of apprehension and insecurity. Encounters with 'difference' are being read not as pleasurable and part of the vitality of the streets, but rather as potentially threatening and dangerous. Changes in contemporary policing which have resulted in forces becoming organizationally and operationally divorced from the public, and a perceived decline in neighbourhood community, are also contributing to people's perceptions of vulnerability in public space.

White middle-class fears of, and desires to be insulated from, 'otherness' have led to white middle-class withdrawal from visible roles within the city (Davis 1990, Sorkin 1992). This has had an effect on the social, physical and aesthetic appearance of public environments. Michael Waltzer (1986) argues that, as a result, a spiral of aban-donment and decay has set in because, as the streets are filled up with social, sex-ual and political deviants, so more ordinary people are fleeing from them, which is leading to the further degeneration of public space. Consequently, he claims, pub-lic space has become a place of fear and loathing where suspicious and hostile encoun-ters are expected. Waltzer (1986) grudgingly concedes that deviant others belong in the urban mix but he argues they can only be tolerated in public space as long as they stay on the margins.

Numerous cities have embraced a range of measures in a bid to halt this per-ceived decline in public space. These include the introduction of CCTV and private security forces into 'public' environments and urban renewal schemes which have replaced 'public' spaces altogether with surrogate 'private' spaces such as shopping malls and festival marketplaces. These measures are leading to a general process of transformation. First, the policing initiatives, which are driven by citizen fear, are designed to filter out 'undesirables' or 'deviants' and so establish or reproduce a white middle-class moral order on the streets. Indeed, Smith (1996b) goes so far as to sug-gest that, in the USA, once-marginalized and disadvantaged groups such as the poor and the homeless are now perceived by the middle classes to be receiving too much 'special assistance', and that exclusionary policies such as 'zero tolerance' of crime and disorder, anti-homelessness laws (see Mitchell 1997) and policing strategies aimed at 'cleaning up' the streets are all part of a backlash or politics of revenge against these social groups. Second, gentrification (see also Chapter 7) and the development of new privatized spaces of consumption are serving to homogenize and domestic-ate public spaces by reducing and controlling diversity in order to make these envir-onments safe for the wealthy middle classes (Smith 1992b). It is a process which has been termed the disneyfying of space (Sorkin 1992).

Critics of this trend argue that state planners and corporate enterprises are pro-ducing carefully monitored and controlled environments 'that are based on desires for security rather than interaction, for entertainment rather than (perhaps divisive) politics' (Mitchell 1995: 119). There is increased willingness on behalf of the white middle classes to surrender privacy and to tolerate interventions that restrict the civil rights and liberties of 'others' in return for a belief that their personal security is in

---

Box 6.4: The People's Park

An example of the sort of oppositional and irreconcilable visions of public space articulated by Waltzer (1986) and Berman (1986) is found in Mitchell's (1995) study of
the People's Park in Berkeley, USA. Here the University of California wanted to redevelop the park which was one of the few areas in the San Francisco Bay area that
was a relatively safe place for the homeless to congregate without being harassed.
The developers wanted to remove what they regarded as unsightly homeless
people and political protestors because they were driving away other users, and to
create the park as a controlled, orderly and safe space into which only an 'appropriate' public would be allowed. These redevelopment plans were opposed by activists
and homeless people. For these groups, the park represented a truly public space
because it was an unconstrained space that was open to anyone and where
political movements could organize, and where the homeless were free to live.
In the eyes of the campaigners, homeless people had the right to be included or
represented within 'the public' and their use of the park constituted a legitimate and necessary use of public space. The risk of disorder and conflict as a result
of allowing free and unmediated interactions was understood to be valuable and
manageable.

---

some way enhanced (Surette 1985). Through such measures, it is argued, 'public'
space is being drained of its vitality and meaning. Free expression is being prohib-
ited. The democratic mix of the street is being undermined by the exclusion of so-
called undesirables such as youth, the homeless, those with mental ill-health and
so on (see Box 6.4). This, in turn, is putting their legitimacy as members of public in
doubt (Mitchell 1995).

Don Mitchell (1995) points out that the elision of shopping malls and guarded
communities with notions of public space hides from us the extent to which the public realm is being privatized and commodified. These spaces are now becoming a
site of struggle between a particular kind of public interest (the police) and private
interests masquerading as 'public' interests (e.g. private security industries) (Boyer
1993, 1996).

Pessimists (Sorkin 1992, Sennett 1993) are going so far as to proclaim the death
of the street and the end of truly public space. For such writers, the maintenance
of 'public' space requires its frequent use by a wide mixture of people. Marshall
Berman (1986), for example, is opposed to those like Waltzer (1986) who support
the privatization and commodification of 'public' space on the grounds that this is
what is necessary to return the middle classes to the street. Rather, Berman (1986)
challenges Waltzer's (1986) privileging of middle-class lifestyles and the emphasis
on the significance of their voice in determining the nature of 'public' space. For

Berman (1986), public space fails not when it is full of so-called 'deviants' but when they are absent. He writes, '[T]here isn't much point in having public space, unless these problematical people are free to come into the centre of the scene . . . The glory of modern public space is that it can pull together all the different sorts of people who are there. It can both compel and empower these people to see each other, not through a glass darkly but face to face' (Berman 1986: 482).

In contrast to Waltzer's (1986) single-minded space, Berman argues for what he terms open-minded space. This is space which is open to encounters between people of different classes, races, ages, religions, ideologies, cultures and stances towards life. He wants public space to be planned to attract all these different populations and to be discreetly policed.

Berman (1986) observes that 'public' space has always been a site of encounter and conflict between different groups. The central issue for him is to find ways of resolving these differences in ways which preserve 'public' space by sustaining its shared use rather than diminishing and destroying it by exclusion. He states that 'No doubt there would be all sorts of dissonance and conflict and trouble in this [open-minded] space', but, he argues, 'that would be exactly what we'd be after. In a genuinely open space, all of a city's loose ends can hang out, all of a society's inner contradictions can express and unfold themselves' (Berman 1986: 484).

This view is also evident in the work of Richard Sennett. He claims that 'The public realm should be gritty and disturbing rather than pleasant' and that disorder and painful events are important 'because they force us to engage with "otherness", to go beyond one's own defined boundaries of self, and are thus central to civilised and civilising social life' (Sennett 1996: 131–2). For Berman, Sennett and others too, fear and anxiety are not something to be avoided, rather they are merely the flip side of the stimulation and challenge associated with cosmopolitanism (Robins 1995).

Other writers, however, have sounded a note of caution in the face of this debate between what appear to be two oppositional and irreconcilable visions of public space. Loretta Lees (1998), for example, argues that geographers are in danger of think-ing of spaces such as the street either as spaces of political expression and celebra-tion or as a sites of repression and control, and that to do so is to overemphasize the extent to which public space is being 'controlled' and to underemphasize its con-testation. Likewise, Sharon Zukin (1995) also warns against focusing on doomsday scenarios which prophesy the death of public space. She regards theme parks and shopping malls as public spaces even though they are privately owned and managed, pointing out the fuzziness and permeability of boundaries between the concepts of 'public' and 'private' held by urban citizens. For Zukin (1995), public spaces are always taking new forms, and different interests are continually interacting within them and struggling for influence. The current debates about different visions of pub-lic space are just another part of an endless story through which citizens contribute to defining what is, or is not, 'public' space and who and what is, or is not, accept-able within it.

■ **Summary**

- A middle-class desire to be insulated from 'otherness' is, some claim, leading to a middle-class withdrawal from public life and a decline in public space.

- In response, cities are embarking on a range of measures which are aimed at halting this decay by creating safe and ordered 'public' environments.

- Pessimists complain that these measures are privatizing and commodifying public space and killing the democratic mix and vitality of the streets.

- Critics argue that truly public space should be gritty and disturbing because for a space to be 'public' it must, by definition, involve encounters between strangers that will produce conflict and dissonance.

- Others warn that this debate may be overemphasizing the extent to which public space is being 'controlled' and underemphasizing its contestation.

■ Further Reading

- Important sources which explore the general debate about the nature of public space include papers by Waltzer and Berman in *Dissent*, vol. 33, 1990, some of the work of Mitchell, especially his articles in *Annals of the Association of American Geographers*, 1995, 85, 108–33; *Urban Geography*, 1996, **17**, 158–78; *Antipode*, 1997, **29**, 303–35 and books such as Fyfe, N. (ed.) (1998) *Images of the Street*, Routledge, London; Smith, N. (1996b) *The New Urban Frontier: Gentrification and the Revanchist City*, Routledge, London, and Sorkin, M. (ed.) (1992) *Variations on a Theme Park: the New American City and the End of Public Space*, Hill and Wang, New York. A more positive view of the street is presented in Zukin, S. (1995) *The Culture of Cities*, Blackwell, Oxford.

- The moral order and policing of the street are the subject of a number of books including Baumgartner, M.P. (1988) *The Moral Order of a Suburb*, Oxford University Press, New York; Davis, M. (1990) *City of Quartz: Excavating the Future in Los Angeles*, Verso, London, and Lyon, D. (1994) *The Electronic Eye: The Rise of the Surveillance Society*, Polity Press, Cambridge. These issues are also explored in Fyfe, N. and Bannister, J. (1996) 'City watching: closed circuit television surveillance in public spaces', *Area*, **28**, 37–47; Fyfe, N. and Bannister, J. (1998) ' "The eyes upon the street":

closed-circuit television, surveillance and the city', in Fyfe, N. (ed.) *Images of the Street*, Routledge, London, and Graham, S. (1998) 'Spaces of surveillant simulation: new technologies, digital representations and material geographies', *Environment and Planning D: Society and Space*, **16**, 483–504.

■ Exercises

1. For a week keep a diary of where you go and when. Reading your diary, think about your use of space. What patterns can you observe? Are there places to which you do not go or where you are fearful of going? Why? What sort of precautionary strategies, if any, do you use? How does your use of space compare with that of your friends?

2. Mark onto a base map of your local area all the CCTV cameras that you can find in your neighbourhood/campus. How many are there, what sort of spaces do they cover? How do you think these surveillant systems affect the production of space? What is your position in the debate between the crime reduction and civil rights lobbies?

■ Essay Titles

1. To what extent do you agree with Mike Davis's (1990) argument that crime control programmes driven by citizen fear are destroying accessible public space?

2. Critically assess Richard's Sennett's (1993) argument that '... the public realm should be gritty and disturbing, not pleasant'.

3. Consider, with the use of geographical examples, how 'moral orders' are produced and contested in specific places.

7 The city

■ 7.1 The city

Millions of people live in cities around the world. They are thus hugely significant spaces. Geographers have focused not only on their size and physical features (architecture, communication networks), but also on the people, goods and information which flow through them and the sort of social relations and spatialities these flows produce.

In particular, the size and density of urban populations mean that cities are places where all different sorts of people are thrown together. This diversity and the juxtapositions it produces offer many positive possibilities but they can also produce tensions and conflicts (see Chapter 6). Likewise, the density of city living also raises issues about the sustainability of this lifestyle and about the relationship between human residents, their environment and other urban dwellers such as animals and wildlife.

This chapter explores the differentiation of urban space. It begins by examining geographical work on ethnic segregation, gentrification, the underclass and sexual dissidents in the city. It then moves on to think about city life. In the 1970s and 1980s cities in the affluent West were generally in decline, being characterized by de-industrialization, population decline and structural unemployment. However, the end of the twentieth and beginning of the twenty-first centuries have witnessed efforts to revalorize city centres by converting their cultural capital (as sites that contain cultural heritage and are sites of cultural production) into economic capital. This chapter considers attempts to reorganize city centres around consumption in which an emphasis is being placed on reviving the act of wandering the streets (flânerie) and on promoting consumer goods and leisure-time activities such as shopping, restaurants and clubs. Part of this process has also involved the re-imaging of cities. From this emphasis on culture the chapter then turns to focus on the relationship between humans and nature within the city, particularly the place of animals in the city and urban environmental politics. It concludes by reflecting on the emergence of virtual cities.

7.2 The heterogeneous city

Cities are characterized by *density* in terms of their populations, buildings and traffic and by *intensity* in terms of the pace of life, social interactions and the range of opportunities they offer. The density and intensity of cities also means that they are above all else sites of *proximity*, the 'place of our meeting with the other' (Barthes 1981), with all the emotions of excitement, frustration or anxiety that this heterogeneity engenders (see Chapter 6).

Yet, within the heterogeneous city it is also possible to differentiate relatively homogeneous areas (see Figures 7.1 and 7.2). Urban sociology and geography have a long

Figure 7.1 The percentage of the population in ethnic London by ward: white. Redrawn from Pile, Brook & Mooney *Unruly Cities?* (1999).

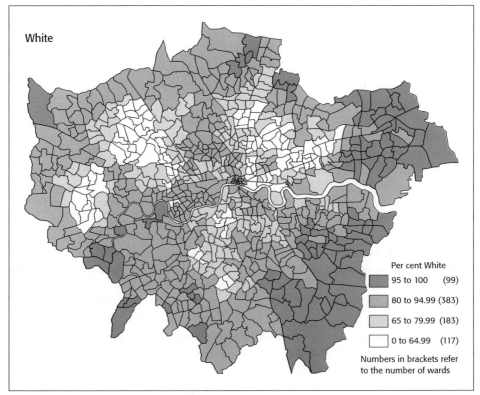

White

Per cent White

95 to 100 (99)

80 to 94.99 (383)

65 to 79.99 (183)

0 to 64.99 (117)

Numbers in brackets refer to the number of wards

Figure 7.2 The percentage of the population in ethnic London by ward: Afro-Caribbean. Redrawn from Pile, Brook & Mooney *Unruly Cities?* (1999).

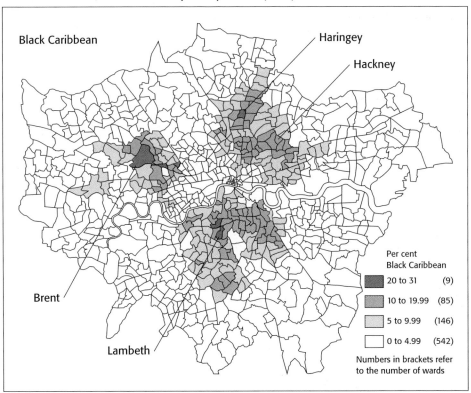

history of focusing on different worlds within the city. The Chicago School of Human Ecology (CSHE) (see Chapter 4) famously mapped and modelled what they regarded as the 'natural' and predictable spatial patterns of ethnic neighbourhoods, the ghetto, areas of 'vice' and the suburbs. This work was very influential, and in the 1960s and 1970s, when positivism was at its height, social geography was pre-occupied with using techniques such as social area analysis and factorial ecology to map the segregation of different housing areas within cities identified on the basis of class or ethnicity.

In the 1970s radical geographers sought to explain and challenge patterns of inequality and material poverty within cities, for example, by focusing on the role of the housing market and building societies in structuring urban space. Sub-sequently, postmodernism has led to a questioning of the focus purely on economics and a recognition of the intertwining of the material and the symbolic. While eco-nomic institutions are obviously fundamental in the (re)production of the material condition of everyday life, cultural representations of particular groups (e.g. race or underclass) in popular discourses are also tangled up with operation of labour, hous-ing markets, etc. In the same way, 'contemporary political struggles around rights and entitlements are often as not struggles that cohere around a politics of identity constituted through processes as much cultural as economic' (Fincher and Jacobs 1998: 3).

The following sections examine some of these debates in relation to ethnic segregation, the underclass, gentrification, and the location of sexual dissidents within the city.

■ 7.2.1 Ethnic segregation

In the 1960s and 1970s social geographers used a range of indices to measure and quantify ethnic segregation (e.g. Peach 1975, Peach *et al.* 1981). These descriptions of spatial patterns were accompanied by vague attempts to explain them which were framed in terms of 'constraint' or 'choice'. This work was heavily criticized by radical geographers and black political activists as 'narrow empiricism' and as 'sociocultural apologism for racial segregation' (Bridge 1982: 83–4 cited in Jackson 1987). The assumptions about 'race' made within this work are at odds with contemporary understandings of identity and difference. 'Race' was regarded as an 'essential' or pre-given category (see Chapter 2). The concentration of particular ethnic minority groups within the city was understood as a problem, but one that would be solved by assimilation, which, it was assumed, would be the inevitable outcome of 'natural' processes of competition and growth (see Chapter 4). This work is now seen as outdated and problematic (Jackson 1987). Some writers still continue to claim that social differences are a product of genetic difference, however – an example being Richard Herrnstein and Charles Murray's (1994) book *The Bell Curve*, in which the authors made claims about the natural or immutable 'cognitive abilities' of different social groups, which purported to demonstrate white superiority. *The Bell Curve* provoked huge controversy in the USA and was widely criticized as racist (see Mitchell 2000).

In the late 1980s researchers began to argue that 'race' is a social construction rather than a 'natural' difference and that, as such, it has no explanatory value in itself. This work examined how people have been classified on the basis of genetic traits and physical differences, such as skin colour, and demonstrated that these social classifications are the product of specific historical circumstances – particularly European colonialism and slavery in the USA – rather than any innate distinctions (Miles 1982) (see Chapter 2 and Chapter 6).

This focus on the social construction of 'race' prompted geographers to think about segregation as a product not of 'natural' ecological processes but rather of white racism. Peter Jackson (1987: 12) defined racism as the 'attempt by a dominant group to exclude a sub-ordinate group from the material and symbolic rewards of status and power on the basis of physical, or cultural traits which are thought to be inherent characteristics of particular social groups'. Racism, he argues, is commonly dismissed as individual prejudice, when in fact it is reproduced, often unintentionally, through the policies and practices of public and private institutions (in relation to housing, education, policing, employment, and so on), and through representations in the media and culture (this cultural racism includes stereotyping of black people as criminal or lazy, Pakistanis as docile or nimble-fingered, the Irish as stupid, etc) (Jackson 1987). This new-found focus on racism led geographers to understand racial segregation as a product of processes of inclusion and exclusion that were both material and symbolic.

In a study of patterns of residential segregation in the UK, Susan Smith (1987: 25) observed that the black population (whom she defined as 'people of South Asian and Afro-Caribbean appearance') were overwhelmingly concentrated in the poorest inner neighbourhoods of cities like London, Birmingham and Manchester. This pattern emerged during the postwar years, when, in response to a labour shortage, people from the Commonwealth countries were encouraged to migrate to the UK to fill these vacancies. This was also a time of a parallel housing shortage, which meant that black workers arriving in Britain were initially forced into poor-quality, inner-city, privately rented accommodation. The government of the time did nothing to address this. It was reluctant to build a concept of race into legislation to combat disadvantage and discrimination and assumed that dispersal would occur 'naturally' as a result of economic development. Instead, this initial pattern of segregation largely continues until today.

Smith (1987) argued that its persistence cannot be understood as chance, or just as a result of income differentials or economic marginality, but rather that it has been sustained through the racist assumptions and practices of individuals and government bodies in a process termed institutional racism. Her work showed that central government policies, including the sale of the public housing stock, housing legislation, the policies and practices of local authorities, the role of financial institutions in providing mortgages, and the actions of estate agents (realtors), have all sustained this pattern of racial segregation by restricting black households to inner-city areas. Smith (1987) concluded that:

> the fact is that irrespective of intentionality, the thrust of national housing policies has been towards racial segregation, the effects of most local institutions have been to protect the housing environment of privileged whites from the entry of blacks, and the outcome is that racial segregation is associated with black people's disadvantage.

Smith (1987) further located the government's laissez-faire approach to racial segregation within broader cultural and political discourses about 'race'. In a study of Hansard which records parliamentary debates she shows how, in the immediate postwar period, black people were constructed as inferior, childlike and backward. In the mid-late 1960s the political rhetoric became more blatantly racist and segregationist. Peter Griffiths fought for the parliamentary seat of Smethwick at the 1964 General Election under the slogan 'If you want a nigger for a neighbour vote Labour.' At this time black people were presented as a threat to white jobs and housing – a fear which culminated in Enoch Powell's infamous 'rivers of blood' speech in which he presented an image of black outsiders swamping white culture and called for a policy of repatriation. (Not too dissimilar sentiments have also been reproduced at the beginning of the twenty-first century in parliamentary debates about asylum seekers entering the UK.) In the 1980s, the emergence of New Right politics brought with it 'new racism', in which social divisions were explained as a product of cultural difference and individual choice rather than white superiority. Smith (1987) suggests that this was merely a more euphemistic or sanitary way of articulating much the same sentiments as Powell.

A further example of the role of racism in producing ethnic segregation within the city is Kay Anderson's (1991) research on Chinatown in Vancouver, Canada, in the period 1880–1980. In this classic study she demonstrated how the racist attitudes of the white council and the determination of white populations to force the Chinese out of particular neighbourhoods through violence led to the emergence in the nineteenth century of a segregated Chinatown, located in a marginal swampy area of the city. This process of segregation was set against a backdrop of ideas about racial difference and the inferior nature of oriental people, which were articulated through the media, scientific publications, and links with the Empire. White settlers regarded themselves as superior to Chinese settlers, viewing the Chinese as a threat to their jobs, trade and land. Both law-making and law-enforcing institutions, as well as mob violence, were used to reproduce white privilege and racial separation. Indeed, efforts were even made to prevent Chinese immigration into Canada, as well as to stop the dispersal of the Chinatown ghetto. Until the mid-1930s Chinatown was negatively stereotyped as a source of diseases, opium dens and other evils.

However, although segregation is often the cause of despair, poverty and danger there can also be a positive and empowering side to it. Ethnic minority groups can remake areas and transform their exclusive nature. In the twentieth century white people have adopted a more romantic vision of the Orient and Vancouver's Chinatown has been marketed as a tourist attraction. This commodifying of the neighbourhood has created the basis for community empowerment and new forms of political action. Notably, the economic clout of Chinatown as an important visitor attraction has led the Vancouver authorities to consult 'community' leaders, enabling the Chinese to be more vocal about their own interests. In this way, the meanings of divided spaces can be recast (Anderson 1991).

In the 1990s the emphasis on the social construction of race and racism also opened up the issue of whiteness (see, for example, Frankenberg 1993a, 1993b, Jackson 1998b, Nayak 1999a, 1999b, Bonnett 2000). In particular, it exposed the way in which white people view themselves as racially and culturally neutral, rather than recognizing their racial and cultural privilege (see Box 7.1) or thinking about what it means to be part of a dominant and normative racial and cultural group (Frankenberg 1993b). In research among white women, Ruth Frankenberg (1993a) asked her informants to recall childhood memories of when they first became aware of 'race'. Although these women had grown up with differing levels of ties with black, Asian and Chicano people, some in anti-racist environments and some in racist contexts, their narratives

---

Box 7.1: White Privilege

White Privilege
Today I got permission to do it in graduate school
That which you have been lynched for,
That which you have been shot for,
That which you have been jailed for,
Sterilised for,
Raped for,
Told you were mad for –
By which I mean
Challenging racism –
Can you believe
The enormity
Of that?

Frankenberg 1993a

---

shared a subtle racism. The form this took included 'educational and economic privileges, verbal assertions of white superiority, the maintenance of all-white neighbourhoods, the invisibility of black and Latina domestic workers, white people's fear of people of colour [even though black people have far more to fear from white people than the other way round] and the "colonial" notion that the cultures of peoples of colour were great only in the past' (Frankenberg 1993a: 77–88). On the basis of these observations, Frankenberg (1993a) classifies three dimensions of whiteness: *structural advantages* – in other words, the privileges white people receive in terms of higher wages, better access to health care, education and the legal system; *standpoint* – the place from which white people view self and society; and *cultural practices*, which are often not recognized by white people as being 'white' but rather are thought of as 'normal'.

Frankenberg (1993a: 78) concludes that racism 'appears not only as an ideology or political orientation chosen or rejected at will; it is also a system and set of ideas embedded in social relations'. She argues that white people can never be outside racism as a social system and that we need to recognize how deeply embedded racism is in all white people's lives. To achieve this Frankenberg (1993a) suggests that white people need to re-examine their personal histories, change their consciousness, and participate in political projects to achieve structural change. At the same time, she also recognizes that whiteness is not a universal experience but rather is inflected by nationhood. In other words, while there may be similarities in terms of what it means to be white in the UK, USA or Australia, there are also differences, because whiteness is produced by particular historical social and political processes (see Chapter 9).

Likewise, white identities are also cross-cut by other differences such as gender, age and sexuality (Frankenberg 1993b).

Whiteness is profoundly encoded in the US city–suburb distinction. The suburbs provide a refuge for the white middle classes from the dirt, crime, disorder and, above all, the heterogeneity of the city. While the city is characterized by different worlds, the suburbs are synonymous with homogeneity, stability, peacefulness and whiteness. The defining qualities of the city are its intensity, proximity and encounters between strangers, whereas the defining qualities of the suburban landscape are its prosperity, and a moral order based on restraint and non-confrontation, in which encounters with difference engender fear and dread (see Chapter 6). Minnie Bruce Pratt (1992: 325) describes how the whiteness of the suburbia in which she grew up, and where she lived with her husband until she came out as a lesbian and moved to an inner-city neighbourhood, moulded her understandings of self and society. She writes: 'I was shaped by my relation to those buildings and by the people in the buildings, by ideas of who should be working in the board of education, of who should be in the bank handling money, of who should have the guns and the keys to the jail, of who should be in the jail; and I was shaped by what I didn't see, or didn't notice on those streets.'

However, Kay Anderson (1998a) cautions in an autocritique that social geographers need to be wary of always thinking in terms of racialized dichotomies such as black–white. She argues that focusing on only one fixed attribute such as 'race' obscures the multiple faces of social power, and that privileging a singular identity as a political 'rallying point' can also produce other exclusions (see Chapter 4). 'Without trying to discredit research that identifies the contribution that race-based oppression makes to patterns of segregation and inequality', Anderson (1998a: 206) highlights the importance of recognizing a 'multiplicity and mobility of subject positionings, including race and class'. Returning to her work on Chinatown, in Vancouver, she identifies the need to refine previous conceptualizations of Chinatowns as having a stably positioned racialized identity in a fixed and antagonistic relationship with a coherent European oppressor. Rather, she argues that Chinatowns are complex sites of difference (see Chapter 2).

According to Anderson (1998a), the early Chinatown in Vancouver was characterized by a 'socio-economic pyramid' headed by a group of men who were some of the wealthiest individuals in the city. These Chinese merchants benefited from the marginalized position of their racialized identity because, without the protection of white unions, and facing immigration tax debts, their fellow Chinatown residents were willing to work long hours for low pay. Indeed, many women worked unpaid for Chinese tailors. These class- and gender-based oppressions are still being reproduced in contemporary Chinatowns in North America. In this way, the Chinese are a complexly differentiated minority group rather than a coherent victim of white oppression.

The category 'Chinese' is also ethnically differentiated. Anderson (1998a) cites the example of Chinatown in New York where non-Cantonese-speaking workers from mainland China are racially 'othered' as inferior foreigners within the Chinese 'community'. Many of these people are illegal immigrants who owe huge debts to the contractors who smuggled them into the country and, as such, they have to work in appalling conditions for a very low wage. Thus Anderson (1998a: 209) argues that 'the race typification of "Chinese" works to obscure other vectors of power that have the enclave as their protection. It follows that to write a more inclusive Chinatown story [or other stories of ethnic segregation] requires one to draw on the reciprocal determinations of (at least) class *and* race, economy *and* culture'.

■ 7.2.2 The underclass

The term '**underclass**' is one in a long history of terms used in North America and Britain to describe those on the bottom rung of the social ladder whose experiences are characterized by persistent intergenerational poverty, dependency on welfare, unstable employment/unemployment, low skills, a limited access to education and the social services, and a high incidence of health problems (physical and mental) (Robinson and Gregson 1992). More controversially, within the USA the underclass is often read as synonymous with African American, and to some extent Hispanic populations, in the inner city. As such, it has become a codeword for 'race' which 'may serve both to promulgate racism and conceal the issue of racial discrimination' (Robinson and Gregson 1992: 41). Likewise, much attention has been drawn to the feminization of poverty, specifically the number of female-headed households and levels of teenage pregnancies in poor urban neighbourhoods (Kelly 1994).

In the USA the underclass is typically spatially concentrated in inner-city 'ghettos' (Cottingham 1982: 3), while in the UK it is elided with specific urban public housing estates (Campbell 1993). These marginal locations have come to symbolize the place of the underclass within the city. The existence of the underclass also represents a powerful critique of urban policies that promote gentrification (see section 7.2.3) and marginalize the urban poor.

At the same time as the underclass has been constructed in and through the space of 'the ghetto', so too these spaces have been constructed through the people who inhabit them. In the context of the UK, Bea Campbell (1993) argues that, because the underclass has been demonized, the city neighbourhoods where these people live have come to symbolize crime and drug abuse at both a local and national level. In the USA guns and gangs also play a part in these representations (see Chapter 6).

The term 'underclass' was first employed by Gunnar Myrdal (1962) to describe those excluded from the labour market as a result of wider structural change. It was adopted more widely in the 1980s when, in contrast to the way it had first been defined, the New Right employed it to highlight what it regarded as a 'dependency culture' and to evoke negative stereotypes of beggars and ghettos. This was a time when rising levels of poverty were increasing the social marginalization of those in US inner cities. Rather than consider that this might be an outcome of laissez-faire

Reaganomics, the Right argued that members of the underclass were victims of their own unwillingness to work, and of their own anti-social conduct, and that their negative attitudes and 'deviant' behaviour were being passed from one generation to another, producing a destructive underclass culture and cycle of intergenerational poverty (Robinson and Gregson 1992). Indeed, some US conservatives, most notably Charles Murray (1984), went so far as to claim that welfare actually promoted dependency and 'deviance' and should be cut further (Robinson and Gregson 1992).

While the Left shared the Right's diagnosis that members of the underclass were trapped in a vicious cycle of intergenerational poverty, it identified a very different cause of their disadvantage. In a book entitled *The Truly Disadvantaged*, Wilson (1987, 1989) argued that the emergence and growth of the underclass was an inevitable consequence firstly of structural economic changes which had created unemployment in the city, and secondly of the severe social policies pursued by the New Right, which had created spatial concentrations of poverty. In Wilson's view, these problems needed to be addressed through social programmes to promote economic growth (Robinson and Gregson 1992). Otherwise, trapped in poverty and with no stake in society, the underclass were being denied the full rights of citizenship (see Chapter 9).

The explanations for the plight of the underclass offered by both the Right and the Left represent highly simplistic accounts of a complex situation. While the Right overemphasizes the role of the individual and underplays structural changes, the Left overlooks the different processes through which individuals may become marginalized and their agency in developing adaptive survival strategies, of which a 'deviant culture' might be considered a part (Robinson and Gregson 1992). For people with little or no access to employment, crime represents one way to acquire the consumer goods so valued within society, which they are otherwise denied (Jordan and Jones 1988).

Despite the debates about the origin and definition of the underclass, Robinson and Gregson (1992) argue that the concept has played an important part in highlighting the fact that social polarization did increase in the USA and UK towards the end of the twentieth century and that, as a consequence, a group of people have become trapped and isolated both socially and spatially from the rest of society. However, they also warn against the danger of placing too much emphasis on the term 'underclass', which they point out actually homogenizes what is a diverse group of people who face different problems and have different responses to their situation.

For women, it may be childcare that keeps them trapped in poverty, whereas for black people it may be racism. Bea Campbell (1993) provides a powerful account of the responses of young British white working-class men to their situation (see also Nayak 1999b). She argues that, with no work, no income, no property and no cars, the only way in which they can produce a masculine identity and earn both the self- and social respect which previous generations of men achieved in the workplace, is through stealing, joy-riding, arson, and car theft (Box 7.2). Through such demonstrations of courage and power they are able to (re)produce hegemonic notions of working-class masculinity. In a similar study located in two US cities, Buffalo in New

---

Box 7.2: Boys will be boys

'I'd leave the house about nine o'clock, when my mam thought I was going to school – I never told her I wasn't. I'd go down to the park at Scotswood Dene with my mates. We messed about…My mam found out when the school board woman came after about eight months. My mam said I had to go to school! Then she took me to school. I felt shown up. My mates were just laughing. I didn't go back.'

Then he became a burglar, and he had a habit. 'We were all bored. We wanted some glue and one of the lads mentioned burgling a house. It was near where we lived. We kicked the front door in and took the video. One of my mates sold it for £100, so we all got £33 each.' That lasted for about two days. 'We just kept burgling. I liked it – till we started getting caught. I liked the money.' Hanging around the pub connected the boys to a network of fences who would sell their stolen goods. The lads sat on the front steps of the pub and did business in the back… 'Me and my mates just had a good laugh, just pinching cars and having a laugh. I got out of my head really. We got glue from the paper shop. I liked the illusions, just seeing things, like trees moving in front of you when they weren't really there.'

When he and his mates began stealing cars he started driving them up the Armstrong Road…

Campbell 1993: 191–2

York State and Jersey City in New Jersey, Fine *et al.* (1997) show how young white men displace their anger and frustration at their structural circumstances by blaming scapegoats such as women, black people and sexual dissidents for the loss of the relatively privileged status of working-class men.

■ 7.2.3 Gentrification

At the end of the twentieth and the beginning of the twenty-first centuries major European and North American cities have been at the centre of the world economy. They have thus benefited from the rapid growth of the service sector (e.g. law, banking, finance) and creative industries (e.g. music, advertising, media, fashion and design). This has produced the emergence of a highly educated, well-paid, creative group of middle-class professionals who want to live in the centre of the city. Such a location offers the advantage of proximity both to the workplace (important for a group who may work long and unsociable hours) and to a wide range of entertainment and leisure opportunities.

In the 1980s middle-class people (mainly single people and dual-income households) began to move into poor or working-class residential urban neighbourhoods, a trend which has continued into the twenty-first century. This has led to the **gentrification** of these post-industrial areas (although gentrification has also occurred in rural areas: see Chapter 8). Older houses have been 'done up' and multi-occupancy dwellings converted into single-occupancy middle-class homes by individuals, while industrial buildings and warehouses have also been transformed into luxury apartments by property developers. Examples of this process include the SoHo area of Manhattan, USA, described by Sharon Zukin (1988) in her book *Loft Living*; the Docklands area of London, UK, where gentrification has spread from the development of warehouse properties on the river Thames into Clerkenwell (Hamnett 1999); and the Marais, around the Bastille, and parts of the Latin Quarter of Central Paris, France (Noin and White 1997).

The process of gentrification brings about changes in both the social and physical make-up of city neighbourhoods (Smith 1987, Hamnett 1991). First, it leads to the economic reordering of property values, commercial opportunities for the construction industry and an expansion of private owner occupation. Second, this process results in the displacement of low-income residents by those of a higher social status, until middle-class residents outnumber the original inhabitants of the neighbourhood. Third, gentrification leads to the transformation of built environments. Property conversions are often characterized by the use of postmodern architectural styles which draw on local themes while also borrowing inspiration from other times and places. Caroline Mills (1988) cites the example of Fairview Slopes in Vancouver, Canada, where industrial materials and colour have been used to reflect the heritage of the area while also being fused with motifs drawn from the Mediterranean (reflecting Fairview's waterfront) and San Francisco (which plays on the notion that Fairview is the San Francisco of Canada), which have further been combined with more classical forms of architecture derived from Boston, USA, and Georgian London. Fourth, changes in the social composition of neighbourhoods and an increase in property prices are often followed by changes in the retail and service landscape (e.g. the opening of quality restaurants and clothes outlets) to meet the needs of gentrifiers, creating a shared middle-class culture and lifestyle around conspicuous consumption (Noin and White 1997) (see Plate 7.1a and b).

Neil Smith (1979b, 1982) has explained gentrification as a movement, not of people back to the city, but of capital back to the city, arguing that the movement of people has only happened because particular investment opportunities have emerged. In what is known as the 'rent gap' thesis, Smith observed that the dilapidated state of many buildings in inner-city neighbourhoods has created a situation where there is a gap between the purchase price of properties and the potential value (because of their central location) at which they can be sold on or rented out once they have been converted or 'done up'. As more buildings in these neighbourhoods are renovated to realize their underutilized potential, so the area becomes an even more desirable place for the middle classes to live, the values of properties soar, and the original inhabitants are displaced because they can no longer afford the rents.

Plate 7.1a and b Gentrification in Berlin has been met with resistance (© Paul White)

Gentrification represents a strategy of **capital accumulation** (or sound financial investment). Smith (1979b: 540) argues that this profit motive is there even if it is not acknowledged by gentrifiers because 'few would consider rehabilitation [of buildings] if a financial loss were to be expected'. This profit motive not only applies to the gentrifiers themselves but also is shared by a whole range of professional agents involved in the process of gentrification, including builders, surveyors, landlords, mortgage lenders, estate agents, government agencies, architects and tenants and even commodity advertisers.

Other writers, however, have criticized the rent gap thesis, arguing that investment opportunities in run-down property are not the only explanation for gentrification. Damaris Rose (1984: 62), for example, writes: 'the terms "gentrification" and "gentrifiers", as commonly used in the literature, are chaotic conceptions which obscure the fact that a multiplicity of processes, rather than a single process, produces changes in the occupation of inner-city neighbourhoods'.

Other motivations for gentrification include changes in the occupational structures of production (Hamnett 1991). Beauregard (1986), for example, claims that a new social class has emerged, the professional middle class, and that gentrification is therefore the material and cultural manifestation of this social group in the landscape. Rose (1984, 1989) queries this, however, pointing out that many gentrifiers are on moderate or low incomes and so are only marginal members of the professional class.

In contrast to explanations for gentrification which focus on production, some writers have focused on lifestyle choice (Ley 1980, Mills 1988). The consumption of housing, like the consumption of other sorts of goods, can play an important role in individuals' identity-formation. Buying a house in a particular neighbourhood can be a strategy to buy into a particular lifestyle and identity because the gentrification of an area often leads to the development of an associated landscape of consumption that includes restaurants, shops, delicatessens, marinas, jogging paths, markets, and so on (Mills 1988). Demographic changes such as the 'baby boom', and the emergence of new household patterns in the face of the decline of the traditional nuclear family (see Chapter 3), are also credited with creating a preference for city living and a demand for alternative forms of housing which are not available in the family-oriented suburbs. In this way, gentrification is not just attributable to the actions of corporations investing in major building and construction projects but is also the product of the more piecemeal actions of individuals buying into particular lifestyles (Hamnett 1984).

Rose (1989) suggests that gentrification is better understood as originating in changes in women's position within the labour market. Her explanation emphasizes gender and sexuality as opposed to the focus of Smith or Beauregard on class. Rose points out that a high proportion of gentrifiers are educated women working in high-income professional occupations and living in dual-earner households. For these women, gentrified neighbourhoods represent a solution to the problems of combining paid and unpaid work. A city-centre location reduces the time-space constraints of their dual role, while gentrified neighbourhoods usually offer more support services (late-night delicatessens, laundry and cleaning services, and so on) than the suburbs, and so represent a way of reducing women's reproductive labour. Accessibility to social networks and 'community' facilities can be just as important in people's choice of neighbourhood as the actual housing unit or apartment. Indeed, Liz Bondi (1991) argues that the construction of gender identities and relations is a key aspect of gentrification, in which she sees gender as a process rather than a category.

■ 7.2.4 Sexual dissidents within the city

In the late nineteenth and early twentieth centuries gay men (who included gay-identified queers, effeminate men known as 'fairies', and men who had sex with men but identified as heterosexual) established a public world in New York, USA, which included gay neighbourhood areas in Greenwich Village, Times Square and Harlem, where beauty contests, drag balls, dances and other social events took place in cafés, restaurants, bathhouses and speakeasies (Chauncey 1995).

More usually, however, the emergence of visible gay neighbourhoods within the heterosexual city is dated from 1967, when a police raid on a gay male bar provoked what became known as the Stonewall riots. This led to the politicization of lesbians and gay men and the emergence of their more visible presence within major North American and European cities.

The Castro district in San Francisco, USA, is perhaps the most famous lesbian and gay neighbourhood (see Figure 7.3). Since the 1950s San Francisco has had a reputation as a city which has a tolerant attitude towards non-heterosexual lifestyles. The Second World War and the period immediately afterwards are credited with playing an important role in the emergence of this subculture. It was in the port of San Francisco that servicemen both departed for and returned from overseas duties and it was also here that dishonourable discharges were carried out. Many of those leaving or being dismissed from the services remained in the city. San Francisco developed a reputation for tolerance and for supporting bohemian ways of life. This, combined with California's liberal state laws on homosexuality, led to it emerging as a lesbian- and gay-friendly city. In turn, this reputation attracted queer migrants fleeing from prejudice and discrimination in more conservative towns, cities and rural areas (see also Chapter 8).

Initially, a handful of bars and clubs acted as spaces for social networks to develop in the city. However, these were subject to police raids and often proved to be transient and unstable environments. In the 1970s the Castro district began to emerge as a gay neighbourhood. The rise of this neighbourhood is often associated with Harvey Milk, a dynamic political activist, who opened a camera shop in the Castro in 1972. Milk was influential in developing neighbourhood campaigns and harnessing a gay political vote, before being assassinated (his story was made into the film *The Life and Times of Harvey Milk*).

Figure 7.3 The Castro District, San Francisco, USA. Redrawn from Castalls (1983).

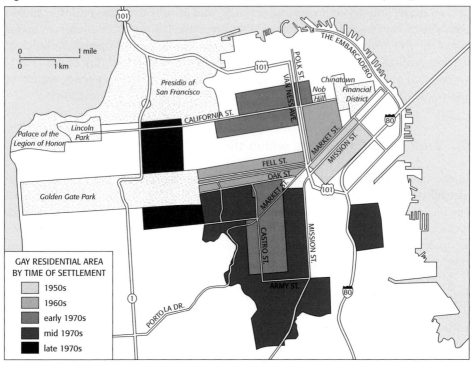

The Castro district developed a reputation as an area of relatively cheap housing which had the potential for renovation, and as a neighbourhood where it was possible to live a gay lifestyle. As more gay men moved into the neighbourhood, bars, clubs, bookstores and other commercial services opened to cater for their needs (Castells 1983, Lauria and Knopp 1985). As a result, gay gentrifiers (mainly men) gradually displaced the long-term poor, minority residents, as well as squeezing out low-income lesbians and gay men (Castells 1983, D'Emilio 1992). This also caused a knock-on effect into the neighbouring Latino Mission district on one side of the Castro and an African American neighbourhood in the Hayes Valley on the other side (Castells 1983). This process of gentrification has caused some debate amongst geographers about the complicity of one oppressed group – gay men (although there is now more of a lesbian presence in the Castro than in the 1970s) – in the perpetuation, through strategies of capital accumulation, of social injustices against other minority groups (see Jackson 1989, Knopp 1992). As Larry Knopp (1998: 159) observes: 'The forging of identities through the economic and political colonisation of territorial spaces (and the related creation of gay-identified places) is much facilitated by class, racial and gender privilege.'

Pink economies based on consumption have also created gay enclaves in many other North American, European and Australasian cities, including, for example, the Marigny neighbourhood in New Orleans, USA (Knopp 1998); the gay village along Canal Street in Manchester, UK (Quilley 1995); Soho/Old Compton Street in London, UK (Binnie 1995); Oxford Street and the surrounding inner-city neighbourhoods of Darlinghurst, Surrey Hills and Paddington in Sydney, Australia (Knopp 1998); and in Amsterdam, Netherlands (Binnie 1995).

Like gay men, lesbians also create their own spaces within cities, although these environments are often less visible to heterosexuals (e.g. Adler and Brenner 1992, Rothenberg 1995, Valentine 1995). Sy Adler and Joanna Brenner (1992) suggest that this is because, like heterosexual women, lesbians have less access to capital than men, and because a fear of male violence deters their willingness to have an obvious presence in the landscape. The influence of feminism has also meant that lesbian 'communities' have tended to be more radical, politicized, and less materially oriented, than those of gay men, which has stymied the development of businesses and bars run for, and by, women.

Rothenberg's study of the Park Slope area of Brooklyn in New York, USA (1995) shows how women tend to create residential rather than commercial spaces. Indeed, the institutional bases of lesbian communities are often made up of non-commercial venues such as support groups, self-defence classes, alternative cafés and co-operative bookstores which are promoted by word of mouth or flyers and are reliant on the energy and enthusiasm of volunteers rather than paid staff. Many of these spaces are shared with other non-commercial users, only being appropriated and transformed into lesbian spaces on specific days at specific times. In this sense, these institutional bases represent a series of spatially concentrated venues that are reasonably fixed in location and regular but not permanent (Valentine 1995). Despite their ephemeral nature, however, these spaces are important locations where lesbian communities are imagined and contested (see Chapter 4).

The visibility of gay men, and, to a lesser extent, lesbians, within major cities reflects the growing confidence of sexual dissidents to assert a claim to sexual citizenship (see also Chapter 9). Indeed, in the late 1990s activists set out not just to establish gay neighbourhood ghettos but to queer the hegemonically straight streets of the whole city. From spectacular celebrations of dissident sexualities in the form of lesbian, gay and bisexual parades and Mardi Gras (Johnson 1997) through to transgressive and disruptive events such as holding 'weddings' and kiss-ins in everyday public spaces, queer activists have both radically appropriated 'public' space and exposed its normative coding as heterosexual (Bell *et al.* 1994, Bell and Valentine 1995a, 1995b, Valentine 1996a) (see also Chapter 6).

However, there are negative sides to this visibility too. Wayne Myslik (1996) suggests that the spatial concentration of lesbian and gay men in particular districts of the city makes it easier for heterosexuals to both control and target them. He notes, for example, that gay men are more likely to be victimized in gay-identified neighbourhoods or cruising areas than on the heterosexual street – an argument further supported by the bombing in 1999 of a lesbian and gay pub, the *Admiral Duncan*, in Soho, London as part of a wider series of hate crimes aimed at a number of different minority groups.

The **commodification** of gay lifestyles as chic cosmopolitanism and the courting of pink pounds, dollars and euros by the heterosexual market are also causing new problems of what Larry Knopp (1998) terms 'managing success'. Lesbian and gay neighbourhoods such as the Castro and Manchester's gay village are increasingly attracting heterosexual visitors. Likewise, spectacles celebrating lesbian and gay sexuality such as the Sydney Mardi Gras (which is broadcast on state television and advertised in prime tourist spots around the Opera House and Circular Quay) are now being marketed for non-gay-identified consumption. Such examples have provoked some anxiety amongst lesbians and gay men that spaces that were previously considered to be the 'property' of sexual dissidents – effectively, collective 'private spaces' as opposed to the heteronormativity of public space – are being invaded and colonized by heterosexuals. This process is feared to be undermining the gay identity of these spaces, so eroding what lesbians and gay men have often taken for granted as safe environments (Whittle 1994).

These examples demonstrate that sexual and spatial identities are mutually constituted. Sexual identities depend to some extent on particular spaces for their production (for example, an individual's sexual identity may be read as lesbian or gay from the space they occupy, or a person may only feel able to 'come out' and identify as gay in a lesbian or gay space). Spatial visibility has thus played a key part in the development of sexual dissidents' rights. In turn, space is also produced through the performance of identities. For example, the performance of lesbian and gay identities can queer environments that are taken for granted as heterosexual, while likewise the performance of heterosexual identities within spaces assumed to be gay can challenge these productions of space too.

So far, this section has focused on lesbian and gay sexuality within the city. Yet, heterosexuality – in terms of sex workers and their clients – has also traditionally been identified with particular city neighbourhoods. Phil Hubbard (1999) points out that the dominant image of prostitution in the West is of women walking the streets, although they actually constitute a minority of those employed in sex work. Both women and men work in saunas, massage parlours, sex clubs, brothels, bars or even from their own homes. Within the sex industry different forms of employment are graded in a hierarchy with street work at the bottom, high-class escort agencies at the top and window work (where prostitutes sit in public view) and stripping/dancing somewhere in between.

Street prostitutes are frequently subject to violence at the hands of their 'punters' and pimps and receive little protection or support from the police, who have a history of regarding them as 'deserving of violence'. Although they regard themselves as skilled at spotting 'dodgy' clients, they only have a split second to make these judgements, and finding 'private' spaces (such as deserted car parks, alleyways and waste ground) to have sex can put prostitutes into situations where they are vulnerable to attack (Hubbard 1999). Off-street sex workers are safer in that they work in familiar and contained environments to which entry is often controlled by a receptionist, bouncer or CCTV but there are other disadvantages too. They are highly constrained and often have to work long shifts with none of the rights or benefits enjoyed by employees in other sorts of occupation, and they are open to financial exploitation as well as violence from those who run the bars, clubs or brothels in which they work.

Street work has traditionally been associated with particular city neighbourhoods that are stigmatized as immoral. For example, the Chicago School of Human Ecology identified certain inner-city areas as 'vice districts', contrasting the sexual morality of these neighbourhoods with the stable and settled residential suburbs (see Chapter 4). The CSHE's work is problematic in terms of contemporary understandings of sexuality and space because it represented space as a passive container or backdrop for sexual relations, rather than recognizing the active role space plays in their constitution (see above). It also represented prostitutes as inherently deviant, rather than showing any understandings of the lifestyles and practices of sex workers (Hubbard 1999).

Subsequent geographies of sex work have moved away from taking such a moralistic stance. Ashworth et al. (1988) contrast erotic entertainment areas in city centres which cater for international tourists such as Zeedijk in Amsterdam, Soho in London and La Pigalle in Paris with street work, which they suggest is usually located outside the centre city in marginalized neighbourhoods that are part-residential and part-commercial. While they suggest that these red-light districts are a product of local demand and supply, other studies show that clients are often drawn from a wide area and that sex workers themselves also prefer to work away from where they live (Hart 1995, 1998).

In contrast to Ashworth et al.'s (1988) account, which understands the emergence of red-light districts in terms of market forces, Symanski (1981: 35) explains their location in terms of moral orders (see Chapter 6), arguing that they are located 'where public opinion, financial interests and those who enforce the laws have pushed prostitution or allowed it to remain'. Understandings of the geographies of prostitution in terms of sexuality, morality, power and the city have been developed by Hubbard (1999). He claims that 'prostitutes are both socially and spatially marginalised by the State and the law in such a way as to maintain and legitimise the moral values of heterosexuality, generally (but not exclusively) maintained in the interests of white, middle class, male subjects' and further demonstrates 'the way that the separation, sequestration and enclosure of prostitutes in specific sites produces and reproduces the distinctions between moral heterosexual subjects and those who embody an immoral and illicit sexuality' (Hubbard 1999: 30). Yet at the same time, Hubbard (1999) draws on Foucauldian notions of power (see Chapter 5), in which power is understood not only as an oppressive force, but also, through the creation of resistances, as an enabling force, to look at how the spatial production of immoral sexuality is contested. For example, prostitutes' rights groups are fighting for greater protection and support – in other words, for sexual citizenship (see Chapter 9).

■   **Summary**

• Cities are characterized by density, intensity and proximity.

• Despite their heterogeneity, it is also possible to differentiate relatively homogeneous areas within cities.

• In the 1960s and 1970s social geographers mapped the segregation of different housing areas within cities identified on the basis of class or ethnicity.

• In the 1980s researchers began to conceptualize 'race' as a social construction and to understand patterns of segregation in terms of institutional racism.

• This opened up the issue of whiteness – particularly the way in which white people view themselves as racially and culturally neutral, rather than recognizing their racial and cultural privilege.

• Contemporary work focuses on unpacking racialized dichotomies such as black–white and recognizing the complex differentiation of racial identities.

• The term 'underclass' is used to describe those who experience intergenerational poverty, welfare dependency and unemployment.

• The underclass is constructed in and through the space of 'the ghetto', while these spaces are constructed through the people which inhabit them.

• Since the 1980s middle-class people have been moving into working-class neighbourhoods, resulting in the gentrification of these post-industrial areas.

• The process of gentrification brings about changes in both the social and physical make-up of city neighbourhoods.

• There are a range of different theoretical explanations for gentrification.

• Gay men are often in the vanguard of gentrification. Lesbians also create their own spaces within cities, although these are often less visible to heterosexuals.

• Geographies of prostitution can be understood in terms of morality and power within the city.

■ 7.3 The flâneur

The everyday act of wandering the city streets has a long history and has been the subject of theoretical exploration. In the nineteenth century the rise of capitalism produced new spaces of consumption such as boulevards, cafés, and arcades in the city. These spaces became home to the flâneur, who, indifferent to the pace of modern life, enjoyed strolling anonymously around the streets in the role of an urban onlooker, voyeuristically taking in the spectacle of city life but not participating in it. Mike Featherstone (1998: 913) explains: 'On the one hand, the flâneur is the idler or waster; on the other hand, he is the observer or detective, the suspicious person who is always looking, noting and classifying . . . The flâneur seeks an immersion in the sensations of the city, he seeks to "bathe in the crowd", to become lost in feelings, to succumb to the pull of random desires and the pleasures of scopophilia.' In this way, the flâneur swings between positions of immersion and detachment, while exhibiting the reflexivity of an artist, writer, or journalist.

Historically, the flâneur is associated with the writings of the poet Charles Baudelaire and was inevitably assumed to be a man. Respectable middle-class women of the nineteenth century were assumed to be at home in the suburbs and not to have the freedom to stroll around the city, although lesbians, the elderly, widows and prostitutes did have more opportunities to be on the streets.

However, Elizabeth Wilson (1992), in her book *The Sphinx in the City* suggests that a female flâneur or a flâneuse was not an impossibility. She argues that nineteenth-century women did have more freedom in the city than their counterparts who lived in villages and towns. Women were often visible in public and liminal public/private spaces like department stores, tearooms, hotels and museums and did therefore have some awareness of the excitement and possibilities of the city. Certainly, they became more visible in it towards the end of the nineteenth century, although they were still constrained by the male gaze and a fear of harassment. In this sense, women were torn between a desire for, and a dread of, city life (Walkovitz 1992). Wilson (1991) argues that, towards the end of the nineteenth century and the beginning of the twentieth century, urban living was important to the development of feminism because it offered women some escape from patriarchal control by providing them with more anonymity, new opportunities to associate with women from wider backgrounds, and the chance to develop their sexualities. In contrast, to many accounts of the flâneur which credit him with a voyeuristic mastery, Wilson (1991) presents him as a more insecure, ambiguous and marginal character. To her he is a passive, feminine figure rather than a symbol of active masculine power.

Other writers, such as Sally Munt (1995) and David Bell and Jon Binnie (1999), have highlighted queer appropriations of street life and spectatorship. Appropriating the street, walking, looking and being looked at, the chance both to 'disappear in a crowd but also to catch glances, to look and be looked at, the chance for a brief encounter' (Bell and Binnie 1998: 131) are all important for sexual dissidents. Drawing on lesbian literature from the 1950s onwards, Munt considers the role of the lesbian flâneur as a sexual adventurer. She writes, 'Swaggering down the street in her butch drag casting her roving eye left and right, the lesbian flâneur signifies a mobilised female sexuality *in control*, not out of control' (Munt 1995: 121). Similarly, Bell and Binnie (1998) focus on the novel *The Dancer from the Dance* by Edmund White, which pre-dates the AIDS epidemic, as an example of the eroticization of the city and its streets by gay men. In this novel the city streets are represented as spaces for the realization of desire and as utopian and democratic spaces. They then go on to consider how the city was also appropriated within queer skinhead culture of the 1990s by reflecting on a series of skinhead novels by Stewart Home. In these books, streets are still eroticized, but here it is a brutal eroticism where the streets are the sites of violence and revolution.

In the late twentieth and early twenty-first centuries the construction and transformation of cultural sites and refurbishment of city centres has created new places in which to shop, and to visit, new sensations and experiences and hence new spaces in which people can stroll (see section 7.4). The contemporary focus on individualization is also creating a new wave of urban flâneurs who are concerned with fashion and the presentation of the self. Indeed, some writers have argued that the contemporary recreational shopper is a flâneur (Falk and Campbell 1997, Featherstone 1998). Shopping malls do not just sell goods to be bought and taken away, they also offer free experiences of sociality and display to be consumed on the spot in the form of aesthetic and highly designed environments containing fountains, mirrors, plants, food halls and entertainments (see section 7.4). Shopping is no longer just about purchasing goods, it is also about the act of strolling, mingling in a crowd, and enjoying the opportunity to consume the sensations and experience of the surroundings. In other words, the contemporary shopper in the mall is caught in the same tension between immersion and detachment as the nineteenth-century flâneur wandering down the boulevards (Falk and Campbell 1997, Featherstone 1998).

However, in celebrating the contemporary flâneur/flâneuse it is important to remember that mobility and the act of seeing, which are so fundamental to the voyeuristic stroller, are not shared by all citizens. People with physical or visual impairments can be denied the freedoms to wander the streets in ways taken for granted by able-bodied citizens because of the disabling nature of many city environments (Hahn 1986, Butler and Bowlby 1997, Gleeson 1998, Parr and Butler 1999; see also Chapter 2).

■   **Summary**

- The everyday act of wandering the city streets has a long history and has been the subject of theoretical exploration.

- The flâneur – an urban onlooker who enjoys taking in the spectacle of the city but not participating in it – is associated with the writings of the poet Charles Baudelaire.

- The flâneur is often assumed to be a man but some writers have highlighted the flâneuse and queer appropriations of street life and spectatorship.

■ 7.4 Landscapes of consumption

In the 1970s and early 1980s North American and European cities that had been developed around industrial production were undermined by economic restructuring and the associated process of de-industrialization. Large areas of city centres became derelict and urban populations experienced long-term structural unemployment, the outcome of which was that many major cities lost their sense of both purpose and identity. Lovatt and O'Connor (1995: 127) comment, 'Ugly grim cities they may have been, but formerly they produced, they made for the world. Now they were just ugly and grim.' In the USA de-industrialized cities were pathologized as dangerous; in the UK there was concern that they were becoming monofunctional places dominated by shops and offices. Poor provision of public transport at night, licensing law constraints, home-centredness, suburbanization and the dominance of the car at the expense of pedestrians were all cited as factors that were eroding an urban way of life (Bianchini 1995).

Despite this decline in major cities, their Central Business Districts still represented to organizations important fixed capital in terms of building and land. They were not about to write this off. In the late 1980s and 1990s a process of revalorizing city centres began, which involved both a re-emphasis on the importance of centrality and an attempt to mobilize culture to lure capital back into the city (Harvey 1989). Cities contain art and cultural heritage (e.g. museums, galleries, buildings) as well as being sites of cultural production, housing creative industries associated with fashion, television, music, food, the arts, tourism, leisure and publishing. Culture was identified as an alternative source of wealth both because it was realized that the cultural capital of particular cities could be converted into economic capital and because cultural industries themselves generate wealth.

A new significance was attached to reorganizing city centres around consumption rather than production, in which the emphasis was on consumer goods and leisure time activities that incorporate high levels of design, style and fashionable cultural imagery (Featherstone 1991). Gentrification and press coverage of bohemian or exotic sub-cultures helped to glamorize the city and to create an image of urban living as an aesthetic or artistic lifestyle (Zukin 1988, 1998).

Whereas in the 1970s cities competed to represent themselves as most in need, in order to win subsidies and government support, now the emphasis is on re-imaging cities in a positive light to sell them in a competitive global marketplace (see section 7.5). Cultural differentiation and vibrancy are playing a key part in this process of re-imaging (Lovatt and O'Connor 1995). As Montgomery (1995a: 143) explains:

Plate 7.2   The 24-hour city (© Becky Kennison)

'Culture is the means by which cities express identity, character, uniqueness, and make positive statements about themselves, who they are, what they do and where they are going.'

City officials who in the 1960s and 1970s would have criticized ethnic or lesbian and gay lifestyles now celebrate these neighbourhoods as examples of the cities' diversity and vitality (see section 7.5). Quilley (1995: 47) claims that Manchester's gay village 'has led the way in re-appropriating the street for pedestrians and flâneurs: for a mode of urban experience that is central to European notions of urbanity in which positive ambience in public space is the result of social (face to face) vibrancy and participation. This is best achieved and experienced through the act of strolling. Gay gentrification in the neighbouring Granby Village development has contributed to this project in the vicinity of Canal Street.'

Promoting a vibrant culture has also meant recognizing the potential of the night-time economy. Without the daytime constraints of work and social obligations, night is the time when most people have the freedom to do what they want (Bianchini 1995). It is a time of play, socialization and encounter. By encouraging daytime businesses to stay open late or even all through the night, Bianchini (1995) observes, cities have the opportunity to effectively double their economy (see Plate 7.2). As well as traditional night-time spaces such as bars, clubs and restaurants, other places such as shops, gyms, galleries and so on are now also opening longer and later, and non-commercial activities like street theatre or social clubs and evening societies are also flourishing.

In European cities what is known as *Animation Culturel* (the word 'animation' is used to mean 'giving life back') emerged in the 1970s as one component in attempts to revitalize cities. This involved the organization of street festivals, street theatre and events such as concerts and art exhibitions in public places both at lunchtime and during the evenings. Between 1977 and 1985 Rome City Council held a pro-gramme of summer cultural events known as *Estate romana*. Four different centres of activity, connected by cheap, late-night bus services, were promoted: City of Film (located in a disused slaughterhouse), City of Sports and Dance (held in an archaeo-logical park), City of Television (in the park of a municipally owned villa) and City of Theatre (in a street of redundant warehouses) (Bianchini 1995). Another ex-ample of *Animation Culturel* is found in Barcelona which holds a large-scale cultural event in the form of a five-day major fiesta known as *La Mercè*. Here arts, music, and other forms of performance, including a parade of giant papier mâché figures, a competition of castellers (groups of people building human castles), a parade of stilt walkers, correfoc (groups of young people dressed as devils and carrying papier mâché beasts who march through streets letting off fireworks), and night-time firework displays take place in open spaces throughout the city (Schuster 1995).

Montgomery (1995b: 104) argues that the theory behind such events is that 'by having people on the streets, in the cafés and moving through the public realm, urban vitality is stimulated'. The intention is to generate flows of people engaged in dif-ferent layers of activities (working, shopping, strolling, socializing, playing) and to reopen the city as a meeting place so that people's friendships and leisure time will be conducted in bars, restaurants and galleries rather than at home. In turn, it is argued, drawing a diverse range of people onto the streets makes everybody feel safer because they believe they can rely on the 'natural surveillance' of other 'eyes on the street' to protect them from 'public' dangers, rather than relying on CCTV and pri-vate security forces (Jacobs 1961). For example, Robert Wassermann, the Assistant to the Police Commissioner, of Boston, USA, claims that: 'We believe that arts activ-ities can generally help reduce street crime. Both in those areas of Boston which have regular street cultural activities and in our theatre districts, there tends to be less crime during those times of the day as the cultural events are ongoing' (cited in Montgomery 1995b: 106).

Temple Bar in Dublin, Eire, provides a good example of the use of culture to revit-alize a city in a process dubbed by Montgomery (1995a: 165) 'urban stewardship'. This district, which was built and developed in the seventeenth and eighteenth cen-turies, is known as the cultural quarter of the city and covers a 218-acre area between O'Connell Bridge to the east, Dame Street to the south and the river Liffey to the north. In the 1970s, despite its central location, Temple Bar had become something of a backwater and its future was uncertain because the State Bus Company (CIE) had declared its intention to develop it as a transportation hub and had begun to buy up property ready for demolition. As a result, property and rental values in the neighbourhood fell. In turn, businesses and activities that could only afford low rents and were willing to accept short tenancy agreements began to move into the area. These included artists' studios, galleries, recording studios, restaurants, second-hand clothes shops, book shops and record stores (Montgomery 1995a).

These cultural industries and businesses began to flourish and so the area developed a reputation for its vitality, being dubbed 'Dublin's Left Bank'. As a consequence, the local traders and community groups then began to organize themselves. A group of individual business people and cultural entrepreneurs formed the Temple Bar 91 group to bid for European Union grants to fund a Temple Bar urban renewal scheme. The intention behind this was to create and sustain the cultural industries; to prevent the loss of Irish talent overseas; to stimulate the night-time as well as the daytime economy; to use place marketing to build on Temple Bar's identity and raise its profile in Eire and beyond; to conserve and renew the architectural identity of the neighbourhood; and to reclaim and redevelop derelict properties. In particular, the intention was to buy the CIE's (the state bus company) property holdings to prevent their sale to another commercial developer. The Temple Bar Development Council (TBDC) was also set up; this produced a manifesto for the creation of a Cultural Enterprise Centre in the Temple Bar neighbourhood. The TBDC's aims were to conserve and improve the environment, to develop recreation and tourism, and to create employment through public and private enterprise (Montgomery 1995a).

The bid to the European Union was eventually successful and the Irish government acted to establish a state-owned trading company called Temple Bar Properties to purchase CIE's and the Dublin Corporation's property portfolios (through compulsory purchase if necessary) and to oversee the development of the area. Tax incentives have been introduced to encourage private property owners to renew and improve buildings and a range of measures have been used to develop major cultural centres to promote the strategic growth of particular cultural activities. These centres include the Irish Film Centre, Temple Bar Photography Gallery, the Temple Bar Gallery and Artist Studio, Children's Arts Centre, and so on (Montgomery 1995a).

The revitalization of cities around culture and consumption has not been without its critics. Property developers and multi-national corporations are accused of creating standardized consumption and entertainment spaces (such as mega malls and multiplex cinemas) within cities, while consumption practices predicated on middle-class lifestyles – particularly gentrification – are blamed for displacing other social groups who cannot afford them. Some critics argue that, through these processes, the uniqueness and cultural diversity of urban identities are being eroded as fast as cities are trying to package and sell themselves around claims of cultural distinctiveness (Zukin 1998).

The desire of city officials and developers to attract the middle classes and middle-aged people with high disposable incomes back into the city has prompted measures designed to make the streets feel safer and so to reverse the downward spiral of abandonment identified in Chapter 6. Part of this has involved efforts to 'clean up' city streets by removing so-called undesirable groups such as the homeless and teenagers. Increasingly, commercial districts, though technically 'public' spaces, are run and policed by private associations or corporations which set the rules about who is allowed in and how they are allowed to behave (see Chapter 6). For example, the New York City Central and Prospect Parks are partly financed and run by private organizations including individual and corporate patrons rather than by the New York City Parks Department (Zukin 1998). Such processes are criticized as undemocratic and exclusionary. The outcome of them is to undermine rather than foster diversity, vibrancy and social justice within the city.

Despite these criticisms of the private sector management of public space, Sharon Zukin (1998) is still upbeat about the ability of different cultural and ethnic groups to forge their own 'urban' lifestyles within their own neighbourhoods. Here, she argues, the interaction and juxtaposition of many different lifestyles and traditions continues to create a 'hybrid' urban culture which resists domination by corporations or the middle class. Zukin writes (1998: 836): 'Newspaper stands owned by members of one immigrant group sell newspapers written in other languages. Store owners stock distinctive ethnic goods that will appeal to several different ethnic groups, and some goods, such as clothing and cosmetics are re-exported to the same or even different countries of origin . . . Here, "transnational" consumers interact and develop their own urban lifestyles.' It is this diversity which Zukin (1998: 837) claims continues to be 'the city's most important product'.

The following subsections focus on three examples of different spaces of culture and consumption within the city: the mall, the restaurant and the club.

■ 7.4.1 The mall

Large shopping malls in, and on the outskirts of, urban areas (such as West Edmonton Mall, Canada, Brent Cross, London, UK, and Faneuil Hall Marketplace in Boston, USA) have developed as a result of a range of forces, including postwar consumer affluence, technological change, growth in car ownership, and the increase in numbers of women in paid employment. Shopping is now the second most important cultural activity in North America after watching TV (Goss 1992). Goods not only sustain our everyday material living standards, they also define our individual and collective identities.

Downtown malls are often designed to reflect a 'modernist nostalgia for authentic community, perceived to exist only in past and distant places' (Goss 1993: 22). An example is Faneuil Hall Marketplace in Boston, where the restored waterfront area combines idealized historical urban community and a street market. The shops are decorated with antique signs and props and street entertainers in costume are on hand (Goss 1992). These malls provide not only shops but also entertainment, food and drink, dance/concerts and, in some cases, funfairs and fashion shows. Jon Goss (1993) points out that the ability of shopping centres to cater to all tastes is actually the result of their organizational intelligence and spatial strategies of control. He writes: 'A sophisticated apparatus researches consumers' personal profiles, their insecurities and desires, and produces a space that comfortably satisfies both individual and mass consumers and manipulates the behaviour of both to not-so-different degrees' (Goss 1993: 40–1).

Most large-scale malls are designed to circulate shoppers in order to maximize the number of goods they will see, though, according to Frank Mort (1989), this also contributes to making them 'new' sexual cruising grounds for some men and women. Seating is provided at 'pause points' so that shoppers can review progress, rearrange their bags and regroup, but these are never too comfortable because they might distract shoppers from consuming or attract undesirable groups (e.g. the elderly, the young, the homeless). There are usually no clocks or windows and the temperature and lighting are carefully controlled to mask the passing of time. Cleaning, security staff and CCTV maintain the cleanliness and order of the space, while water and music are used to calm tensions and promote sales (Goss 1992). As such, malls 'are artificial environments, which, unlike the main street, have no prior reason for existence and no historic rootedness in place' (Goss 1992: 166).

Yet, this very artificiality – what Goss (1992) dubs the 'magic of the mall' – means that many consumers regard them as idealized spaces. Here, they are free from the inconvenience of the weather, pollution and traffic, and able to avoid any confrontation with social difference, fear of crime and the general sense of disorder and incivility which is associated with 'the street' (see Chapter 6). In this sense, Jackson (1998a) argues, malls represent in effect the 'domestication' of the street. However, he claims, 'whereas the idealised form of the public street is a relatively open and democratic space, the shopping centre offers only a parody of participation: where "credit card citizenship" allows the consumer to purchase an identity, engaging vicariously in their chosen lifestyle' without any of the inconveniences or responsibilities of a truly 'public' space (Jackson 1997: 178). Indeed, although malls are a magnet for young people, offering them not only the chance to window shop but to escape the dangers of the streets (Vanderbeck 1999), as a group with low spending power, they often find themselves excluded from these spaces and virtually disenfranchised from city life (Bianchini 1990). This is not to suggest, however, that shoppers and browsers are entirely passive in the face of the dictates of shopping centre managers and security staff; control is also tactically appropriated by consumers.

■   7.4.2 The restaurant

Eating places are increasingly important to the urban landscape. There has been a renaissance in dining out in contemporary Western cities. The diversity of different ethnic restaurants available gives diners the chance to get a taste of other places. For example, London's listings magazine *Time Out* recently implored its readers to 'Give your tongue a holiday and treat yourself to the best meals in the world – all without setting foot outside our fair capital' (Cook and Crang 1996: 131). On the surface, these places offer a way of stimulating a new level of cultural interest in different cultures and places and support for anti-racism. Yet, May (1996) argues that consumption of such foods is often dependent on and helps to reproduce racist imaginative geographies. bell hooks (1992) also warns that, through the process of consuming 'other' places, we often try to contain and represent them within our own frameworks. She writes that this 'commodification of difference promotes paradigms of consumption wherein whatever difference the Other inhabits is eradicated, via exchange, by a consumer cannibalism that not only displaces the Other but denies the significance of that other's history through a process of decontextualisation' (hooks 1992: 31).

Restaurants, however, are not just important for their food but also for the social interactions which take place within them. '[T]he restaurant exists as a feature of the entertainment industry, and is as much concerned with the marketing of emotional moods and desires as with the selling of food. Eating in the public domain has become a mode of demonstrating one's standing and one's distinction by associating oneself with the ready-made ambience of the restaurant itself' (Beardsworth and Keil 1990: 142–3). The behaviour of diners in formal restaurants is very mannered and structured (Finkelstein 1989). These are performative environments where there are often elaborate place ballets in which not only diners are on display but also staff (Crang 1994) (see also Chapter 5). Sharon Zukin (1995) observes, for example, how artists and resting actors are often hired to work in top restaurants on the basis of their performative personalities. She writes: 'Waiters are less important than chefs in creating restaurant food. They are no less significant, however, in creating the experience of dining out. For many people, oblivious of restaurant workers' social background, waiters are actors in the daily drama of urban culture' (Zukin 1995: 154).

While those who know how to behave in such social environments may feel socially comfortable and even take pleasure from demonstrating a shared social knowledge, those who are less familiar with the social etiquettes of particular restaurants can feel 'out of place' in these socially regulated environments, although eating above one's social station can also provide a rare taste (in both the culinary and social senses of the word) of how the 'other half lives' (Valentine 1999e).

Restaurants are the site of all the daily dramas of urban life – everything from births, marriages, adulterous affairs, divorce and death are recorded at city restaurant tables. They represent a halfway house between the communal environment of a 'public' space such as a bar or club and the intimacy of home. As such, they are a 'safe' yet fairly 'private' and 'romantic' environment for women to meet men on first dates. Meals, like other forms of consumption, also provide currency for everyday conversations and even a way for people to talk about their relationships. Through the performative act of sharing a meal, individuals can articulate their identities and competence in public culture, develop contingent knowledges about each other and even assess their compatibility. The twin restaurant characteristics of 'privacy' and 'romance' also mean that they can be important sites for playing out adulterous affairs (Cline 1990). Sally Cline (1990) cites the example of Robin, a secretary, who described how, when she started an affair, food assumed great significance in her life because she and her lover would use restaurants as their meeting place.

The business lunch has long been recognized as an essential part of corporate entertainment. During the early 1980s, not only the 'power lunch', but also the 'power breakfast', was imported from the USA into the UK. A recent British survey of advertising executives found that, on average, they attend or host one lunch per week (Athenaeum, 1996). The restaurant provides a neutral space in which to meet competitors, support services (e.g. lawyers and accountants) and producers or suppliers, all of whom are often concentrated within offices in the centres of cities (Bergman, 1979). The importance of these face-to-face meetings was emphasized by Michael Korda, editor-in-chief of publishers Simon and Schuster, when he claimed that *'the most powerful place in town [for my industry] is the Grill Room of the Four Seasons [a restaurant in New York]'* (*New York Times* 1976, quoted in Bergman 1979: 236). The business lunch is a place not only for negotiating, dealing and hiring and firing, but also for binding people together and creating business networks.

■ 7.4.3 The club

The UK nightclub industry is estimated to be worth approximately £2 billion per year and attracts over a million people per week (Hyder 1995). It therefore offers many regenerative possibilities for decaying inner-city areas. Notably, clubs bring young people into the city and onto the streets, while at the same time also being staffed by other young people, often students, who can be paid low wages and be expected to work long, unsocial shifts.

Most spaces of the city are designed for travelling through rather than socializing in, or are designed in such a way as to minimize unplanned or undesired encounters with strangers (Malbon 1998). Ben Malbon (1998) suggests that in the presence of other people whom we do not know we often project a sense of indifference to them: a disassociation (Malbon 1998). In contrast, he argues that clubs are actually spaces where people enjoy being near to others: they are spaces of identification. A desire for a sense of togetherness with other people is a crucial part of the experience of clubbing. The club can generate affinity between the place and the people in it in the same way that sporting venues and festivals do. Malbon (1999) observes that the ritual of queuing at the door and the knowledges and competencies in terms of bodily practices (the look, etc) which are necessary to gain entry all foster a sense of group identity, albeit one that is fleeting and transitory.

The dance floor of the club itself offers a sensuous mind and body experience. Music in particular plays a powerful role in creating this emotionally charged atmosphere. The sense of conviviality, empathy and unity produced by sharing a space is often boosted by the use of drugs such as Ecstasy. The mixture of dark and light spaces within the club means that, although it is often crowded, it is easy for club-bers to lose sight of others. In this way, the intense but fleeting nature of social con-tact within this space offers a respite from normal rules of interaction and ordering outside of the club (Malbon 1999).

Malbon (1998: 271) defines clubbing as a performance 'where the lights (or dark-ness), the sounds, the possible use of drugs, the practices (and rituals) of dancing and the proximity of the "audience" all add to its intensity'. In the club, he argues, it is possible for young people to lose themselves, to forget the outside world and all its pressures. The space of the club offers a dislocation from the ties of everyday life and the opportunity for clubbers 'to inscribe their own creativities upon a shared space, to create a space of their own making of which they are also the consumers' (Malbon 1998: 280).

■ Summary

- In the face of de-industrialization cities have sought to revive their decaying centres by mobilizing culture to lure capital back into the city.

- Cultural differentiation and the vibrancy of consumption spaces such as malls, restaurants and clubs play a key part in this process of re-imaging.

- Critics argue that such processes are creating standardized consumption spaces within cities, displacing other social groups, and are undemocratic and exclusionary.

■ 7.5 Selling the city

'The practice of selling places entails the various ways in which public and private agencies – local authorities and local entrepreneurs, often working collaboratively – strive to "sell" the image of a particular geographically defined "place", usually a town or city, so as to make it attractive to economic enterprise [especially footloose high-tech industries], tourists and even to inhabitants of the place' (Philo and Kearns 1993: 3). Cities are competing against each other in the scramble to get a share of capital investment (at a time when new technologies have produced an unpre-cedented mobility of capital) and so each needs to carve out a place for itself in the global economy (see Plate 7.3). Its ability to do so is dependent upon it being able to offer something 'different' or 'more' than its counterparts. In this way places are becoming commodities to be packaged, advertised, marketed and consumed, just like any other goods (Philo and Kearns 1993).

But, as the opening quote suggested, this (re)imaging of cities is not just an economic process but also a social one, which aims to rebuild civic pride and achieve

Plate 7.3   Manchester: at the heart of the global economy?

social consensus for changes. As such, it is a 'subtle form of socialisation designed to convince local people, many of whom will be disadvantaged and potentially disaffected, that they are important cogs in a successful community and that all sorts of "good things" are really done on their behalf' (Philo and Kearns 1993: 3). David Harvey (1989) has dubbed the use of events and celebrations to sell cities, not only on a national or global scale but also to local people, as the phenomenon of 'bread and circuses', in the sense that local people are encouraged to enjoy a taste of fun for the day and to forget the problems of their everyday lives.

The selling of cities often involves promoting traditions, lifestyles, and the arts, which are supposed to be local or 'authentic', even though in practice these representations often use cultural motifs that are only loosely associated with the place, or they play upon or manipulate pride in local history to mobilize particular moments, or anniversaries. These images of past industrial prosperity or heroic imagery around particular events are often put together to create marketable pastiches of culture and history which are both decontextualized and superficial.

All cities play upon their uniqueness and cosmopolitanism. In marketing Manchester, UK, its City Council has attempted to portray it as culturally diverse, using the gay village along with other cultural quarters, such as Chinatown, as proof of the city's cosmopolitan and progressive credentials (Quilley 1995). The gay village in particular has been central to the re-imaging of Manchester and its attempt to represent itself as a post-industrial, service-based economy with an international reputation for its clubs, nightlife and European-style street ambience. Maps of the village are produced by the City Council tourist office, which identify its boundaries, entertainments and institutions.

However, despite the emphasis on uniqueness, Philo and Kearns (1993) point out the somewhat ironic universal vocabulary of central, bigger, better and more beautiful which appears in the imaging of most cities. They write, 'the practice of selling places may even generate sameness and blandness despite its appearance of

bringing geographical difference into the fold of contemporary economic and political discourse' (Philo and Kearns 1993: 21).

Not surprisingly, conflicts over cultural representations often arise when local people consider place marketeers' representations to be unfaithful or unwanted, when
they intentionally or unintentionally obscure 'other' groups or their histories, or when
they go against the experiences and understandings of local people, who may contrast the ideals of the place marketeers with the reality of what is actually offered
to locals in terms of employment and social opportunities (Philo and Kearns 1993).

Glasgow, Scotland, provides a good example of the marketing of a city which
has caused local disquiet. Glasgow was once a famous industrial city built on the
river Clyde and known as the 'second city of empire'. Following de-industrialization
and decline in the 1970s, the city authorities have attempted to create a new economic identity for the city. These economic regeneration initiatives were started in
the early 1980s by Glasgow District Council. Since the 1990s the Glasgow Development Agency has attempted to reposition the city by promoting its success at winning investment in face of globalization. The language of its marketing emphasizes
positive change, with slogans such as 'Glasgow's Miles Better', the 'Cinderella City',
'Phoenix from the Ashes', and so on (McInroy and Boyle 1996). McInroy and Boyle
(1996) also provide an example of a newspaper article in the local *Evening Times*
entitled 'Fat City: Glasgow is billions better as investment money rolls in'. This provided a map of the 'New Glasgow' in which buildings were marked as evidence of
Glasgow's rejuvenation. The Glasgow District Council was so pleased with this representation that it had hundreds of copies printed off to give to other journalists,
visitors and tourists.

At the same time 'the local state has also been active in manufacturing other cultural identities for the city' (McInroy and Boyle 1996: 74). In particular, Glasgow
has been marketed as a city that is committed to the arts. Since the early 1980s an
annual arts festival, the Mayfest, has been launched, and the Burrell Collection, a
new Royal Concert Hall and the Gallery of Modern Art have all been opened. In
1990 Glasgow was European City of Culture and in 1999 it was named British City
of Architecture. Mass participation spectacles such as the Glasgow Garden Festival
have also been used to distract the local people from the economic and social problems of the city.

However, rather than legitimating local economic development strategies, civic
boosterism in Glasgow actually antagonized local people and was met with opposition. In 1990 the Workers' City was launched by a group of 40 left-wing activists to
contest the way the City Council was handling the European City of Culture initiative (McInroy and Boyle 1996). They opposed the way the leader of the Council was,
in their opinion, forcing through place promotion, regardless of the views of other
councillors, civil servants and the public. They were also very unhappy about the
sanitized and self-important image of the city that was being marketed. Although
this group did not share a universal vision of Glasgow, they were united in their
criticism that the marketeers were ignoring the city's heritage, which was strongly

bound up with working-class industrial struggles and a history of municipal socialism. They regarded the efforts to promote the city as friendly to global capital rather
than to market its socialist history and identity as, in effect, selling the soul of the
city. The activists organized protests, petitions, meetings and marches, and put their
case through letter-writing campaigns to the press, and contributions to various TV
and radio programmes (McInroy and Boyle 1996). Although they recognized the
difficulties of competition in a global world, the Workers' City activists argued that
Glasgow needed to work through its own identity and that the process of cultural
transition needed to happen at a slower pace so that the city could find a new image
with which it could be comfortable.

■ Summary

- In the competition to attract inward investment cities are becoming
  commodities to be packaged, advertised, marketed and consumed.

- Promotional campaigns aim to create new economic and cultural
  identities for cities and to distract local people from local economic and
  social problems.

- Conflicts over cultural representations often arise when local people
  consider the place marketeers' representations to be unfaithful or
  unwanted.

■ 7.6 Nature in the city

Western thought has always positioned the rural and urban in opposition to each
other (see also Chapter 8). All green and open spaces are assumed to be in the countryside, whereas the city is imagined as an overcrowded, polluted, concrete jungle:
the very antithesis of nature and sustainability. The contrast between the two spaces
has prompted urban-to-rural migration and the flight of people to the green suburbs. While a lot of attention has been paid to the way the city is penetrating the
country in the form of urban sprawl, there is little consideration of the opposite process: how nature pervades the city.

Yet the urban/rural dichotomy is another example of the dangers of binary thinking (see also discussions of man/woman in Chapter 2 and public/private in Chapter 3 and Chapter 6). There is no clear separation between country and city. Open
spaces are embedded in the city rather than separate from it, a variety of species,
plants and animals inhabit or move within and through urban environments. Indeed,
the urban and the rural are part of the same ecosystems so that the city can affect
nature, for example through air or water pollution (Hinchliffe 1999).

7.6.1 Open spaces

What Jacqueline Burgess (1998: 115) describes as 'soft urban landscapes' – parks, gardens, waterways, commonland, woodland – make up important in-between spaces within the city. It is estimated that around 11 per cent of London is actually open space (McLaren 1992).

Public space has always been an integral feature of cities. In Victorian Britain, concern about industrial urbanism and the risks to health and morals of separating working people from nature and fresh air led to the development of urban parks such as Marylebourne Park (now known as Regents Park), Victoria Park, Kennington Park and Battersea Park, which were based on landscaped English country estates. Campaigns were also established to save common lands, like Hampstead Heath and Wimbledon Common, which were threatened by urban expansion (Bunce 1994).

A similar nineteenth-century movement to develop open spaces took place in North American cities. Central Park in New York, which is based on large meadows and areas of woodland, took ten years to develop. Other examples comprise Golden Gate Park in San Francisco, Prospect Park in Brooklyn, Fairmount Park in Philadelphia, and Franklin Park in Boston. In European cities, efforts to develop high-quality open spaces include Bijlmermeer in Amsterdam, the Netherlands, a predominantly high-rise estate which is located in a naturalistic woodland and wetland setting (McLaren 1992).

Urban parks are highly valued landscapes. They represent sources of pleasure, leisure, an escape from the concrete of the streets, and everyday sensuous encounters with 'nature': a chance to touch, see, smell and hear the 'natural' world (Harrison et al. 1987, Burgess et al. 1988). Contrary to stereotypical assumptions that urban residents are cut off from these landscapes and have to seek them out through special trips to distant parks or rural landscapes, Burgess et al.'s (1988) study emphasized the everyday nature of city residents' contact with the living world as part of their routine journeys and activities in familiar environments round the home or on the way to work.

Open spaces such as parks and gardens are appreciated not only for their so-called 'natural' qualities but also as spaces for social encounters. The elderly like to sit in parks and watch the world go by, and parents value them as non-materialistic environments where their children can enjoy controlled adventure and exploration while they meet up with and talk to other adults. Extended families and 'communities' also use urban open spaces as gathering points for games and picnics. In this way, these environments can become imbued with personal meanings and memories of 'community' life or childhood (Burgess et al. 1988).

Despite the fact of the importance of open spaces to urban residents, they have been subject to development pressures and have suffered neglect and disinvestment. In the UK a national initiative to revive these environments, involving the Countryside Commission, local authorities and the private sector, has set out to

---

Box 7.3: The pleasures of nature in the city

Richard: 'When I'm depressed I like to sit, not walk. And there is one little bench on top of the hill at Shrewsbury Park [in Eltham, London] that looks right out over the Thames Valley...and you can sit there and just look at the horizon and feel quite happy. And then I can walk into the wood down below and watch the squirrels which relaxes me. A little bit of wildlife around as well. It's marvellous.'

Viv: 'I could lie about for hours if I didn't have kids, if I had nobody to be responsible for...In Hall Place, they've got them sort of sprucy trees. They smell like the trees in Corfu. And you laid down, sort of in between them. I've done it with Lynn [daughter] lots of times. You very rarely get people come that close. It's lovely. I can lay there for ages...Close your eyes and you can smell the trees and that. If the sun's shining, you can imagine you're somewhere else.'

Michael: 'My strongest memory in moving this summer was waking up one morning and the whole area was covered in multi-coloured poppies!...Coming from an inner urban area to suddenly find this beautiful, spectacular background. That was quite a feeling.'

Burgess *et al*. 1998, 460–1

---

create 12 'community' forests on the edge of major cities (Burgess 1998). Attempts to restore open spaces are not always concerned with aesthetics and pleasure; they can also be used as potential means for regenerating urban 'community' (see Chapter 4). In 1976 the Bronx Frontier Development Corporation was established in the derelict areas of South Bronx in New York, USA. This grass-roots group initiated the development of over 50 community gardens, some of which have had commercial success in selling herbs to restaurants (Bunce 1994).

However, not all urban open spaces are appreciated; some are regarded as monotonous or sterile environments where there is nothing to do, or as dirty places where there are problems of litter, waste-dumping and dog faeces. There is also what Burgess *et al.* (1988: 464) have termed the 'dark side' of open space (see Box 7.3). Parks and common land are settings where people feel afraid of crime, despite the fact that, statistically, relatively few attacks take place in these environments. Perceptions of danger are often associated with enclosed or remote environments (see also Chapter 6). These features are an intrinsic part of many woodland areas. Participants in Burgess *et al.*'s (1988) study described their fears that woods offer the opportunity for attackers to hide, that vegetation might inhibit potential escape routes, and that, because it is difficult to see very far ahead, these can be isolating environments (Burgess 1998). Indeed, in folk stories and fairy tales woods themselves often symbolize danger.

All the women's groups which took part in Burgess *et al.*'s (1988) project expressed a fear of being attacked by a man or men, whom they described as 'maniacs, weirdos and nutters', and recounted experiences of flashing that often went unreported to the police. Participants also expressed concerns about children's safety, racially motivated attacks and anti-social behaviour such as glue sniffing and vandalism (Burgess *et al.* 1988). These fears are compounded by two contemporary processes. First, there has been a decline in the management and social control of public parks and open spaces, largely because of public spending cuts which have resulted in the loss of park keepers and wardens from these environments. Second, increased surveillance and exclusionary practices on the streets and in semi-public spaces such as shopping malls have pushed so-called 'undesirable' groups (such as the homeless and teenagers) into parks and open spaces, which represent the only places where they are free to hang out (see Chapter 6).

7.6.2 Animals

City animals are usually imagined to be pets (such as cats and dogs), while the countryside is seen as the realm of livestock animals (pigs, cows, sheep, etc) and the wilderness is the space of wild animals (such as bears and wolves) (Philo and Wilbert 2000). Yet, in practice, this neat classification does not hold up. Zoos represent an example of humans domesticating and containing wild animals within the city. Many zoos have their origins in imperialism when a representative range of wild species was brought back from the colonies to be classified and displayed (Anderson 1995, 1997, 1998b, 2000). Today, however, zoos are reinventing themselves under the guise of animal conservation.

A whole range of animals also come into the city of their own accord and forge their own living space within it (Philo and Wilbert 2000). Some of these keep their distance from humans by occupying marginal locations such as sewers or waste ground – indeed, we may even be unaware that they are there. Of those animals who make regular appearances in the city, some (such as hedgehogs and urban foxes or badgers) are valued by humans and are therefore fed or encouraged, whereas others (such as pigeons and rats) are coded as pests and vermin.

These animals threaten to disturb human spatial orderings and are often conceptualized in terms of metaphors of contagion and pollution (Philo and Wilbert 2000). The nature–culture boundary is maintained and policed by environmental health officers and animal welfare organizations who often carry out extermination programmes. For example, in London, pigeons, although a tourist attraction in Trafalgar Square, are also regarded as a visual eyesore; their excrement is blamed for damaging buildings, and they are assumed to carry diseases which are a threat to human health and so secret culls often take place at night.

Griffiths *et al.* (2000: 61) explain that: 'Those animals which transgress the boundary between civilisation and nature, or between public and private, which do not stay in their allotted space, are commonly sources of abjection [see Chapter 2], engendering feelings of discomfort or even nausea which we try to distance from the self, the group and associated spaces (but which we can never banish from the psyche). This is clearly the case with cockroaches and rats which invade public and domestic space, emerging from where they "belong", out of sight on a stratum below civilised life, and eliding with other cultures in racist discourse to symbolise racialised "others" '.

They go onto examine human–animal relations in the city by studying the place of feral cats in Hull, UK. Their research explores the extent to which these animals are accepted as having a legitimate place in the city by focusing on the ways that some humans try to engage with them and feed them, while others are antagonistic towards them, claiming that they need to be redomesticated and returned to households. Griffiths *et al.* (2000) conclude that, just as the heterogeneity of human life within the city creates conflicts about who belongs where and who is 'in place' or 'out of place' (see Chapter 6), so the same is true of humans' engagement with the 'natural' world. Wild nature is desired but at the same time feared because it signals a loss of human control over the environment. In particular, some animals are understood to disturb the urban order and to threaten humans' precarious control over nature within the city, hence our desire to domesticate or exterminate them (Griffiths *et al.* 2000).

■ 7.6.3 Urban environmental politics

Conservation is traditionally considered a country issue. However, since the late 1980s the conservation movement and a network of wildlife organizations (such as the Royal Society for the Protection of Birds, and Friends of the Earth) have begun to pay attention to the conservation of urban nature in the UK. An inventory of natural habitats and wildlife species in London by the Greater London Council identified over 2000 sites for wildlife and conservation projects; these included locations such as railway embankments and derelict or abandoned areas (Bunce 1994). The County Trusts for Nature Conservation have also acquired urban sites for nature and wildlife reserves (Harrison *et al.* 1987).

Carolyn Harrison and Jacquelin Burgess's (1994) study of plans to develop Rainham Marshes in London provides a good example of some of the issues encountered by urban conservation movements. Rainham Marshes was classified as a Site of Special Scientific Interest (SSSI) because of its distinctive fauna and flora. This designation should have protected it from any major development. However, in 1989 a consortium of developers sought permission to build a £2.4 billion theme park-style entertainment centre and film studio on the marshes, which offered to bring jobs and tourism to the area. Conservationists were opposed to this plan but local residents were more ambivalent because the mud flats and marshes were poorly managed, often being used for fly tipping and by motorbikers.

Plate 7.4 Cyclists reclaim the city streets, Bastille, France (© Paul White)

In reworking their proposal, the developers then highlighted the pollution and degradation of the marshes and proposed instead a nature and leisure park with a managed wetland in what they argued would be a sustainable development. In this way they played upon the local residents' preference for a very particular kind of 'nature' in which they re-imagined 'nature' and represented themselves as environmental stewards. In 1990 Havering Borough Council granted planning permission for the development.

It is not only specific nature sites which are at risk in the city. Urban environmental campaigns have also sought to raise awareness about the poor quality of the air in cities where nitrogen dioxide, carbon monoxide and ozone guidelines are frequently exceeded. Large cities also have distinctive local climatic effects. The combination of cities absorbing incoming radiation and the output of waste heat and energy creates urban heat islands and temperature inversions trap and concentrate pollution in the city (McLaren 1992). Energy conservation measures to improve the efficiency of ageing housing stocks with double glazing and insulation, and urban green spaces which can help to absorb pollution and lower temperatures are both regarded as potentially important ways of protecting the environment. Urban environmental activists also have taken over the streets of major cities, such as Paris, on bikes and rollerblades (see Plate 7.4) to promote traffic calming schemes, and the need for restrictions on vehicles in order to cut emissions, and to advocate more green forms of transport.

Cities generate large amounts of waste. Urban open spaces and rivers suffer from pollution by waste, metals, pesticides, litter (90 tons a day collected off the streets of London) and illegal dumping. There are also large tracts of derelict land in many cities that were once the site of waste disposal facilities, power stations, chemical works or gasworks and are now too contaminated to be used for other purposes. Although some cities are mounting recycling and urban redevelopment schemes the proportions of household waste that are recycled remain low. In London, UK, the figure is only 2–3 per cent; this compares unfavourably with Portland, Oregon, where the figure is 22 per cent, while in the Netherlands cities recycle up to 50 per cent of their aluminium, paper and glass waste (McLaren 1992).

In the face of such environmental problems, urban environmental campaigners are promoting a programme to create sustainable cities that includes a focus on energy efficiency, traffic calming, effective decontamination and the use of derelict land, waste recycling and the development of more accessible higher-quality open spaces and wildlife resources (McClaren 1992, Elkin, McClaren and Hillman 1991).

■ **Summary**

- Western thought has positioned the country and city in opposition to each other, yet open spaces are embedded in the city rather than separate from it, and a variety of species, plants and animals inhabit urban environments.

- Urban open spaces such as parks and gardens are appreciated for their 'natural' qualities and as spaces for social encounters.

- Animals occupy an ambiguous place in the city: while humans engage with some, others are perceived to disturb human spatial orderings.

- Conservation is traditionally considered a country issue but urban environmental campaigns focus on the need to promote sustainable cities.

■ **7.7 Virtual cities**

The earliest example of a virtual urban space was Habitat, a graphical computer environment initiated in the 1980s, which allowed people both to see representations of themselves and to interact. Its creators, Moringstar and Farmer, described it as a system that could support a population of thousands of users in a single shared cyberspace. It presented its users with 'a real time animated view into an on-line simulated world in which the users could communicate, play games, go on adventures, fall in love, get married, get divorced, wage wars, protest against them and experiment with self government (1991: 273, cited in Ostwald 1997: 139).

The Habitat world was broken up into different regions. The people were depicted as silhouettes of jagged black lines infilled with colour, termed Avatars. They had torsos, heads, arms and legs and could, by moving a joystick, be made to turn and walk in four directions as well as pick up and use objects. The Avatars could move between regions, encounter others (up to 20 000 computers could simultaneously access Habitat), and even talk to each other, by typing words which would appear in a balloon above their heads. Through these interactions communities developed as Avatars bought land, held meetings and developed newspapers and traditions. 'Habitat became a model urban community with hundreds and eventually thousands of people participating' (Ostwald 1997: 140). But it was not a utopia which lasted long. Crime began to occur and, with murders taking place and gangs roaming the streets, Habitat elected a sheriff, guns were banned and this virtual urban space eventually stabilized. However, by this time the Internet was emerging and Habitat was eventually shut down. Although Habitat was the first virtual urban space, subsequently hundreds of similar attempts to parody community and city life have been set up in on-line spaces. One such example is Geocities, a commercial site which describes itself as a 'community' and invites members to become residents of this space by establishing their own web pages within particular Geocity neighbourhoods (see Bassett and Wilbert 1999).

As well as offering new opportunities for creating virtual worlds, the Internet has also been credited with creating the electronic flâneur who browses on-line space. In contrast to the slow strolling of the flâneur described in Section 7.3, who saunters around the streets of the city, the electronic flâneur strolling around on-line worlds is not restricted by the limitations of the human body (see also Chapter 2) but can jump from one on-line space to another. Indeed, the term 'surfing' suggests a sense of mobility, of riding a wave and changing scene (Featherstone 1998).

While contemporary technology has until now been largely text-based, Mike Featherstone (1998) suggests that the development of 3-D programmes means that we can now move through datasets constructed to simulate buildings and streets. Electronic flâneurs can therefore not just browse the Net but also immerse themselves in parallel worlds. Indeed, eventually they will be able to enjoy full sensory involvement and interactions with digital entities being operated by other computer users. This notion of a parallel universe is central to William Gibson's (1986) vision of cyberspace as a data city in his novel *Neuromancer* (Grafton, London).

Featherstone (1998: 922) describes these simulated environments thus: 'In effect, there can be a high degree of replication of bodily presence in environments and interactions with others. One could, for example, stroll through a simulation of a Parisian arcade of the 1830s, and take in the sensation given off at street level.' He also goes on to point out that, like the Net-surfer, the electronic flâneur strolling the virtual streets of a simulated city is not confined to the streets in which they are walking, but can jump from one virtual space to another. He explains: 'While one can have a simulation of the "thickness" of everyday embodied existence one need not bump one's head when one walks into a wall, one need not grow tired at the prospect of a long walk home when one is lost in a strange quarter of the city, one can first jump out of the situation, or zoom out of the local, so that the simulated city appears below like a three dimensional map. Hence it is possible to experience the emotional excitement (free from the physical threats found in dangerous cities) and aesthetic sensations of the street-level stroller, but also that of the detached city planner' (Featherstone 1998: 922–3).

■ Summary

- Technology enables us to move through datasets constructed to simulate buildings, streets and cities.

- Eventually we will be able to enjoy full sensory involvement and interactions with digital entities operated by other PC users in simulated worlds.

- The electronic flâneur strolling in on-line city worlds is not restricted by the limitations of a body but can jump from one on-line space to another.

■ Further Reading

- The city is the subject of a number of specialist books, notably Fincher, R. and Jacobs, J.M. (eds) (1998) *Cities of Difference*, Guilford Press, London; Westwood, S. and Williams, J. (eds) (1997) *Imagining Cities: Scripts, Signs and Memory*, Routledge, London, and the Understanding Cities series published by the Open University Press: Allen, J., Massey, D. and Pryke, M. (eds) (1999) *Unsettling Cities*; Massey, D., Allen, J. and Pile, S. (eds) (1999) *City Worlds*, and Pile, S., Brook, C. and Mooney, G. (eds) (1999) *Unruly Cities?* Useful journals in which to look for articles on the city include *Urban Studies, Urban Geography*, the *International Journal of Urban and Regional Research*, and *Planning Practice and Research*.

- Good examples of articles about different aspects of urban differentiation include the following. **Ethnic segregation:** Jackson, P. (ed.) (1987) *Race and Racism: Essays in Social Geography*, Allen & Unwin, London; Anderson, K. (1991) *Vancouver's Chinatown: Racial Discourse in Canada, 1875–1980*, McGill-Queen's University Press, Montreal; Bonnett, A. (2000) *White Identities: Historical and International Perspectives*, Longman, Harlow. The **underclass:** Gregson, N. and Robinson, F. (1992) 'The "underclass": a class apart?' *Critical Social Policy*, 38–51. **Gentrification:** see the work of Neil Smith in *Antipode*, 1979, **11**, 139–55, *Journal of the American Planners' Association*, 1979, **45**, 538–48 and *Economic Geography* 1982, **58**, 139–55. **Sexual dissidents: Knopp, L. (1998) 'Sexuality and urban space: gay male identity politics in the United States, the United Kingdom and Australia', in Fincher, R. and Jacobs, J. (eds) *Cities of Difference*, Guilford Press, London, and Hubbard, P. (1999) *Sex and the City: Geographies of Prostitution in the Urban West*, Ashgate, Aldershot. In terms of fiction Tom Wolfe's (1987) *Bonfire of the Vanities*, Farrar, Strauss and Giroux, New York, and Jonathan Raban's (1990) *Hunting Mr Heartbreak*, Collin and Harvill, London, both capture the different worlds of the rich and poor who occupy the same city of New York, USA, while Armistead Maupin's (1980) *Tales of the City*, Corgi, London, is a wonderful description of sexual adventures in the city.

- The journal *Planning Practice and Research*, 1995, vol. 10 contains an excellent collection of papers on the city as a landscape of consumption. A good summary of work on the commodification and marketing of cities is Kearns, G. and Philo, C. (eds) (1993) *Selling Places: The City as Cultural, Capital, Past and Present*, Pergamon, Oxford.

- Burgess and Harrison have produced a number of important papers on nature in the city. See, for example: Burgess, J., Harrison, C.M. and Limb, M. (1988) 'People, parks and the urban green: a study of popular meanings and values for open spaces in the city', *Urban Studies*, 25, 455–73; and Harrison, C., Limb, M. and Burgess, J. (1987) 'Nature in the city: popular values for a living world', *Journal of Environmental Management*, 25, 347–62. The best starting point for work on animals is Philo, C. and Wilbert, C. (2000) *Animal Spaces, Beastly Places*, Routledge, London.

■ **Exercises**

1. Take on the role of the flâneur and go for a stroll around the streets of your city, taking in but not participating in the spectacle of city life. In the style of a journalist write an account of your observations and the sensations you experienced.

2. Collect four pieces of different sorts of promotional material (such as a tourist guide, an advertisement for a local company, a promotional image produced by the local development corporation, etc) about the city you live in. Write an interpretative account of the way your city is being represented in these materials. Identify themes highlighted in different ways by the organizations that have produced the materials you have looked at. What in the images is being promoted or given prominence? What is being hidden or suppressed?

3. Choose a consumption space (e.g. a shopping mall or a restaurant or a club). Carry out participation observation in this space at different times. Think about how this space is being produced, how it is being used and the social relations taking place within it.

■ **Essay Titles**

1. Using examples, outline and evaluate the contribution that race-based oppression makes to patterns of segregation and inequality in contemporary cities.
2. Critically evaluate the different explanations for gentrification.
3. To what extent do you agree that the revitalization of cities around consumption is eroding their uniqueness and cultural diversity?
4. Nature is both feared and revered. Critically assess this statement as an explanation for human attitudes to wildlife in the city.

8 The rural

8.1 The rural

The story of how 'the rural' has been constructed in the discourse of social science is a story of a continual struggle to define what is meant by the 'rural' and to establish the extent to which it is the same as, or distinctive from, the 'urban'. The very process of attempting to distinguish between these two opposites has given meanings to them both (Murdoch and Pratt 1993). In particular, there has been an emphasis within geographical work on imagining an opposition between rural space which is understood to have a rural society, and urban space which is viewed as having an urban society (Cloke 1999). The rural has therefore been conceptualized as a distinct, bounded space, while rural society has been distinguished from life in the city in two main ways. First, it is believed to have a strong sense of 'community' (everyone knowing everyone else). Second, rural life has been imagined as closer to nature than urban life (as less competitive, as less predicated on material possessions and status and having a slower pace of life) (Bell 1992).

However, this debate about the relationship between the town and country is increasingly becoming outdated (Mormont 1990). Rather than focusing on how to define the identity of 'the rural' and 'the urban', rural geographers have now begun to challenge the universality of these concepts and have instead started to understand the rural as a social construction, unpacking the ways in which rurality is culturally constructed, by people living both in the country and elsewhere, through discourses and language deployed, emphasizing the production and contestation of meanings. An important aspect of this work has been a recognition that there are 'multiple' cultural constructions of rurality and meanings ascribed to living in the countryside (Philo 1992).

Rural space is often represented as a 'natural' or pure environment in opposition to the pace and activity of the city. As such, the countryside is often seen as a refuge from the oppressive aspects of city life. There is a long history of radical organizations attempting to establish alternative ways of living in the utopian space of the country, while more generally, the countryside is also appreciated as a leisure space which can offer a range of pleasures and forms of relaxation. Yet, at the same time the rural landscape also remains an important site of production and a workspace. Given the many different uses to which rural space is put, it is not surprising that conflicts often break out between those pursuing different activities and that there are huge pressures on the landscape.

This chapter begins by exploring the way that rural society has been conceptualized, focusing on the notion of 'community' (here there are strong links with Chapter 4). It then goes on to examine the countryside as a social construct, looking at the dominant national meanings of the rural in England, the USA and New Zealand. In the following section on 'other' rurals, the chapter examines how the meanings of the country are negotiated and contested by the rural poor, children, lesbians and gay men, ethnic minorities and New Age Travellers. The attention then shifts to examining rural space, firstly as a utopian environment, secondly as people's playground and thirdly as a space of production. The final section considers conflicts between these different uses and the extent to which nature is under threat from them.

■ 8.2 Rural society: community

Claims of an ethos of co-operation and mutual aid have characterized numerous studies of rural life (Newby 1986, Bell 1992). In other words, rural space is often defined as synonymous with a particular form of social relations: *gemeinschaft*, 'community'. In contrast, urban areas are imagined to be characterized by individualistic, impersonal, anonymous – *gesellschaft* – relationships (see Chapter 4). Little (1984, 1986), for example, cites a social survey carried out in two Wiltshire, UK, villages which found that 46 per cent of those questioned identified a sense of 'community' as the most important quality of village life. Bell's (1992: 66, see Box 8.1) study of the people of Childerley, UK, found that the residents distinguished country life from city life on the basis that there is far more 'community' in the country (similar claims are evident in other countries too; see, for example, Williams and Kaltenborn's (1999) discussion of cottage dwellers' community spirit in Wisconsin, USA). Particular gender relations are fundamental to this imagining of rural community. It is women who are the linchpin of community life. Their work underpins many of the village

---

Box 8.1: The rural idyll

'Country life to me is – well, if you look out over there, there's quite a nice view. Now, I can go out over there every day and every day you see something just slightly different. Now there's blooms just starting to get on the trees, getting ready to burst. And all over there, I know all those woods and fields by name like the back of my hand.'

Older villager, Childerley

'The bit I really enjoy is walking around in the evening – taking the sheepdogs, going around looking at the sheep standing in amongst the beef, and looking at good cattle. And that's something...The beauty in the countryside is something too, just walking around in the autumn mist. Once your eyes have been opened to beauty, you can see it everywhere, can't you?'

Local farmer, Childerley

'Peaceful. It's a whole new "ball game". You know, there's a quietness about genuine country people that sort of just plod along. The no-hassle of life... People have got time, time for living, time to talk, which I think is smashing. I mean even in our little country shop, they've got time to serve somebody rather than expect them to rush around and get it all themselves and get 'em out as quickly as possible.'

Women resident for over two decades

Bell 1992: 68–9

---

activities which are said to epitomize 'community', such as fêtes, the church, the parish council, mother and toddler groups, volunteer work, and so on (Little and Austin 1996). In this sense, family and community are inseparable, though it is also important to note that a lack of childcare, employment opportunities and transport mean that some women can experience so-called idyllic rural communities as isolating and oppressive (see Little 1986 and also section 8.4.2).

The popular understanding of the countryside as an appropriate 'family' environment in which to bring up children is one of the reasons why more people in Britain are choosing to leave the city for the countryside. As Miller (1997) explains, 'A new generation is fleeing the city in quest of rural bliss and most of those streaming from London will tell you that they are doing it for the children. As part of our [UK] national mythology, we hold the country to be a good thing for children.'

In moving to the countryside, migrants from the city with their urban-oriented lifestyles have been accused of threatening the very rural society which they were seeking when they moved to a rural space. Numerous studies of rural 'communities' in the UK have observed particularly clear divisions between villagers considered 'insiders' (long-term residents) and those who are perceived to lead more urban-oriented lifestyles, 'outsiders' (Newby *et al.* 1978). Within different 'communities', however, villagers identify and classify each other into different groups, and impose membership of social groups onto each other in subtly different ways. Sarah Harper (1987), for example, suggested that rural dwellers could be categorized according to the extent to which the village was the focus of their lives. Those for whom it was important in all aspects of their lives (e.g. work, home, social networks, memories) she described as 'centred' residents; whereas others she categorized as partially-centred or non-centred. Similarly, Bell (1992) in a study of an English village named Childerley applied four general rules: *localism* (length of residence in the village); *ruralism* (length of time living in the country or working in rural occupations); *countryism* (participation in country ways, e.g. riding, hunting, walking, local history, botany and having knowledge or understanding of country ways); and *communalism* (participation in community activities such as the parish council, fêtes, church, and so on) to determine the extent to which people could be considered to 'belong' in the village.

The relationship between what can loosely be described as 'incomers' and 'outsiders' is often conflictual and in many cases antagonistic. Writing in the late 1960s and early 1970s, Pahl (1965, 1970) argued that mobile middle-class immigrants were destroying village communities by conducting their lives (work, social activities, etc) outside the village, thus exposing class divisions between working- and middle-class residents. He claimed that: 'The middle-class people come into rural areas in search of a meaningful community and by their presence help to destroy whatever community was there. That is not to say that the middle-class people change or influence the working class. They simply make them aware of a national class division, thus polarising the local society' (Pahl 1965: 18).

More contemporary studies have argued that a process of rural gentrification (see Chapter 7) is taking place. Jo Little (1987: 186) explains: 'The prevailing trend . . . has been the movement of middle-class migrants . . . into villages, attracted initially by cheaper houses but, more recently, by an idyllic vision of a healthy, peaceful, natural way of life. Such processes, which are by no means uniform, have led in extreme cases to the gentrification of villages and the wholesale replacement of one population by another.' This sort of pattern is particularly evident in the south of England where what Paul Cloke and Nigel Thrift (1987) term 'the service class' are colonizing more and more rural areas. They are attracted to the country because the housing stock is seen as desirable, having positional character, while also being located reasonably near 'theatres of consumption' (Cloke and Thrift 1987: 327). In turn, the house price inflation this has triggered in southern England has attracted more service-class migrants who want to move to areas where property values are rising, in the belief that they can be assured of future profits on any house which they buy (Cloke and Thrift 1987).

Some of these rural gentrification studies have traced the way that middle-class incomers come to dominate housing and labour markets and often try to mould areas politically and culturally to fit their own vision of rural life (e.g. Thrift 1987, Phillips 1993). In doing so, they have highlighted connections between village conservation policies and middle-class interests (Phillips 1993). This can spark clashes with farming and landowning interests. For example, middle-class incomers often become involved in community and political institutions where they can use their professional skills to oppose farmers' plans to sell off land for development. At the same time as praising rural life, and claiming a distinct lifestyle and pattern of social relations, incomers also fear that they will not be able to maintain this 'community', or that it will not live up to their imagined ideal. In this way, their fears echo Raymond Williams' (1973) claim that the ideal of rural society has always gone hand in hand with a sense that it is in decline – a Golden Age which is ebbing away.

Although much of the academic literature on rural society has focused on conflicts between incomers and long-term residents – which are often expressed in terms of class – Paul Cloke and Nigel Thrift (1987), and Martin Phillips (1993) have argued that these conflicts are usually not quite this simple. Gentrification is not one universal process but rather is a term which conceals a multiplicity of processes which may be at work (see for example Rose 1984, Bondi 1991, Phillips 1993 and Chapter 7). An emphasis on simple cases of middle-class groups replacing working-class inhabitants can create a false dichotomy that obscures the complexity of class relationships and the fact that, in many cases, rural gentrification may involve one middle-class group replacing another (Phillips 1993). Cloke and Thrift (1987) point out that incomers (the middle class or what are termed service-class groups), are themselves fractured or cross-cut by other divisions, identifying five fault lines: public/private sector, gender, lifecycle, consumption practices and type of locality. As a result, instead of social conflict in rural areas being between two class positions (middle class and working class, incomers and long-term residents) it is actually more often the product of antagonisms between middle-class groups. In other words, it can be the result of intra-class conflicts rather than inter-class tensions (Cloke and Thrift 1987). The importance of 'difference' within rural society is discussed in section 8.4.

■ 8.2.2 Country and the city: a false opposition?

However, the imagining of an opposition between rural space, which is understood to have a rural society – epitomized by a strong sense of 'community' (albeit one under threat from internal conflicts) – and urban space, which is viewed as having a more anonymous, individualistic society, is increasingly becoming outdated (Mormont 1990). Differences between the country and city are being blurred. The UK census for the last two decades has shown that most rural areas have experienced population growth (Cloke 1999). This choice has been facilitated by greater opportunities for paid work in non-metropolitan areas as a result of the expansion of the service sector, and even, in some cases, manufacturing employment. The development of information and communications technologies which enable home-working – or what North Americans term 'telecottaging' – and improved personal mobility which allows workers to commute further from urban workplaces, have also enabled more people to choose to live in rural environments where there is little or no actual employment or economic activity. As a result, the traditional urban–rural distinction has been challenged. People no longer have to live and work in one place, but can work in the city and live in the country – or vice versa. New economic uses of the country also mean that rural space and rural society no longer necessarily go together. This leads Marc Mormont (1990) to conclude that there is no single rural space, but rather that multiple social spaces can exist and overlap within a specific geographical area.

Understandings of a clear rural–urban distinction are being further blurred by geographers' increasing sensitivity to the fact that rural areas do not have a unitary character but rather vary widely in terms of their scale, distance from cities, nature–culture relations and their use (e.g. some are close to metropolitan areas, while others are very remote, marginal environments; some are landscapes of agriculture, others are landscapes of pleasure, etc). These variations are evident not only within nations (e.g. the Peak District in the UK is very different from southern rural England; or the US Midwest from Californian desert) but also between nations. The Australian 'outback' or North American 'wilderness' are very different environments from more domesticated European rural landscapes (Cloke 1999). This diversity suggests that the general picture of changes in the economic uses of the countryside presented above conceals an even more complex pattern in terms of the changing relationship between society and space.

Yet, while the traditional opposition between rural space and urban space is being blurred by the above processes, and by geographers' recognition of the diversity of rural landscapes, differences between the meanings ascribed to 'the country' and 'the city' are still important (Cloke 1999). Rather than focusing on how to define the identity of 'the rural' and 'the urban' (for example in terms of economic function or specific social relationships such as community), rural geographers have now begun to understand the rural as a social construction, or structure of meaning. Cloke (1999: 260), for example, observes that '[I]n contemporary society the social and cultural views which are *thought* to be attached to rurality provide clearer grounds for differentiating between urban and rural than do the differences manifest in geographic space.' As a result, some contemporary rural geography has focused on unpacking the ways in which rurality is culturally constructed, by people living both in the country and elsewhere, through discourses, emphasizing the production and contestation of meanings. An important aspect of this work has been a recognition that there are 'multiple' cultural constructions of rurality and meanings ascribed to living in the countryside (Philo 1992). The next two sections focus first on how different countrysides are being culturally constructed, emphasizing the ways in which they are commodified; and then on the way that dominant imaginings of rurality obscure 'other' meanings and experiences of the rural.

✗■   **Summary**

- Rural space has been understood to have a particular form of rural society: community.

- Urbanites who move to rural space in pursuit of rural society have been accused of undermining the very community which they sought by migrating.

- Studies of rural society have focused on conflicts between incomers and long-term residents, which are often expressed in terms of class.

- The middle class is fractured by other divisions. Social conflict in rural areas is therefore often the product of antagonisms between middle-class groups.

- The debate about rural space and rural society is increasingly outdated. Differences between the country and city are being blurred.

- However, differences between the meanings ascribed to 'the country' and 'the city' are still important. Rather than rurality being defined in terms of function or particular society it is now understood as a social construct.

8.3 Meanings and commodification of the landscape

The discussion in section 8.2.1 about community was very Anglo-centric, reflecting the emphasis put on this within Britain, and particularly England. Yet, as section 8.2.2 highlighted, rural areas do not have a unitary character but, rather, there are wide variations both within, and between, nations. Different countries ascribe different dominant national meanings to the rural landscape, which are generally positive.

In turn, these cultural meanings are increasingly employed by the media, advertising and other forms of popular culture to sell a range of products or places which may be associated with the country but may also be aspatial or even, in the past, associated with the city. In this way, the meanings of the rural landscape are no longer tied to a particular type of space but rather are being appropriated and consumed in a range of different contexts. In other words, the countryside is being commodified. For example, it is marketed as a desirable or exclusive place to live, rustic traditions are used to sell fabric and furnishings, while themes of nature, history/nostalgia, outdoor fun and adventure are all used to market rural environments themselves as tourist attractions (Cloke 1993, Wilson 1992). Nature, too, is being commodified, with multi-national retail chains like Nature Company creating and selling products that idealize the natural world (e.g. fossils, plant kits, cuddly animals, bird-watching guides, coffee table books, etc) (Smith 1996a).

The following subsections look at the different dominant national meanings of the rural in England, North America and New Zealand, and the different ways in which these environments are commodified.

■ 8.3.1 The English rural idyll

The dominant imagining of the English countryside is of 'a green and pleasant land' distinguished by pretty rustic villages, winding, narrow lanes and beautiful rolling hills and fields (see Plate 8.1) (Bunce 1994). Within this landscape, the dominant imagining is of a peaceful, tranquil, close-knit and timeless or unchanging community, characterized by harmonious social relationships, which are regarded somehow as more 'authentic' and sincere than the falseness and competitiveness of urban relationships (Newby 1979, Short 1991, Little and Austin 1996). This is a romantic vision based on nostalgia for a past way of life which is 'remembered' as purer, simpler, more innocent and closer to nature (Short 1991, Bell 1992). In this imagining the rural is seen to offer stability, a sense of belonging and an escape from the congested,

Plate 8.1   The English rural idyll (© Brenda Prince/Format)

unhealthy environment of the city and the misery of a disordered and socially unstable urban society (Bunce 1994) – this despite the fact that many people experience
rural life very differently (see, for example, section 8.4).

It is also a representation which is tightly bound up with notions of heterosexual family life and articulates and reproduces particular class (Cloke and Thrift
1990) and patriarchal gender relations (Davidoff and Hall 1987, Little 1987).
Paul Cloke and Nigel Thrift (1990), for example, highlight what they see as the
growing significance among the new middle classes of home domain, 'familism' and
domesticity in rural villages. In particular, the rural is imagined as an idyllic setting for family life because it is regarded as a safer space than the city in which to
bring up children, and a space where children can have more freedom to explore
the environment than their urban peers (Little and Austin 1996, Valentine 1997b).
'The phrases "better for the children" and "good for the family" are conversational
cowslips for the village. Like that famous English country flower, they are touch-
stones of what is right and good about country living' (Bell 1994: 93). Such representations of rural life are prevalent in English literature, lifestyle magazines
and children's toys (Jones 1997). Examples of autobiographical novels which recall
idyllic rural childhoods include Flora Thompson's *The Country Child*, and Laurie
Lee's *Cider with Rosie*.

This cultural construction – the English rural idyll – is actively mobilized and
reproduced thorough the marketing and commodification of the countryside for
urban consumption on multiple scales from the local to the national (Thrift 1987,
Winstanley 1989, Cloke and Milbourne 1992). The countryside has become a positional good – 'something which is limited in supply and whose consumption is reliant
upon a person's position in society' (Phillips 1993: 126). There is a desire, particularly among the middle classes, not simply to visit the rural but to possess it by buying into this way of living. The village environment is marketed as an exclusive place
to live, and as having a particular rural lifestyle (associated with 'community'). The
emphasis on nostalgia within the English rural idyll means that traditional rural culture in the form of crafts, clothing and designs, countryside customs and pastimes
(such as village festivals), buildings and even entire villages are also packaged and
marketed to tourists (Cloke and Milbourne 1992). It is an image of England which
is readily embraced by other nations (Bunce 1994).

8.3.2 Taming the American wilderness

In contrast to the UK, where the countryside is constructed as a beautiful landscape
with symbolic national meanings, the meanings of rural America are tied up with
the wilderness, nostalgia for pioneering settlers and with backwardness (see Wilson
1992).

When migrants arrived in New England and Pennsylvania, USA, from Europe
in the seventeenth and eighteenth centuries, they initially set out to re-create

village-style communities but this way of living quickly broke down. Many of the migrants had fled religious or political intolerance or wanted to escape the straitjacket of hierarchical European society (Bunce 1994). They were seeking new opportunities and the chance to make a fortune. The emphasis was therefore very much on individualism rather than 'community'. Whereas land had been scarce – being dominated by elite landowners – in the Europe which they had left, there was a huge expanse of land available to be taken from indigenous people in the USA which was free or at least cheap (Bunce 1994). The migrants were keen to escape any form of institutional restriction and to be as independent as possible in order to maximize the potential opportunities open to them. As a result, the initial village-style communities were quickly replaced by a pattern (except in the southern plantations) of widespread individual family farms. It is this vision of independent, hard-working pioneer farmers carving out a living in the wilderness, as the backbone of the nation and the root of its strength, which has come to dominate US national ideology (Bunce 1994).

Whereas the English 'rural idyll' is an imagining of the countryside which reflects a nostalgia for a sentimentalized, domesticated and picturesque rural landscape, characterized by tranquillity and a particular social order of village life, rural America does not have the same romantic and emotional connotations. This is partly because of the scale and diversity of the North American landscape, but also because it is associated with a hard-working life, the pioneers' struggle to tame the wilderness and a simple way of living (Bunce 1994). These meanings have been commodified through products such as jeans and cigarettes. Although nostalgia within the US imagining of the rural is associated with hardship rather than tranquillity, pioneer culture in the form of crafts, clothing and farms, etc has, like the English idyll, been marketed as a tourist attraction.

The independence and insularity of rural inhabitants has also fostered cultural myths in the USA about rural incest, inbreeding, backwardness and even cannibalism. Landscapes that appear idyllic and self-contained are feared to hide malignancy, decay and chaos, in contrast to the civilized and sophisticated lifestyles of urban inhabitants. These cultural stereotypes of rural dwellers who are alienated from contemporary urban life are epitomized by poor rural white people based in the southern USA (particularly Southern Appalachia) who are popularly labelled 'hicks', 'hillbillies', 'rednecks', 'yokels' or 'mountain men' (Creed and Ching 1997, Bell 2000). Such representations are exploited in horror films in which innocent urban newcomers or visitors find themselves engaged in a struggle for survival with what Glenn (1995: 46, cited in Bell 1997: 97) describes as 'imbecilic unwashed mountain geeks' who prey on the urbanites who stumble into their territory.

The isolation and deserted spaces of American rural landscapes provide the perfect backdrop against which to stage these encounters between urban travellers and monstrous rural inhabitants (Bell 1997). David Bell (1997: 98) picks out examples of a range of movies, such as *Two Thousand Maniacs*, *Texas Chainsaw massacre*,

Friday the 13th, and *Deliverance*, in which '[i]n a kind of anti-frontier myth, they [urban travellers or campers] find themselves inept at civilising nature; it becomes uncontrollable, alien, terrifying and ultimately murderous'. The motive for the violence of the rural dwellers in these films is often that of the country exacting revenge on the city (Bell 1997, 2000). The rapacious greed of urban capitalists is blamed for destroying the livelihoods of, and marginalizing, rural residents. For example, in *Texas Chainsaw Massacre* rural people suffering as a result of the mechanization of slaughterhouses which has destroyed local jobs, resort to murder and cannibalism against a backdrop of decaying farm buildings and discarded farm equipment (Bell 1997).

8.3.3 New Zealand adventure

The rural landscape of contemporary New Zealand is being commodified through its tourist industry, which is attempting to create a new image or role for the country. In contrast to the American wilderness which is imagined as a bleak frontier, the New Zealand countryside is being represented as a beautiful, untouched wilderness: a paradise (Cloke and Perkins 1998).

Yet, this is not an idyllic landscape which is just to be to be gazed upon or admired; rather, the tourist marketing (magazine and television advertisements and films) stresses participation through adventurous activities (such as trekking on foot, by horse or four-wheel-drive vehicles, white and black water rafting, climbing, skiing, hang gliding, bungee jumping, abseiling, caving, and so on). The emphasis is on doing and being rather than just seeing, and the environment is marketed as akin to a giant adventure theme park (Cloke and Perkins 1988).

The New Zealand countryside, which includes dramatic cave, river and mountain scenery, is constructed through these activities not only as beautiful, but also as an exotic and dangerous wilderness which can be tamed through a combination of personal courage, technological expertise, and the support of skilled guides. In the process of conquering the environment, individuals are also presented with opportunities for self-discovery and personal growth. The final stage of this effort to triumph over nature is, Paul Cloke and Perkins (1998) argue, the capturing or appropriation of the scenery through photography.

This commodification of the New Zealand landscape based on excitement and adventure therefore contrasts strongly with the commodification of the English countryside where the emphasis is on a nostalgia for a peaceful, more tranquil, past way of life. It is important to recognize, however, that this contemporary representation of New Zealand also submerges myths and meanings ascribed by indigenous people to the landscape, and the whole history of the struggles over land ownership between Maoris and Pakeha settlers (Cloke and Perkins 1998).

■ **Summary**

- There are different dominant national meanings of the rural landscape.
- Cultural meanings of rurality are employed by the media, advertising and other forms of popular culture to sell products and places.
- Thus the meanings of rurality are no longer tied to a particular type of space but are being appropriated and consumed in different contexts. The countryside is being commodified.

■ 8.4 'Other' rurals

In a review essay on Colin Ward's book *The Child in the Country*, Chris Philo (1992: 200) commented on rural geographers' general lack of sensitivity to what he termed 'other human groupings'. He observed that, at the time, many contributions to the literature on rural geography tended to portray people as ' "Mr Averages": as being men in employment, earning enough to live, white and probably English, straight and somehow without sexuality, able in body and sound in mind, and devoid of any quirks of (say) religious belief or political affiliation'. Although this quotation perhaps overstates the case, since Philo (1992) himself also noted that rural geographers have recognized some differences, particularly between incomers and long-term residents (see section 8.2), the general point about the subdiscipline's failure to acknowledge how rural environments are experienced and given meaning by lesbians, gay men and bisexuals, children and the elderly, ethnic minorities, and so on, was well made. Philo (1992: 201) concluded his review by calling for rural geography to recognize that 'the social life of rural areas is . . . fractured along numerous lines of difference constitutive of overlapping and multiple forms of otherness, all of which are surely deserving of careful study by geographers'. This challenge has subsequently been met by a range of studies which emphasize that the experiences of rural 'others', and the meanings they ascribe to living in the country do not necessarily fit in with the dominant imaginings or idyllization of rural life set out in section 8.3. This section explores how different meanings of the country are negotiated and contested by the rural poor, children, lesbians and gay men, ethnic minorities and New Age Travellers.

■ 8.4.1 Rural poverty

Rural people suffer deprivation in the form of a lack of work, affordable housing, and public transport. Rates of pay in rural occupations such as agriculture and tourism are low and work is often seasonal, so there is a problem not just of unemployment but also of underemployment. Indeed, Cloke *et al.* (1994) suggest that the proportion of those in UK rural areas living on the margins of poverty (measured by mean income and levels of state benefits) may be as high as 39.2 per cent. Yet, the rural poor are an invisible social group.

Paul Cloke (1997) argues that it is the idyllization of the rural lifestyle – which is predicated on middle-class imaginings of 'family', 'community', good health and so on (see section 8.3.1 above) – which contributes to concealing UK rural poverty. He suggests that in the eyes of policy makers the benefits of rural life itself are assumed to outweigh or mitigate experiences of deprivation. Likewise, he attributes the willingness of rural people living in poverty to tolerate their position to the fact that they have low expectations because they too prioritize family, health and the pleasures of being in the country over material aspirations, and regard asking for help from the state as a stigma.

In the USA rural people living in poverty are not invisible, in the same way as they are in the UK, because there is an established poverty line, a state-defined measure of poverty – albeit one which is criticized for being set at too low a level and for making unrealistic assumptions about the costs of essential items (Cloke 1997). Yet, the rural poor are still eclipsed by the urban poor because of US imaginings of rural life. The legacy of the pioneers is that hard times are regarded as 'natural' in rural areas, intrinsic to the pioneer lifestyle. Hardship is assumed to be part of the price, but also the thrill, of taming nature (Cloke 1997). Rural people are therefore understood to be the 'deserving poor', in contrast to their urban counterparts, who are characterized as dysfunctional, dependent on welfare and therefore as the 'undeserving poor'. The consequence of this binary categorization of those in poverty is that the US public debate and attention generally focus on the 'undeserving poor', with the consequence that rural poverty is forgotten (Cloke 1997). This is despite the fact that, statistically, the poverty rate is consistently higher in non-metropolitan than in urban areas (Rural Sociological Society Task Force 1993 cited in Cloke 1997: 263).

Rural poverty is also disregarded because it is strongly associated with particular regions of the USA, notably Appalachia and the South, even though the problem is more widespread. Appalachia has a strong regional identity predicated on supportive kin and community networks, unique customs and language, and a subsistence lifestyle which rejects many modern economic values. As a result of the independent or contained world in which they live, cultural myths have developed about the Appalachian people, who are stigmatized as 'backward', incestuous, and so on (see section 8.3.2). This 'othering' of a regional identity serves to explain or justify rural poverty (Cloke 1997) and enables the national government to avoid addressing its real causes or acknowledging that black poverty is concentrated in the South.

■ 8.4.2 Rural children

The academic literature on children (Ward 1990, Hart 1979, Wood 1982, Shoard 1980) suggests that young people prefer to play in 'natural' spaces, where they are free to create their own activities and adapt their surroundings to their games, rather than in adult-defined play spaces where their activities are constrained. The countryside is seen to offer more possibilities for this sort of imaginative play than the city because rural environments are not as ordered and regimented by adults as urban environments. There are claimed to be more open or abandoned and derelict spaces in the country than the city, which children can adapt to their own ends, away from the surveillant gaze of grown-ups (Jones 2000). Marion Shoard (1980) claims that five characteristics make rural environments such a rich place for children to play: (1) the number of props available to play in or with (grass, trees, ditches, etc); (2) the freedom of movement the country offers; (3) it is a separate space where they can escape from the parental home; (4) rural animals and wildlife provide a source of fun; (5) the countryside is a site of the unknown, full of possible surprises and scrapes. Rural locations are therefore frequently represented as idyllic places for children to grow up. It is an image reproduced through fictional accounts of childhood such as *Swallows and Amazons* and *Cider with Rosie* in which the joys of nature and the country are celebrated and where the characters embark on adventures together independently from adults.

In bringing such fun and entertainment into children's lives, the rural environment is also imagined to enhance their physical and spiritual health. There is a whole discourse around the countryside as a place of fresh air, health and healing for children (Jones 1997). For parents, rural environments have further appeal as a place for their children to grow up, because the country is presumed to shelter young people from the commercial pressures of the fashion industry and peer group pressures to engage in drugs, underage sex, bullying, and violent crime, and so to provide a more innocent, less worldly and purer experience of childhood than that offered by the city (Jones 1997, Valentine 1997b).

However, studies of children and young people's actual experiences of the countryside present a rather different picture. The mechanization of agriculture, the erosion of residential areas into green belt sites, and landowners enforcing laws against trespass to keep children off their property have all contributed towards marginalizing young people within the rural landscape. Indeed, most land in rural areas is privately owned, so many of the disputes which occur between adults and children in the city (see Chapter 6) over their access to, and use of, space (Valentine 1996b) are also replicated in rural areas (Davis and Ridge 1997). Likewise, parental fears about the vulnerability of children to stranger-dangers (notably walkers, tourists from urban areas, etc), traffic and rural demons such as travellers and gypsies (Valentine 1997b) are also contributing to the contraction of rural children's spatial ranges. (Nevertheless, 'community' is often drawn upon to justify adult beliefs that, despite these dangers, the country is still a safer environment for children than the city.)

Colin Ward writes (1990: 100) that, though we imagine that rural children have unlimited access to open space, in practice '[t]oday's rural landscape has fewer children and fewer places for them'. Ward (1990) further suggests that the contemporary loss of rural hedgerows, ponds, streams and access to woods, documented by Shoard (1980) in *The Theft of the Countryside*, has had – as she herself also argued – as significant an impact on rural children as it has on wildlife. In these terms, the rural environment is represented as a space that was once a golden land of opportunities for children but is now in decline.

Yet rather than wanting to play in woods, fields or by rivers and lakes, the landscapes assumed by adults to be desirable for children, Matthews *et al.* (2000) and Francine Watkins (1998) found that children and teenagers preferred to hang around marginal spaces such as outside shops, phone boxes, in car parks, bus shelters, the park and building sites. In young people's worlds the social is more important than the natural environment. Children hanker after the excitements of the city, such as the cinema, leisure centres, sports venues and shopping malls, rather than the rural idyll. Living in remote or poorly serviced communities, they are dependent on the vagaries of public transport services or adults (usually mothers) with cars to reach urban entertainments (Little 1986). They therefore experience their lack of mobility and isolation perhaps more acutely than adults, though feminist research shows that women can also experience rural life as isolating – see Little (1986) and Little and Austin (1996) (see Box 8.2).

Box 8.2: Rural isolation

'If you don't have your own car it is a bit of a nightmare really. I used to have a bike and I used to have to cycle everywhere because the buses stop at quarter to five in the evening and they cost a fortune. So if I wanted to go out in the evenings to see my mates I used to have to cycle there.'

Teenage girl, Little Hatton

'I'll be out of here by the time I am 18. I want to go to college. It is okay for old people because they don't want to go out much whereas with my age group there is nothing to do because if you want to be able to go out to clubs and movies and do anything you can't because there is nothing round here.'

Teenage boy, Little Hatton

'There isn't anything for young people in the village. Little Hatton is the sort of place I'd like to live in when I'm middle-aged but not in my 20s. It's like all the people on Warren View are married with 2 kids, his and hers Sierras [family car] and everything in the village is for them.'

Teenage girl, Little Hatton

Watkins 1998: 213–15

Children can also experience the close-knit nature of rural 'community' life as more negative and claustrophobic than adults (though see also section 8.4.3 on lesbian and gay experiences of the rural). One feature of small town and village rural areas is the 'public' nature of people's lives. Although children in urban areas are also supervised by parents, teachers, neighbours and friends, the scale and anonymity of urban environments mean that it is still relatively easy for urban youth to escape the gaze of people they know. In small towns and villages – especially where the population is fairly stable, with generations of the same families living there – the young people may be known to most of the town via the school, the church, community groups, service providers (e.g. community health and medical centres), shops, businesses, and so on. This lack of freedom is both oppressive and potentially dangerous. For example, Hillier, Harrison and Bowditch (1999) point out that young people in remote rural areas of Australia often do not practise safe sex because it is impossible for them to buy contraceptives (it is not as easy to get to an anonymous chemist in the Australian outback as it is in the town/city!) or to seek medical services and sexual health counselling without being seen by someone they know and so news of their sexual activity spreading round the 'community'.

Contrary, then, to adult imaginings of rural environments as idyllic places to grow up, children and young people can feel smothered – unable to carve out a space for themselves independently of adults – disaffected, alienated from the 'adult community' and powerless or excluded from channels of representations (James 1990).

■ 8.4.3 Lesbians and gay men

Lesbians and gay men echo children's and teenagers' complaints that there is nothing to do and nowhere to go in rural environments. There is a lack of structural services and facilities (pubs, clubs, support groups, safe sex information, and so on) and basic resources (such as lesbian and gay newspapers, magazines and books) to support non-heterosexual lifestyles in most rural communities (Bell and Valentine 1995c, Kramer 1995). There is some evidence, though, that lesbians and gay men can develop spatially disparate communities without propinquity through telephone helplines, newspapers (Wilson 2000) and the Net, which can offset some of these problems of isolation and a lack of information (see also Chapter 4).

Rural communities are not necessarily exposed to the progress made by lesbians' and gay men's movements in urban areas, and police forces in rural locations are commonly less sensitized than those in the city to dealing with sexual dissidents, which can result in homophobic or ignorant policing of lesbian and gay lifestyles (Bell and Valentine 1995c). The extent of homophobia in rural areas is evident in Hillier *et al.*'s (1999) paper about young people's sexuality in Australian small towns. Not surprisingly, many lesbians and gay men living in the country are reluctant to 'come out' because they are afraid of encountering homophobia in close-knit, family-oriented 'communities' (Wilson 2000). This fear also discourages lesbians and gay men from contacting each other in case they are 'outed' by association (D'Augelli and Hart 1987). Consequently, they have few opportunities to express their sexual identity (see also the experience of rural youth in section 8.4.2). Jerry Lee Kramer's (1995) account of gay men's lives in North Dakota, USA, paints a bleak picture of episodic sexual encounters on the highway or trips to local town adult bookstores. It is worth noting, however, that same-sex activity between those who do *not* identify as lesbian, gay or bisexual may be more common in remote rural areas because the very isolated nature of some landscapes, combined with physical work and homosocial bonds in male-only occupations such as cattle running and lumbering, can overcome moral codes against men having sex with men (Bell and Valentine 1995c).

The picture, however, of lesbian and gay life in the country is generally as bleak as the title – 'Get thee to a big city' – of Kath Weston's (1995) paper on lesbian and gay rural-to-urban migration suggests. Weston (1995) argues that the anonymity offered by urban environments makes them a better place in which to live a lesbian or gay lifestyle than within claustrophobic rural society. Indeed, she argues that the symbolic contrast offered by the urban/rural dichotomy is crucial to making sense of lesbian and gay identities, being central to the organization of many 'coming out' stories in which sexual dissidents migrate from the country to the city to escape prejudice and to forge their own identities.

Despite the everyday realities of lesbian and gay lives in rural communities, rural landscapes feature in lesbian and gay imaginations (Ingram 2000). Body-landscape metaphors are common in homoerotic verse (Woods 1987), while the rural is eroticized as a location for sexual encounters in lesbian and gay fiction, and in tourist guides promoting same-sex accommodation in the countryside (Bell and Valentine 1995c). In such imaginings the rural is presented as a utopian space for same-sex love away from the restrictions of the law (see, for example, Vidal's *The City and The Pillar*, or E.M. Forster's novel *Maurice*). The possibilities rural environments offer for sexual dissidents to establish new ways of living and new communities in space away from the masculinist and homophobic city have been particularly evident in lesbian feminist representations of the country (Valentine 1997c, Wilson 2000). These are discussed in detail in section 8.5.

8.4.4 Non-white ethnic minorities

The dominant imagining of the UK rural idyll is of the countryside as a white landscape. It draws upon and reproduces a nostalgic myth of a white Anglo-Saxon past, neglecting the significant number of non-white people who have lived in, and continue to live in and use the countryside. Julian Agyeman (1995) draws upon diverse historical examples, from the North Africans stationed in Britain on Hadrian's Wall as part of the Roman empire to more contemporary examples, such as British Asians employed in Yorkshire and Lancashire cotton mills, to highlight the contribution of non-white people to the British rural landscape. Today, despite its veneer of whiteness, significant numbers of black people live in the British countryside. Jay (1992) estimates that there are 26 200–36 600 non-white people in the south-west peninsula (counties of Cornwall, Devon, Somerset and Dorset).

The word 'country' has a double meaning, signifying both 'rural' and also 'nation' (see also Chapter 9). Andrew Howkins (1986) argues that the countryside is there-fore also seen as the essence of England. It is idealized or naturalized as a pure space, a signifier of a white national identity, in contrast to the city, which is associated with racial degeneration and immorality. As a result, white people have a sense of attachment to the rural which may not be shared by black and Asian people. Julian Agyeman and Rachel Spooner (1997) point out, for example, how imaginings of the rural idyll are used in National Front and British National Party literature, groups which also recruit heavily from rural areas.

A Countryside Commission document *Enjoying the countryside: policies for people* suggested that economic factors – a disproportionate number of people from ethnic minorities live in the inner city, from which it is difficult, time-consuming and expensive to get away – and a fear of racism deter people of colour from visiting the countryside. Other reports also document examples of institutionalized racism, racial harassment and bigotry (see also Chapter 7) in rural areas (Jay 1992). Just as lesbians and gay men feel isolated amidst the heterosexuality of rural life (section 8.4.3), it is difficult for black and Asian people to develop and sustain a positive identity in the face of racism if they are isolated from other non-white people. People of colour feel alienated, disenfranchised and unsafe in this white landscape, antici-pating abuse from white people, even when they have not had a bad experience them-selves (Malik 1992). The black photographer Ingrid Pollard has used photographs with captions such as 'Feeling I don't belong, walks through leafy glades with a base-ball bat by my side' to challenge her own sense of alienation from the countryside (see Figure 8.2).

The marginalization of black people from the British landscape has also been chal-lenged by the Black Environmental Network. This was formed in 1988, setting up regional forums between environmental organizations and ethnic minority com-munities in a bid not only to promote access to the countryside but also to establish a broader environmental agenda (Agyeman and Spooner 1997).

■ 8.4.5 Travellers

There is a tradition of travelling fairs in the British countryside which began in the 1960s and 1970s to provide a focus for counter-cultural movements. Those with an interest in crafts, alternative technologies and ways of living, and those who rejected mainstream society, began to coalesce around these traditional horse fairs, adopting

Plate 8.2   'Feeling like I don't belong, walks through leafy glades with a baseball bat by my side' (© Ingrid Pollard)

semi-nomadic lifestyles, living in buses and tipis. A festival circuit, including Glastonbury and other rural venues, gradually emerged, which was advertised through alternative book and wholefood stores, magazines and by word of mouth. These festivals were conceived of as utopian spaces where people could be free from the routine and values of 'normal' society. They were also a form of opposition to commercial pop festivals (Hetherington 1996). In the face of rising unemployment in the 1980s and disenchantment with materialistic society, the numbers travelling and the popularity of the festivals grew (Sibley 1997).

The phrase New Age Travellers (NATs), used to describe those leading this sort of life, is an umbrella term for a range of people whom Halfacree (1996: 47) lists as the 'Brew Crew', 'cosmic hippies', '24 hour party people', 'back-to-the-landers'. What they share are a number of characteristics – real or imagined – which contrast with conservative visions of rural society. NATs show 'lower levels of materialism than the settled community which is reflected in their lack of concern for obtaining regular full-time paid employment and they are less respectful of private property, especially land, and the other key institutional supports stressed by the authoritarian right, such as the nuclear family and establishment traditions' (Halfacree 1996: 47).

While they sometimes do seasonal work or sell things for money at festivals, there is also a barter economy and gift exchange between travellers. An emphasis is placed on reskilling – developing alternative magazines, craft skills, proficiency in holistic medicine and therapies, musical and entertainment talents, and so on – rather than conventional educational and employment skills (Hetherington 1992).

In the 1980s the NATs began to be conceptualized by the British government as a threatening minority. These fears tapped into a history of prejudice and hostility towards gypsies and travellers. But whereas gypsies have a place within the national imagining of the rural idyll – the Romany traditions which are associated with brightly painted caravans and their own language being romanticized as ethnically exotic (Sibley 1992) – NATs do not. Instead, through subverting 'the spatiality of sedentary society', they were seen to 'threaten to disrupt the privatised and ordered form that the commodified "post-productivist" countryside is assuming' (Halfacree 1996: 44).

Cat-and-mouse games with the police around south-west England came to a head at the 'Battle of the Beanfield' in 1985. An injunction was put in place to stop some travellers going within five miles of Stonehenge to celebrate the summer solstice. Stonehenge is an ancient site of stones which is associated with pagan traditions and ancient wisdoms. It is venerated as a place of worship and renewal. This is symbolized by the position of the sun on the slaughter stone at the solstice. As a site of alternative character and mystery, Stonehenge symbolically represents the values and beliefs which NATs regard as being marginalized in industrial society (Hetherington 1996). The site is fenced and roped off from the public, being maintained as a 'museum without walls' (Hetherington 1996: 165) by English Heritage, who manage the site, and the National Trust, who own the surrounding land. Whereas it is managed in this way as an ordered and static place, sedimented in the past, and available to be picked over only by archaeologists, for NATs it represents a site of celebration and transgression in the present. In 1985 the efforts of the police to prevent NATs reaching the site resulted in a violent conflict between police and travellers and, ultimately, the demonization and criminalization of NATs (Sibley 1997).

In the ensuing popular debate the NATs' alternative lifestyles, cultures and spatialities were regarded as trespassing or violating the English rural idyll (Halfacree 1996). In parliamentary debates and the popular media they were represented as a 'convoy of pollution' (Rojek 1988) – typified by drugs, criminality, rubbish, risk and uncertainty – invading small rural communities, bringing chaos and anarchy to the peaceful and stable order of rural society and desecrating the beauty of the countryside. In other words, somewhat ironically, NATs were vilified not because they moved on, but because of fears that they might stay and contaminate idyllic rural areas (Hetherington 1992)

While the Conservative government of the time believed in the free market, it also believed that this must be upheld by a strong state and thus, in the face of what it regarded as an 'army of marauders, hell-bent on destroying the peace, quiet and seclusion of country life' (Halfacree 1996: 62), the government saw a need to 'uphold the *spatial* authority of the state' (Halfacree 1996: 48).

In 1994 *The Criminal Justice and Public Order Act* was passed. This was a very broad Act. For example, it expanded the obscenity law to include computer pornography; it introduced tougher laws against racially inflammatory material; it ended the right to silence and it also created new public order offences which were targeted at NATs. These included expanded powers to remove trespassers from land; to enable local authorities to direct unauthorized campers to leave land; to abolish the responsibility of local authorities for providing permanent sites for travellers; and to create a new offence of 'aggravated trespass' (Fyfe 1995).

Vera Chouniard (1994: 430) argues that the law 'is actively involved in shaping people's territories and access to diverse spaces, expressing multiple forms of empowerment and exclusion over how spaces can be used by whom'. In this case, the combination of the trespass law changes and the end of permanent sites for travellers effectively outlawed nomadic and semi-nomadic minorities, putting gypsies and travellers in a vulnerable position. It also criminalized a very diverse group of people in the form of disparate protesters who take direct action in rural areas (such as hunt saboteurs, and environmental protesters campaigning against road building), for the simple reason that they move around the countryside (Fyfe 1995, Sibley 1997). These measures are deeply geographic. They represent a strategy for the purification of rural space by blatantly trying to exclude from rural environments, through criminalizing the use of space for certain activities, those regarded as undesirable 'others', because this group are seen to have no place within the straitjacket of the English rural idyll (Sibley 1997).

■ Summary

- Society in rural areas is fractured along numerous lines of difference. Yet, rural geographers have often been guilty of focusing on the white middle classes, ignoring the experiences of 'other human groupings'.

- The meanings rural 'others', such as the poor, children, lesbians and gay men, and travelling communities, ascribe to living in the country do not necessarily fit in with the dominant imaginings or idyllization of rural life described in section 8.3.

- Geographers are beginning to be sensitive to the way rurality is negotiated and contested by different groups.

■ 8.5 Rural space: a utopian environment

Rural space is epitomized as 'natural', pure and tranquil, and therefore is traditionally understood to have a particular form of society which is regarded as being closer to nature than the sort of society which emerges in urban space. Rural society is imagined to be simple, one which is less competitive, less predicated on material possessions and status, and which has a slower pace of life than that found in urban areas. As a result, the countryside or wilderness is seen to offer a refuge from the 'base instincts' of the consumerist and oppressive city (Short 1991: 31). For this reason, it is often imagined as a utopian space. The country is considered to be a place where it is possible to put into practice another way of life because it offers social and physical space for people to develop alternative models of social and economic living (Mormont 1987a), and because the spirituality of nature is seen to provide opportunities for personal growth and individual healing.

The New Age movement covers a whole range of practices and ideologies (notably Eastern ideologies) which are focused on the importance of spirituality and a desire to find an authentic mode of being and which place great emphasis on sacred space. This cultural movement has been described as being at the root of many contemporary environmental movements. It can be traced back to the eighteenth century, but really gained momentum from counter-cultural activity in the 1960s and has recently been revitalized through the revival of pagan and shamanic ideology (Heelas 1996). Although not a rural movement *per se*, the quest for self-spirituality, rejection of capitalist institutions and reverence for nature mean that many New Age activities and services (e.g. psychic teachings, sexuality workshops, healing practices, and so on) are located in the countryside (Moore 1997).

Likewise, sections 8.5.1 and 8.5.2 outline two more detailed examples of different groups – lesbian separatists and the mythopoetic men's movement – for whom

the countryside and wilderness are seen to offer a utopian space for developing alternative ways of living and personal growth.

8.5.1 Lesbian lands

In the 1970s radical feminists identified heterosexuality as the root of all women's
(lesbian, bisexual and heterosexual) oppression. The only way to avoid (re)produ-
cing patriarchy and for women to construct a new society beyond men's influence
was for women to establish their own communities that were separated or spatially
distanced from heterosexual society. Although some women-only spaces were established in urban areas (e.g. Toronto), the aim of separatism was understood to be
best fulfilled in rural areas. The spatial isolation of the country meant that it was
easier for women to be self-sufficient and therefore purer in their practices than in
the city. Essentialist notions about women's closeness to nature because of their men-
strual cycles and reproductive role also meant that the countryside was identified as
a women's space (see Chapter 2). In contrast, the man-made city was blamed for
draining women's energy (Bell and Valentine 1995c, Valentine 1997c).

In the USA a circuit of women's farms, known as lesbian lands, was established
(see Chapter 9). The control of rural space was seen as vital to give women the freedom to create new ways of living and new ways of relating to the environment.
The communities were based on non-hierarchical lines and effort was put into building new forms of living space (see also Chapter 3). The women did not want to
have to go back to, or rely in any way on, patriarchal society, so they re-learned
old skills such as firemaking, herbal medicine and survival skills, while also
developing a women's culture in terms of language, music and books. Their belief
in women's closeness to nature meant there was a strong spiritual dimension to these
rural communities. The women celebrated the full moon, equinox and solstice, and
practised goddess worship, witchcraft and other women-centred traditions which symbolized their resistance to patriarchy. Their intention was not only to build self-
contained communities but that these communities 'would eventually be built into
a strong state of mind and that might even be powerful enough, through its example, to divert the country and the world from their dangerous course' (Faderman
1991: 217) (see Box 8.3 and also Chapter 9).

However, tensions also arose within and between lesbian lands, particularly over
issues such as boy children and even male animals. While some settlements took
a non-essentialist view of identity and so allowed male children and animals to
remain in lesbian lands, others excluded them in an attempt to create a pure women-
only space. The emphasis on escaping patriarchy meant that many lesbian lands
promoted the residents' shared identities as women over their differences. Class and
issues over ownership and co-ownership of sizeable investments in terms of land,
dwellings, buildings, wells and fences were common sources of dispute, as were sexual
jealousies and relationship breakdowns which caused divisions and exclusions.

Box 8.3: Utopian space

'We view our maintaining lesbian space and protecting these acres from the rape of man and his chemical as a political act of active resistance. Struggling with each other to work through our patriarchal conditioning, and attempting to work and live together in harmony with each other and nature.'

Resident of Wisconsin womyn's land co-operative

'Best of all, there is time and space for a total renewal of ourselves, our connections to each other and our earth'.

Resident of Wisconsin womyn's land co-operative

'We do have a basic ritual. We open the circle with a blessing and a purification, similar to Dianic. We begin with the oldest to the youngest. Lots of the time we use the salt water purification. Each one cleanses the other woman and takes away her negative energy.'

Resident of the Pagoda

'Each spring we were directed to do a medicine walk upon our boundaries. We walk 129 acres up and down the sides of the mountain ... This encircles us and protects ourselves, our animals, our community, our children from any interference.'

Resident of Arco Idris

'The male child issue has been the most painful one. We've all chosen to live without men, yet there are women now who are talking about having babies. This one has created not the most anger between us, but the most pain. We really have tried very hard not to go at each other.'

Resident of Maud's Land

Cheney 1985: 132, 132, 113, 31

Claims of racism and a lack of tolerance of disabled women were other fissures of difference that split the fragile unity of some utopian communities. Black women and Jewish women felt marginalized by the inherent whiteness of most of the lesbian lands, and the white women's lack of awareness of the specificities of oppression. The emphasis on the body and shared physical commitment to the land through physical labour meant that many disabled women felt unable to participate and considered that the communities did not respond to their needs (Valentine 1997c). Some lesbian lands did, however, attempt to make independent space for disabled lesbians or to create non-racist environments specifically for women of colour.

These rural utopian communities often faced hostility and even violence from wider rural society. In the summer of 1993 a radical lesbian feminist retreat known as 'Camp Sister Spirit' was established in the small rural community of Ovett, Mississippi, USA. The local population strongly opposed it, the local newspaper drawing a stark contrast between the idyllic community of Ovett, which it described as: 'little churches with old graveyards. Freshly tilled soil, cows and more cows. Thick, green grass . . . some of the most beautiful land on earth' (Greene 1997: 34) and the camp. This was characterized as dirty, chaotic and the women as aggressive, man-hating and as a threat to the wives and daughters of Ovett (see also section 8.4.5). The newspaper dismissed it as: 'primitive at best. A couple of travel trailers here, junked coaches there and working women in T-shirt and blue jeans, one of them with a knife strapped to her side' (Greene 1997: 34) and described it as a haven for 'every stray lesbian in the country' (Greene 1997: 21).

The women were threatened not only through such articles but also through more direct forms of harassment (e.g. a puppy was shot and left on their mailbox), which escalated to threats that the KKK will 'burn a cross on you' (Greene 1997: 21). A group known as Mississippi for Family Values was set up with support from the radical right to oppose Camp Sister Spirit. To promote its argument that the idyllic rural community of Ovett was being invaded and polluted by the camp this organization used a film from the Christian Coalition and Traditional Values Coalition, representing lesbian and gay sexuality as a sin and an abomination. Eventually, the dispute reached a climax when it was reported in the national media and the FBI became involved. The case then went to court, where the judge ruled in favour of the women, allowing them to continue to use this space as a retreat (Greene 1997).

8.5.2 The mythopoetic men's movement

The mythopoetic men's movement emerged in the USA in the 1980s. It developed from a network of men's groups and journals. The movement identifies contemporary society as feminized and focuses on a quest to recover the 'deeper male'. Some commentators have argued that its origins lie in social and economic restructuring which has created new uncertainties about gender roles and identities, eroding the traditional authority of men at both home and work, threatening in particular the manhood of white men in their 20s–40s. The men's movement is therefore seen as a backlash against men's sense of powerlessness and an attempt to restate male power (Kimmel and Kaufman 1994). Alistair Bonnett (1996) is critical of this explanation, however. Pointing to the middle-class nature of the movement, and the absence of a parallel working-class movement, he suggests these men are not powerless but indeed have the ability to wield power, for example by obtaining resources for their project. Rather, Bonnett suggests that the men's movement is a response to the fact that middle-class men have contradictory experiences of power, experiencing both power and a loss of power because of feminism.

The men's movement draws on practices from non-Western (especially native American) cultures: group drumming, wilderness retreats, initiation rites, and so on, which, like lesbian separatism, are commonly acted out in a wilderness context. Bonnett (1996: 276) provides the example of the Clearwater Men's Group, Wisconsin, USA:

> We would celebrate a ritual of our own making, a ritual of striving, ascending, and joyous welcoming. We broke from the circle to plant our staffs in a double row, forming an aisle running fifty feet up a steep hillside. At the top of the hill we began to drum, building quickly to a throbbing boogie. Then one by one, beginning with the eldest, we descended to the bottom of the hill, entered the path between the staffs and walked, ran or danced our way to the top to be hugged, held and hoisted into the air by our brothers' strong arms. As the youngest man reached the hilltop, the drums built up to a new climax and we raised a triumphant chant 'WE ARE MEN!' (Pierson 1992: 114, in Bonnett 1996: 276).

Like lesbian separatism, the men's movement has a strong wilderness philosophy, encouraging a spiritual and practical reverence for the landscape. Just as lesbian lands were created by women wanting to escape the patriarchal city, the men's movement understands the wilderness as offering men liberation from the claustrophobic, alienating world of feminine society, where luxuries, fashion, and so on, are blamed for producing the 'soft boy' (Bonnett 1996). While the lesbian separatists romanticized witchcraft and other women-centred traditions, the men's movement values what it regards as the authentic wisdom of indigenous peoples such as Native Americans. Bonnett (1996: 281) observes that 'the forms of primitivism it makes use of draw from an established and well-supplied reservoir of imperial fantasies of "tribal" peoples as metropolitan "civilisation's" Natural Other' (see also Chapter 2, section 2.4.1.2). To the men's movement these tribal societies are regarded as being 'at one' with nature, and as being as unchanging or timeless as nature itself. Consequently, the gender inequalities within these societies are understood to represent a more primeval and 'natural' form of gender relations (Bonnett 1996).

Like the lesbian separatist movement, the men's movement is often regarded as essentialist and reactionary. Yet, Bonnett suggests that it is not quite that simple. While the men's movement may be essentialist because it emphasizes 'natural' and eternal differences between men and women, it is also anti-essentialist because of the emphasis it puts on men showing emotion and its strong anti-violence stance. Likewise, while it is reactionary and even racist because of the way it appropriates other cultures and draws on racist colonial fantasies, it is also anti-racist or transgressive because it subverts whiteness through promoting white people's identifications with non-white people and transgresses notions of a fixed white culture, for example, in the way that the men mix up a variety of cultural forms and traditions and rework them in their rituals and ceremonies.

■ **Summary**

- Rural space is traditionally understood to have a particular form of society, which is regarded as being closer to nature, less competitive and materialistic than that which emerges in urban space.

- As a result, the countryside is often imagined to be a utopian space where it is possible to develop alternative ways of living and achieve spiritual growth and healing.

- Lesbian separatists and the mythopoetic men's movement are examples of those who imagine the countryside/wilderness to be a utopian space.

■ 8.6 Society's playground

The countryside is not only valued by particular social movements for the utopian possibilities it offers for developing alternative ways of living and spirituality, but is also appreciated more widely by the urban population as a space which can offer temporary release from the everyday pressures of city life. The countryside is what Michael Bunce (1994: 111) terms 'the people's playground' – a leisure or consumption space which offers a range of pleasures and forms of relaxation from the opportunity to enjoy the scenery to more participatory outdoor activities.

Until the mid-nineteenth century the countryside was principally enjoyed as a landscape of leisure by the wealthy, who liked to indulge in recreational pursuits such as hunting on country estates. In the eighteenth century, however, it had been popular for the middle classes to take tours of the UK countryside. The growth of maps and guidebooks detailing places of interest, such as country houses, and writers, artists and poets (such as Wordsworth) who had begun to romanticize nature and picturesque landscapes, had all encouraged people to begin to appreciate, and take an interest in, 'nature' and 'wild scenery' (Bunce 1994). Prior to the development of the railways the working classes' appreciation of the countryside was limited to the open spaces to which they could walk from the city. It was a popular activity because many of those working in the new factories were ex-rural migrants and because there was a general lack of open space in the industrial cities of the nineteenth century, which were characterized by squalor and overcrowding (Bunce 1994).

In turn, the English middle-class understanding of picturesque scenery and the romanticization of nature translated to North America. Here, it stimulated a trend of travel in search of wild and spectacular landscapes which offered sensory and spiritual pleasures. Bunce (1994) reports that by the mid-nineteenth century the middle classes were taking eastern seaboard carriage journeys as far inland as Niagara Falls, and steamboats up the Hudson River to Albany, into the Catskills, the Adirondacks and the White Mountains. More localized trips to natural features in and around cities were also popular. Painting scenery became a common hobby (Bunce 1994).

In the late nineteenth century concern over public health and the harsh conditions in factories led to social reform in the UK (inspired by trade unions and philanthropists) and the establishment of half-day working on Saturdays, a day off on Sundays and bank holidays. This definition of leisure time gave the working classes more opportunity to escape to the countryside, and, combined with the development of railways and cheaper fares, inspired the growth of day trips to seaside resorts and the Lake and Peak Districts. Country rambling became a hugely popular and well-organized movement, with clubs arranging excursions (Bunce 1994).

In the USA again there was a parallel trend, with the rise of the railways promoting summer excursions from eastern seaboard cities into the countryside for day trips and, more significantly, into remoter, spectacular scenery, which led to the emergence of a 'cult of the wilderness' (Bunce 1994: 118). Bunce (1994) argues, however, that the USA was slower to establish a set leisure time, in part due to a strong cultural emphasis on hard work which was a product of the pioneer tradition (see section 8.3.2). Here, use of countryside for recreation was again inspired by the middle classes, who developed outdoor sports and country clubs.

By the beginning of the twentieth century increased mobility, leisure time and affluence had combined to establish the countryside's role as a physical and mental escape from the city and as a space for recreation and sport. It is a process that was dramatically expanded by the car. Bunce observes that in the USA the car was marketed at the beginning of the twentieth century mainly as a tool of pleasure. Advertisements stressed the joys of the open road and depicted vehicles set against country scenery. Tour guides, and picnicking and camping equipment for the car soon followed. It opened up access to national parks and, in doing so, changed 'the wilderness from a gallery reserved for a discriminating few to a playground where all might absorb what they could' (Schmitt 1969: 155 cited in Bunce 1994: 119).

As the century progressed, mass car ownership and the development of road and highway networks (set against a background of the shortening of the working week, an increase in vacation time, and improvements in the wealth and health of the masses) increased the possibilities for casual trips from the city and for people to establish rural retreats as weekend or second homes. Outdoor recreation became hugely popular across Europe, North America and Scandinavia. In turn, this led to areas being set aside by public and private agencies so that these demands for access to the countryside could be satisfied and managed (Bunce 1994).

The following subsections look more closely at some of the ways that rural environments are used by society as a playground.

▨ 8.6.1 Outdoor activities

Together, the country sports of hunting, shooting and fishing – the original preserve of the landed gentry in the UK – remain the most popular countryside sports in the UK and USA (with fishing the most popular of all). Within Britain, shooting and, to a lesser extent, hunting with horse and hounds are still fairly exclusive activities, riddled with class connotations and snobbery (see also Chapter 2, section 2.3.1.2). In contrast, in the USA and Canada, hunting for sport – moose, rabbits, etc. – and fishing tend to be blue-collar rather than upper/middle-class activities. Wildlife recreation equipment (such as guns, off-road vehicles, boats and camouflage clothing) is a big consumer industry (Cordell and Hendee 1982).

Hiking, backpacking, camping and canoeing are the most popular wilderness activities in North America, Australia and New Zealand. A quarter of all those who participate in outdoor recreation in Canada go wilderness tripping (Jackson 1986). These uses of the landscape blur the boundaries between a range of experiences: survival, subduing nature, adventure, communing with nature and the sensory or aesthetic pleasures of the outdoor scenery (Bunce 1994).

In addition to these traditional outdoor activities the latter part of the twentieth and early twenty-first centuries have witnessed a growth in a much wider range of sporting pursuits. Bunce (1994: 127) categorizes these as (1) terrain-based activities such as skiing, and climbing (see Plate 8.3); (2) water activities in rivers, lakes, canals

Plate 8.3 There has been a rapid growth in outdoor pursuits (© Andrew Milne)

and coastal areas, such as windsurfing, kayaking, sailing and yachting (see Laurier 1999); (3) open space activities such as golf, horse-riding, llama trekking; (4) *relaxation activities*: these are less active, more informal and even unintentional sorts of things like picnics, leisurely strolls (as opposed to hikes), sunbathing, swimming, ball games, and so on.

Generally, British uses of the countryside are more sedate than those in the USA, Canada and New Zealand (Bunce 1994). Camping is more likely to take place in sites with facilities than in wilderness areas, and activities like pony trekking, walking and just driving for pleasure to visit nostalgic sites such as country houses, picturesque villages, castles and sites of interest, and craft shops are still more popular uses of the countryside than most of the adventurous participatory activities outlined above.

■ 8.6.2 Second homes

In Norway and Sweden there is a huge tradition of owning a second home in the countryside (Williams and Kaltenborn 1999, Nordin 1993, Jarlöv 1999). This dates back to a time when the rural subsistence economy meant that extensive use was made of remote resources and outlying areas. As a consequence, people established small cabins and summer farms in the higher pastures, where they stayed when they were looking after the livestock. Later, as the economy changed, these were converted to second homes. It is now a tradition for people to build cabins (see Plate 8.4) close to the district from which their family originates (Nordin 1993). This trend has been promoted by the wealthy middle classes developing rural retreats to enable them to pursue extended salmon farming and large game hunting; and by larger companies which have established cottage communities near cities. Initially, these were for their employees to use, but they are now commonly for corporate entertainment (Williams and Kaltenborn 1999).

Second homes in Norway are seen to provide an escape from the everyday stresses of living and working in the city. The average user spends six weeks a year in the cottage, including eight days in harsh midwinter (Williams and Kaltenborn 1999). Cottages are commonly located so that each is relatively private and spacious, with attractive views and opportunities for a range of outdoor activities such as hiking, skiing, fishing, hunting and picking berries and mushrooms (Jaakson 1986, Williams and Kaltenborn 1999). The fact that many Norwegian cottages have no electricity or running water and may only be accessible by road in the summer means that they also offer an escape from modern life altogether. Residents commonly have to 'step back in time' while in the cottage by fetching firewood and water, and adjusting their lifestyles to the rhythms of natural daylight and darkness. Great importance is placed on experiencing the elements and on the sense of being 'at one with nature' (Williams and Kaltenborn 1999). The cottage, then, is a place where life is lived 'differently' and where the meanings of work and leisure converge (Jaakson

Plate 8.4   Norwegian second home (© Trude and Andy Hodson)

1986). For example, activities such as maintaining the property, chopping wood, fishing or harvesting berries are chores, but they can also be relaxing, giving residents time and space to contemplate their existence and enjoy a simpler way and slower pace of life (Williams and Kaltenborn 1999) (see Box 8.4).

Cottages in Norway are usually kept for life, in contrast to the fact that their owners' 'full-time, permanent' homes may be changed several times during the lifecourse. They are also handed down from generation to generation. Indeed, the cottage often binds the family together because all the members' activities have to be interwoven (e.g. fetching water or wood for each other), in contrast to the city home, where individual family members can lead segmented or individualized lives (Williams and Kaltenborn 1999) (see Chapter 3). In this sense, second homes may provide more of a sense of identity and continuity (in terms of rootedness, 'emotional home', etc) than a 'full-time, permanent' home. This perception can also be reproduced through an emphasis on specific cottage rituals (e.g. annual hunts of ptarmigan, a type of mountain grouse).

This Norwegian tradition of owning a second home also reinforces a sense of national identity (see Chapter 9). Cottage activities such as enjoying the peace and quiet of nature, walking, cross-country skiing, and outdoor rituals like hunting and gathering berries are seen as symbols of 'Norwegian-ness' (Erikson 1997). As a result, the contemporary development of new cottages with modern conveniences such as

---

Box 8.4: Norwegian cabins

'A lot is important, particularly getting close to nature, feeling the wind and weather, having a few people around you, and living as one with nature and the rhythms there.'

'[I] feel comfortable in my own company. The cabin and the mountains give me peace of mind. I often bring with me things to the cabin which I don't feel I have time to deal with at home. At the cabin I can permit myself just "to be" without doing anything. Just sit and enjoy the view.'

'Cabin life is tradition. It has always been my favourite existence. Away from crowds, time pressures and the regulations of civilisation and technical things. Experience, enjoy and protect "untrammelled" nature. Live a simple life materially and in agreement with a few, chosen people. Physical – and, not the least, psychological – balance is very quickly restored during a stay at the cabin.'

Williams and Kaltenborn 1999: 222, 225

---

electricity and water, which are attracting different types of users, is being accused of threatening and unsettling the Norwegian national identity (Williams and Kaltenborn 1999).

■ 8.6.3 Four-wheel-drive vehicles and the wilderness

Four-wheel-drive vehicles have opened up the wilderness and remote places, allowing people to visit extreme environments such as the Australian outback (Plate 8.5). Peter Bishop (1996: 257) describes a typical scenario:

> It is the height of summer in the immense arid region of Central Australia. A smart four-wheel drive (4WD) vehicle has stopped. Inside is a white Australian family. With the engine still running the air conditioner keeps the temperature pleasant, almost cool. The man switches on the radio and, after briefly listening to an Aboriginal radio programme broadcast from Alice Springs, nearly 1000km away, tunes directly into a cricket match from Barbados. Both broadcasts are relayed by AUSSAT, the Australian communication satellite hovering in space overhead. Meanwhile the woman is using a state-of-the-art electronic navigational device incorporating a global positioning system to guide her family to the area where they planned to set up camp. In the backseat the youngest of the two children is playing a game on the powerbook personal computer which their father has brought along to finish off some work whilst on their ecotour.

4WDs have grown in popularity since the 1970s to become a common feature of the affluent suburbs of the Western world. In the popular imagination they are

Plate 8.5   4WD vehicles have opened up the wilderness (© Giles Wiggs)

linked to the military, adventures and expeditions. Media advertising for 4WDs draws on this repertoire of images to connect them to particular rural landscapes. In the USA, advertisements emphasize the ethos of the frontier, the rugged independence of the pioneers, the urban cowboy; in the UK the imagery employed represents them enduring bad weather or in the context of the empire and overseas exploration; whereas in Australia 4WDs are portrayed in the outback (Bishop 1996). In these ways, these vehicles are seen as a crucial way of connecting the city/suburbs with the wilds of the country. In these representations the remote environments such as the outback are once again seen as restorative or healthy locations where it is possible to escape contemporary problems. The outback – characterized by masculine fantasies of heroic struggles with nature – also has a long-standing place in the Australian geographical imagination. Think of the flying doctors and *Crocodile Dundee*. Just as the second home epitomizes Norwegian-ness (see section 8.6.2) so the 'image of the outback has long functioned as a potent symbol of Australian identity both for Australians and for outsiders' (Bishop 1996: 261). This landscape imagery helps to legitimize the purchase of 4WDs – Australia leads the world in terms of their consumption.

Paradoxically, however, while 4WDs are lauded for the opportunity which they offer urban inhabitants to escape civilization and experience extreme landscapes, by making the outback so accessible they threaten the very concept of a 'remote' environment. Indeed, Australian wilderness landscape regions have actually been divided up according to their 4WD accessibility. Environmentalists are concerned about their potential intrusion into every corner of the outback, the environmental impact of off-road driving and their general environmental degradation (Edington and Edington 1986, Davidson 1982). Even 4WDs on city streets and in suburban garages are viewed as a symbol of consumer inauthenticity and as a sign of disregard for the environment (Bishop 1996). Underlying all these debates too are the complex issues of indigenous people's claims for land rights and self-determination, and an obvious clash between the high-tech 4WDs owners' use of the outback and the low-tech ecospirituality of aboriginal beliefs and land use practices (see Chapter 9). All this has led to the introduction of measures (such as zoning, trail improvement, education initiatives and 4WD clubs) designed to curtail 4WD use and to promote environmental awareness and responsibility among drivers (Bishop 1996, Woodward 1995).

■ **Summary**

- The rural environment is a leisure/consumption space which offers everything from the opportunity to enjoy the scenery to participatory outdoor activities.

- Until the mid-nineteenth century it was only enjoyed in this way by the wealthy. But the growth of transport, leisure time and affluence combined in the twentieth century to turn the countryside into society's playground.

- It is used, for example, for a variety of activities, for relaxation, as a second home and by 4WD owners.

■ 8.7 The rural as a space of production

Sections 8.5 and 8.6 drew in different ways on a particular discourse about nature, namely that it is something to be respected and protected. 'The wilderness is represented as a sacred space, and as a transitional space between the wilderness and the city the countryside becomes a place in which humans can experience and commune with nature' (Woods 1998a: 1221). This discourse provides a motivation for tourism, outdoor recreation, and second homes (discussed above) in which the rural environment is a space of consumption. But there are other discourses around nature too. One of the most important is that of the human stewardship of nature. In these terms the countryside is conceptualized as a space of 'tamed wilderness' (Short 1991) to be controlled and managed in the service of human need. It is this discourse – where the rural is regarded as a space of production rather than consumption – which is used to justify its exploitation by humans for commercial and military purposes (Woods 1998a).

■ 8.7.1 Agriculture

'Agriculture is a "natural" production and provisioning process at the same time that its practices are "unnatural"' (Buttel 1998: 1152). In particular, the twentieth century witnessed unprecedented technological change in agricultural production, inspired by advances in science and industry (Goodman and Redclift 1991). Farming has literally become big business. Large scale multi-national and pharmaceutical companies are now involved in agricultural research and development. Consequently, an unprecedented armoury of chemicals and mechanical aids is available to enhance agricultural production, while biotechnologies offer further possibilities for plant and animal improvement.

This advanced technological husbandry of the rural environment has in turn brought unparalleled improvements in the variety, quality, predictability, convenience, cheapness and safety of the foods on the dinner tables of the Western world. Overproduction in the West means that we are able to treat food just like any other commodity. For the majority of people (or at least those who are above the poverty line), this efficient stewardship of the rural environment means that food can play a routine or mundane role in our everyday life – it is no longer a matter of life and death, at least in the sense that malnutrition and starvation are rare in the affluent West (Buttel 1998).

The industrial approach to agriculture and food production has, however, effectively severed the relationship between food and the natural world. The rhythms of nature, seasons and climates, for example, are no longer necessarily important. Technological advances mean that virtually all foods are available all year round. Likewise, global commodity chains have severed the face-to-face links between producers and consumers, while the intensive use of chemicals and pesticides and biotechnologies, such as genetic engineering, have raised questions about the extent to which many 'foods' can be described as 'natural' (see also Chapter 2). Despite the industrial character of the agri-business food manufacturers and retailers continue to make discursive appeals to 'the natural' through their advertising, product labelling and marketing. 'The natural is thus becoming contested terrain in grocery stores and shopping malls as well as in parliaments, regulatory agencies and university laboratories' (Buttel 1998: 1155).

In the context of these contemporary anxieties about the extent to which the agricultural industry is exploiting nature, conservation is increasingly being promoted by policymakers as a potentially important element of farmers' work. In this sense, there is a desire to place renewed emphasis on the role of farmers as stewards of the environment, in which the productive use of land can be combined with nature conservation in agri-environmental schemes (McHenry 1998). McHenry (1998) points out, however, that farmers are more likely to take on this role when they can see practical or financial benefits – such as a market return on conservation from diversified enterprises such as tourism – rather than moral benefits in terms of protecting the environment for the greater good of the human race. They are also more willing to become involved in the conservation of attractive species (e.g. barn owls) which do not cause any conflicts with their productive endeavours than to participate in schemes to conserve animals which they regard as agricultural pests. Their understanding of nature is thus conceptualized in terms of its effect on and interaction with farming (McHenry 1998).

■ 8.7.2 The armed forces

The armed forces occupy large areas of land. During the Second World War a US army battalion required 4000 acres for practising manoeuvres; today a battalion uses up to 80 000 acres (Seagar 1993). Most military training and manoeuvres take place in remote landscapes. The British army, for example, owns land in National Parks and Areas of Outstanding Natural Beauty such as the Otterburn Training Area in the Northumberland National Park and on the Dorset coast (Woodward 1998), while as much as 10 per cent of Hawaii, USA, is owned or controlled by the military (Seager 1993). Rachel Woodward (1998) argues that the armed forces' occupation of the countryside is legitimized both through the demarcation of space with barbed wire, flags and notices into military and civilian areas, and through discourses which depict the countryside as the rightful home of the army.

First, recruitment and training brochures usually show camouflaged soldiers and tanks in the fields and woodland. Urban warfare is very rarely mentioned or depicted. In this way the army is represented as the rightful occupier of territory. British wartime recruiting and propaganda materials have also famously intertwined discourses of rurality and national identity, representing the Britain which must be defended in terms of idealized visions of rural life (see Chapter 9). Second,

the army – like farming – portrays itself as the steward or guardian of the environment. Firing ranges, for example, are presented as a conservation measure because
the land they cover is untouched by people. The British army even has a training
video called *Training Green* which promotes the idea that any environmental destruction is minimized and that the military are careful to avoid disturbing wildlife, and
camp and dispose of rubbish responsibly.

Finally, encountering the natural environment is seen as an important means
through which the army transforms civilians into soldiers. Remote landscapes are a
workspace for the army. Outdoor survival exercises and endurance tests are seen as
important rites of passage for recruits. Here, soldiers are taught to view the landscape not as a space of pleasure but a space of protection in terms of the possibilities it offers for camouflage and fighting positions (Woodward 1998).

■ Summary

- A common rural discourse is that humans have stewardship of nature.
  In this representation the rural environment is conceptualized as a 'tamed
  wilderness' to be controlled and managed in the service of human needs.

- This view of the rural as a space of production rather than consumption
  (see above) is used to justify its exploitation for commercial and military
  purposes

■ 8.8 Rural conflicts: nature under threat?

The previous sections in this chapter have demonstrated that the rural environment
is a space which is idyllized and commodified: it is a space of consumption, a place
in which to live, a place to explore, a lifestyle, a way of spending money, a place in
which to relax, a sports facility, a public park, a tourist attraction, a therapeutic
resource, a nature reserve, a living museum, and so on. It is also a commercial and
military space: a space of production, a landscape to be exploited to meet human
needs, a workspace, a space to be managed and protected, a training ground, and
so on. In fulfilling these twin roles as a landscape of consumption and of production, the rural environment is used by a wide range of different groups of people.
As a result of this scale and diversity of uses and users, it is not surprising that there
are huge pressures on rural environments. Conflicts frequently erupt between different users (Mormont 1987b, Murdoch and Marsden 1994), who draw on a range
of different discourses about 'nature' and the 'rural' to justify their positions. The
following sections outline just some of the many tensions evident in contemporary
uses of the rural landscape, which are leading to demands for the protection of nature,
its amenity value and public accessibility.

▦ 8.8.1 Access disputes: landowners v recreational users

In Britain land is heavily demarcated between owners and non-owners. As a result, many non-agricultural users of rural environments are subject to control and exclusion (Ravenscroft 1992, 1998). In particular, the rights of ramblers to have access to the countryside (notably uplands and open country) have been the subject of confrontation and conflict between recreational users of the landscape and landowners. In the inter-war years a mass trespass of Kinder Scout in the Peak District, UK, took place as part of the campaign for the 'right to roam' – a tactic that was repeated in the late 1990s as part of the Forbidden Britain campaigns (Ravenscroft 1999).

Neil Ravenscroft (1999) argues that these disputes are about citizenship, in which land-based rights have been privileged over citizens' rights to roam. One response of the government has been to promote initiatives such as the Countryside Stewardship Scheme, in which farmers receive designated land payments for opening up some areas for conservation and public enjoyment. Bishop and Phillips (1993: 335) argue that these schemes 'allow farmers to identify relevant environmental services and goods which they can provide and the opportunity to market these and promote their role as custodians and managers of the countryside'. Here landowners are cast as willingly giving up some of their rights for the greater good of the citizenry, taking on the role of benevolent stewards offering safe and sanitized access to particular sites and lessons in countryside conduct (Ravenscroft 1999).

Yet, Ravenscroft argues, rather than providing the 'freedom to roam', the Countryside Stewardship Scheme and similar projects merely sacrifice specific parcels of land to meet the demands for recreational uses of the countryside in order to divert attention from the wider land ownership issues. These initiatives, he argues, offer a privatized notion of citizenship (see Chapter 9) based on the effective consumption of rights, as opposed to a rights-based notion of citizenship sought by the 'right to roam' campaigners.

▦ 8.8.2 Agricultural industry and military v environmental campaigners

The industrialization of agro-food production (see section 8.7.1) (Goodman and Watts 1994) is causing alarm about the potential risks it poses both to the environment and to human bodies, including, for example: (1) risks to humans from pesticides and chemicals in food, water and the environment; (2) doubts as to the naturalness and safety of food produced using biotechnologies; (3) public health risks from rural pollution involving farm effluent, and so on.

- *Pesticides*: A huge array of chemicals and pesticides are used in contemporary agricultural production. Critics of these practices argue that the monitoring of pesticides is limited, their potential effects on ecosystems and diversity have been under-researched, and their effects in small doses on our food and water are virtually impossible to prove. Ward *et al.* (1998) point out that most regulation of pesticides is concentrated on prior approval. In other words, those chemicals available for sale are commonly assumed to be safe because dangerous or contaminating pesticides would not have been allowed to reach the marketplace. Yet, as they point out, this ignores the fact that scientific understandings of pesticides are partial and it is difficult to predict their impacts on the environment on the basis of laboratory testing – not least, because scientists tend to assume that pesticides will be used by competent technicians and responsible farmers (Ward *et al.* 1998). But when farmers choose which chemicals to use they are driven by their anxieties about achieving effective crop protection rather than concerns about the environment. Despite spraying instructions, the pressures of work can mean they may spray in less than ideal conditions (e.g. wind and rain), which can result in the chemicals being spread more widely than intended (Ward 1995, 1996, Ward *et al.* 1998). Advisory systems also militate against a reduction in pesticide use. It is agrochemical merchants selling pesticides who give most crop protection advice to farmers. Their need to maintain their sales figures is a deterrent to their recommending fewer treatments or lower dose rates to farmers (Ward *et al.* 1998). Chemical usage is inadequately policed and monitored in any event. Visits to individual farms are very rare and often only in response to complaints from the public.

- *Biotechnologies*: Concerns about the naturalness and safety of food produced using biotechnologies are evident in debates about the use of the BST growth hormone by dairy producers in the USA. This hormone occurs naturally in pituitary glands and has been proven to increase milk production per cow and feed efficiency. However, it is not possible to obtain it from slaughtered cattle in large enough quantities for commercial use in live dairy herds, though it can also be produced through genetic engineering. A biotechnology coalition (of scientists, administrators, etc) has promoted this genetically engineered version, recombinant BST, as a revolutionary advance in agricultural productivity, arguing that it is safe for human consumption because it is a 'naturally' occurring product which can be safely digested within the human stomach and has no ill-effects (Buttel 1998). Opponents have challenged the 'naturalness' and safety of BST milk. They argue it is not 'natural' because the hormone has to be artificially injected into animals, it contains high levels of IGF-1, and it leads to the burn-out of cows. For these critics, this is an unnecessary intervention which is hooking farmers on technology they do not need and threatens the social and pastoral landscape through its potential to wipe out small-scale family farms in the dairy states of the North-East and Upper Midwest (Buttel 1998).

- *Pollution*: Over 50 per cent of US water pollution is attributed to waste from the livestock industry (Seager 1993). When slurry and silage effluent is accidentally discharged it can cause widespread pollution, killing fish and damaging ecosystems, while smaller gradual leakages, runoff and seepage also cause deoxygenation and nutrient enrichment, damaging river fish and plant communities (Ward *et al.* 1998). Ward *et al.* (1998) observe that, just as pesticide regulations assume that farmers are competent managers and technicians, so too advice and regulations relating to effluent management draw on the same assumptions. Yet, farmers are often indifferent to or incompetent in this respect, prioritizing milk production and husbandry over managing effluent systems. In many cases they simply have too many animals to dispose of the effluent they produce, with the result that low-volume seepage from yards and storage areas causes widespread damage in rural environments (Seager *et al.* 1992, Ward *et al.* 1998). Regulation also places the emphasis on the behaviour of individual farmers when, in fact, '[r]iver quality depends ultimately on the mix and nature of water and land uses within the drainage catchment' (Ward *et al.* 1998: 1171). Only an overall catchment management approach can effectively reduce this sort of pollution.

Farmers, and the agro-industry more generally, are reluctant to acknowledge damage to the rural environment and their role in causing it (McHenry 1998). This has created a history of tensions between farmers and conservation and environmental groups. Farmers are resistant to regulations and pollution controls, regarding them as 'anti-farmer' or 'anti-rural'. As a group they often feel aggrieved and misunderstood, believing, for example, that they are often unfairly stereotyped as ignorant or destructive when they are struggling for economic survival (McHenry 1998).

Like the agro-industry, the military is also charged with threatening and polluting rural environments and their inhabitants. Despite the claims made by the army (see section 8.6.4) about its role as steward of nature and its concern for the countryside, the military is responsible for environmental destruction around the world. This damage occurs not only during wartime but also in peacetime. The army consumes huge amounts of raw materials; uses animals for scientific experimentation; damages landscapes during its manoeuvres, weapons testing and training exercises; and dumps large amounts of waste – including radioactive materials from its nuclear production facilities – as well as a wide range of chemicals, fuels, oils, ammunition, etc, which are used in its everyday activities.

In a damning account of the impact of the military on the environment Seager (1993) points out that US military sites are some of the most dangerous and toxic locations in the world, citing the US Army's Rocky Mountain Arsenal in Denver, Colorado, as just one example. She draws on research which shows that the Pentagon alone generates more toxic waste than the five largest US chemical corporations combined, and that a third of all the hazardous waste generated in the USA is produced by the military. This and other evidence is used to make her case that the armed forces, along with other traditional institutions of power, such as transnational corporations and governments, are inextricably linked to contemporary environmental crises. Joni Seager (1993: 14) argues that: 'Militaries are privileged environmental vandals. Their daily operations are typically beyond the reach of civil law, and they are protected from public and government scrutiny, even in "democracies". When military bureaucrats are challenged or asked to explain themselves, they hide behind the "national security" cloak of secrecy and silence.'

The practices and the power of the agro-industry and the military are challenged by a range of environmental movements. In tandem with left-wing, feminist, black civil rights and student movements, dozens of environmental action groups were founded in the 1960s and 1970s. These included organizations like Greenpeace, the World Wildlife Fund, and Friends of the Earth. In the subsequent decades many of these pressure groups have been professionalized, establishing headquarters, using advertising agencies and hiring political lobbyists and lawyers to advance their cause (Seager 1993). Green parties have also reshaped the political map of Europe, becoming a viable electoral force, winning seats in national elections (e.g. in Germany, Sweden, Austria, the Netherlands, Belgium, Finland, Italy, and so on) as well as for the European Parliament. In response, politicians in the UK, the USA and Australia have been quick to try to promote their green credentials. Not surprisingly, the professionalization of the environmental movement has triggered conflicts over tactics, styles, campaigns, leadership, finance and publicity strategies in which accusations of sexism and racism have been evident (Seager 1993).

A dissatisfaction with these mainstream environmental movements has, in turn, spawned alternative forms of activism, including deep ecology and eco-feminism (see, for example, Griffin 1978, Pepper 1993, Warren 1994). 'Both movements posit the need for affirmation of a "deeper" human relationship with the earth – a relationship that at its best comprises elements of mysticism, awe, and an appreciation of the "sacred" in nature. Both movements couch their environmentalism in "woman-identified" terms, and deep ecologists are the only environmentalists other than ecofeminists to explicitly assert an affinity with women's culture and feminist politics' (Seager 1993: 223). The earth as Mother visions of both groups have, however, been criticized as essentialist (see Chapter 2) and dangerous because they draw on stereotypes of women as nurturing, emotional and so on, assume universal women's values, ignore differences between women, and leave unchallenged patriarchal dualisms such as male/female, culture/nature, rational/irrational (Seager 1993).

At the same time, hundreds of small-scale, often local or community-based groups of unpaid volunteers make up grassroots movements which also continue to challenge the environmental impact of the agro-industry and military. Women's traditional domestic role, which involves managing the home, in which they commonly take decisions about resources, and responsibility for the families' food and health, leads Seager (1993) to argue that it is often women who are the first to become environmentalists within their communities. One such example is women's participation in Landcare – groups of people in rural areas of Australia who are organizing around their shared concerns about land degradation and land management practices (Liepins 1998).

The diversity of campaigners' responses (and the conflicts between them) to the threats posed to humans and the environment by agro-industry and the military, and the frequent reluctance of farmers, the services or governments to be held accountable for this damage, demonstrate that there is a strong political dimension to conflict over the environment, in which the power of particular groups to dominate each other, as well as dominate 'nature', is at stake (Redclift 1992).

■ 8.8.3 Tourists *v* residents

The commodification of the rural landscape for tourism and leisure is a source of many conflicts. One such example is Mordue's (1999) study of Goathland, a village in the North York Moors National Park, UK. This is a traditional picturesque village which contains all the trappings associated with the English rural idyll (see section 8.3.1) – it is a small, peaceful community, pretty, steeped in nostalgia, set in beautiful countryside. It has always been popular with retirees and especially walkers, who have used it as a base to explore the surrounding countryside. Such visitors have been generally unobtrusive, spending all day out of the village in the countryside.

Recently, however, Goathland has become the location for the filming of a popular television series, *Heartbeat*, a nostalgic programme, set in the 1960s, about a young policeman who has moved up from London to live in a Yorkshire village. As a result, Goathland has become a honeypot for day-trippers. Rather than spending their time in the surrounding countryside as the walkers do, these visitors have a strong physical presence in the village, causing congestion and overcrowding on the streets and generally being intrusive (playing music, football, etc). The village car park has already been expanded in response to the increased traffic pressures and there is talk of establishing a further car parking area. However, this has caused NIMBY disputes (see Chapter 4) among the locals, who are concerned about its possible location. Residents are also up in arms about the creeping commodification of the village. Some have even switched their homes around, moving the lounge from a front room to a back room to achieve some measure of privacy from the tourist gaze. An increase in car crime and petty theft has been blamed on urban criminals. The residents fear that their village is being overwhelmed by tourists and that, in the process, these consumers are destroying the very qualities of the 'rural idyll' – tranquillity, beauty, community – which they expect to find there.

Elsewhere, similar pressures on the countryside, which have caused congestion at popular sites and demands for new roads and car parking areas, have led to the establishment of parks (with car parking, and set facilities for walking, fishing, etc) on the edge of metropolitan areas in order to relieve the pressure on the so-called 'natural countryside' (Bunce 1994).

Other tensions are often evident between recreational users of the countryside and those participating in environmental conservation and education programmes

(Bunce 1994). Clashes between different groups of recreational users often flare up in rural areas too – notably, hikers and horse riders resent the noise, pollution and threat to people and the environment posed by off-road motorized vehicle users (e.g. motor bikes, trail bikes, four-wheel-drive vehicles) (Bishop 1996).

■ 8.8.4 Hunters *v* animal rights

The cultivation of the wilderness by humans led to the domestication of particular animals. Some have been used to work closely alongside humans, assisting in our stewardship of nature (e.g. horses and dogs), others are controlled by humans and kept for food (e.g. livestock). Wild animals which do not contribute to humans' productive activities are often perceived to be a threat to crops or domesticated animals and therefore to be controlled through hunting with guns, dogs or traps.

These activities are often justified as part of the 'natural' order in which every creature (except humans) has a 'natural' predator which keeps its population in check. In these terms, humans are responsible for controlling the animal population as part of our role as stewards of the environment (Woods 1998a). Hunting is further legitimated by recourse to arguments that the 'chase' offers animals a fair chance to escape – only the old or the weak are culled. In the UK the rituals (hunt meetings, dog shows, hunt balls, and so on) that accompany hunting have, in turn, become a symbol and celebration of the rural way of life and 'community', part of the rural heritage or tradition. In other words, country people understand animals and people to hold different places within the system.

In contrast, for some people who go to the country in search of nature and the rural idyll 'animals are imagined in an almost wholly aesthetic sense, divorced from the dirty noxious aspects of agriculture and the harsh predatory order of nature. Rather animals are expected to act as "props" for an imagined rural idyll, such that the countryside becomes a space of consumption or spectacle in which there is an essential distance between the animals who "perform" and the people who watch or listen. Furthermore, the presence of wild animals is judged to increase the "authenticity" of the natural experience, such that they become creatures to be protected, not controlled, and activities such as hunting are recast as being "against nature" ' (Woods 1998a: 1221).

These opponents of hunting contest rural dwellers' representation of what is natural by drawing on images which enforce the idea that animals are at home in the countryside, that they are part of the realm of nature (Woods 1998a). Rather, they argue that hunting is a cruel and implicitly 'unnatural' recreational activity which, Joni Seager (1993) speculates, is forcing as many as 100 species a day into extinction.

Related to the anti-hunting lobby is a broader discourse of animal rights in which animals are understood as having value independently of humans. In this sense, animal rights is the latest in a long line of campaigns against various forms of social injustice in which the oppression of animals parallels the oppression of other groups such as women and ethnic minorities (Gold 1999). Seager (1993), for example, talks about factory farming and experimentation in terms of women's experiences of gynaecology and pornography, while she also observes that hunting, like rape, is a form of male violence.

The contemporary animal protection movement comprises a loose network of groups (often very localized in their activities) campaigning against fur farms, the use of animals in laboratories, live exports, hunting, and so on. It has a broad constituency, attracting the support of a wide spectrum of the population, though it also has a more radical hard core whose grass-roots direct actions and 'terrorism' have attracted the attention of the police and security services (Garner 1993).

■ Summary

- The countryside is a landscape of consumption and production. The scale and diversity of the uses to which it is put mean that huge pressures are placed on rural environments.

- As a result, conflicts frequently erupt between different users of the countryside who draw on different discourses about 'nature' and the 'rural' to justify their positions.

■ Further Reading

- The *Journal of Rural Studies* is home to most of the important papers about rural geographies (other useful journals include *Sociologica Ruralis, Rural Sociology*). Among those that have played an important part in shaping understandings about rural society are: Cloke, P. and Thrift, N. (1987) 'Intra-class conflict in rural areas', *Journal of Rural Studies*, 3, 321–33; Little, J. and Austin, P. (1996) 'Women and the rural idyll', *Journal of Rural Studies*, 12, 101–11; Philo, C. (1992) 'Neglected rural geographies: a review', *Journal of Rural Studies*, 8, 193–207, and Woods M. (1998b) 'Researching rural conflicts: hunting, local politics and action networks', *Journal of Rural Studies*, 3, 321–40. In addition to these journal articles a good place to start is the edited book: Cloke, P. and Little, J. (1997) *Contested Countryside: Otherness, Marginality and Rurality*, Routledge, London.

- A number of single-authored and edited books deal with the diverse meanings, uses of, and pressures on rural space. In particular see: Bunce, M. (1994) *The Countryside Ideal: Anglo-American Images of Landscape*, Routledge, London; Shoard, M. (1980) *The Theft of the Countryside*, Maurice Temple Smith, London; Crouch, D. (ed.) (1999) *Leisure/Tourism Geographies: Practices and*

Geographical Knowledge, Routledge, London, and Marsden, T., Lowe, P. and Whatmore, S. (eds) (1990) *Rural Restructuring*, David Fulton, London.

- Good sources on environmentalism and eco-feminism include: Pepper, D. (1993) *Eco-socialism: From Deep Ecology to Social Justice*, Routledge, London; Seager, J. (1993) *Earth Follies: Feminism, Politics and the Environment*, Earthscan Publications Ltd, London, and Warren, K. (ed.) (1994) *Ecological Feminism*, Routledge, London.

■ Exercises

1. Read some extracts from fictional accounts of childhood set within the countryside such as *Swallows and Amazons* or *Cider with Rosie*. How is rural life represented in these stories? From your academic reading and possible personal experience how do you think this fiction compares with children's actual experiences of rural life?
2. Look through some magazines and newspapers and collect as many cuttings as you can that feature disputes over the meanings or use of the countryside. What are these disputes about and who are the main protagonists? In each case analyse how the journalists and those quoted in the stories mobilize different discourses about 'nature' to make their case.

■ Essay Titles

1. To what extent do you agree with Mormont's (1990) claim that the rural–urban dichotomy is a false opposition that is increasingly becoming outdated?
2. Using examples, critically evaluate the attempts of middle-class rural dwellers to mould their communities to fit their own vision of rural life.
3. Justify Philo's (1992) call for rural geography to pay more attention to 'otherness'.
4. Critically evaluate the argument that rural utopian movements are both essentialist and reactionary.

9 The nation

9.1 The nation

Geographies of the nation have focused on defining and exploring the production and contestation of national identities, nationalism and citizenship. Around the world an increasing number of nation states are under threat from nationalist movements which threaten to dissolve them from below. Likewise, processes of globalization, and the emergence of supranational and transnational organizations are alleged to have made the boundaries of nation states increasingly irrelevant and therefore to be threatening the sovereignty of nation states from above. The growing importance of social identifications such as gender, ethnicity, sexuality and disability also stand accused of undermining citizens' attachments and loyalty to the state. As a result of these trends, and a recognition of the interconnectedness and interdependence of nations, various international regimes of governance and new notions of global citizenship are emerging.

This chapter begins by examining geographical understandings of nations and national identities and explores some of the forms of nationalism which both facilitate and threaten contemporary nation states. It then goes onto explore sexual citizenship and disaporic citizenship as examples of two forms of social identification that some writers have claimed are replacing national identity as a master narrative. The focus then switches onto processes of globalization and how people's lives around the world are being woven together in increasingly complex ways by the global economy. The final section outlines some of the new spaces which may be opening up for different forms of politics to develop. It explores the possibilities of technological and ecological citizenship and reflects on cosmopolitanism as a potential transnational structure of political action to replace the present structure of international laws mediated between sovereign states. The chapter concludes by arguing that globalization and post-modernism are causing a shift away from a homogenous and universal concept of citizenship based on membership of a nation state. Rather, it suggests that the identity of the post-modern citizen is an ensemble of overlapping and interconnected forms of citizenship.

■ 9.2 The nation and national identities

The nation can be defined as '[a] community of people whose members are bound together by a sense of solidarity rooted in an historic attachment to a homeland and a common culture' (Johnston *et al.* 2000: 532). For Anderson (1983) nations are more specifically 'Imagined Communities'(see Chapter 4). He explains that nations are '[I]magined because the members of even the smallest nation will never know their fellow-members, meet them, or even hear of them, yet in the minds of each they carry the image of their communion' (Anderson 1983: 15). He further argues that nations are imagined communities because fellow citizens often have a deep sense of comradeship or identity with each other, even though, in reality, there may be differences, exploitation and inequality between them. In this sense, nations are often assumed to represent the primary form of belonging or identification for their populations, taking precedence over other axes of identification such as class, race, gender, sexuality, religious belief, and so on. Indeed, ideologies of national solidarity and identity have often been used to justify the subordination of class and gender interests by nationalist movements (McClintock 1995, Sharp 1996).

According to Anderson (1983) the development of technology, capitalism and linguistics played an important part in forging and reproducing national identities in Europe during the eighteenth and nineteenth centuries. During this period the emergence of print capitalism enabled the mass production of books and newspapers, and linguistic diversity to flourish. In particular, old sacred languages such as Latin and Greek, which until then had been the preserve of the ruling elites of Europe, were demoted in favour of a multitude of vernaculars that became stabilized print-languages. These developments, by enhancing communication between people within specific geographical territories, contributed to fostering a sense among them that they shared interests and concerns and so belonged to a nation. Anderson (1983: 84) explains, 'The lexicographic revolution in Europe . . . created, and gradually spread, the conviction that languages (in Europe at least) were, so to speak, the personal property of quite specific groups – their daily speakers and readers – and moreover that these groups, imagined as communities, were entitled to their autonomous place in a fraternity of equals.'

The nation as imagined community is predicated, however, not only on spoken and written language but also on a wider range of cultural fictions (Bennington 1994).

As Sharp explains (1996: 98), 'Each drawing of maps of nation state territory, each playing of the national anthem or laying of wreaths at war memorials, every spectatorship of national sports events and so on represents . . . [a] daily affirmation of national identification.' In particular, banal acts such as the singing of a national anthem or the victory of a national sports team represent an experience of simultaneity. At such moments, people who are unknown to each other sing the same words or celebrate the same event, aware that others whom they do not know and cannot see or hear are doing so too. In such ways, members of a nation often experience a sense of unison with their fellow citizens even though nothing connects them at all but an imagined sound or image (Anderson 1983). As such, '[a] nation does not express itself through its culture: it is culture that produces "the nation"' (Donald 1993: 167).

The repetitive performance of symbols and stories of the nation congeal over time to produce the appearance that the nation is a 'natural' and timeless spatial entity, a product of the inherent relationship between a group of people and a particular place. Yet a flick through atlases published at different moments of the twentieth century – during which time the map of Europe was redrawn several times, with nations being created and broken up – clearly demonstrates that the nation is not a pre-given spatial entity but rather is a social invention, one whose boundaries and very existence can be both fluid and contested. Yet, the seemingly timeless rituals of traditions and institutions often obscure the production of national identity to such an extent that it is taken for granted as 'natural' or inherent (Sharp 1996).

National identities are not only produced through processes which foster a sense of sameness but also those which emphasize a differentiation from others. As Anderson (1983: 16) points out, the nation is always imagined as limited in reach by finite boundaries, '[n]o nation imagines itself as coterminous with mankind [sic]'. Nations must, therefore, have something to define themselves against. As Parker et al. (1992: 5) explain, 'nationality is a relational term whose identity derives from its inherence in a system of differences. In the same way [as] "man" and "woman" define themselves reciprocally . . . national identity is determined not on the basis of its own intrinsic properties but as a function of what it (presumably) is not.' Indeed, 'the idea of the nation is dependent upon an extensive appeal to ethnicity, frequently constructed against the threat of the "alien", and his or her "foreign" tastes, cultures and habits' (Chambers 1993: 153).

In a now classic book, Orientalism, the Palestinian social theorist Edward Said (1978) reflects on the fact that the Orient has been an object of fascination to the West for centuries. He considers Western visitors' visions of the Orient (e.g. as exotic, mysterious, decadent, corrupt, barbarous) as evidenced by paintings, photographs and writings (academic work, poetry, novels, and so on). By focusing on this discourse of the 'mystic Orient', Said (1978) shows how the traditions of thought and imagery which give the Orient a reality are in fact a European invention, a product of the European imagination. He argues that, by containing and representing the

Orient through this dominant framework, European culture has gained in strength and identity, setting itself off against the Orient in a series of asymmetrical relationships in which it is always seen in a favourable light (i.e. as civilized, ordered, etc) (Clifford 1988). In this way, '[B]y dramatising the distance and difference between what is close to . . . and what is far away' (Said 1978: 55) imaginative geographies not only produce images of the 'Other' but of the 'Self' too.

Box 9.1 draws on extracts from e-mails written by children involved in an Interlink project, funded by the British Council and developed and managed by Copeland Wilson & Associates Ltd (a specialist producer of educational learning materials in New Zealand). This brought children from Britain and New Zealand together on line to explore their impressions of each other's countries and lifestyles (detailed in Holloway and Valentine forthcoming; see also Hengst's 1997 study of British, German and Turkish children's understandings of their own and others' national identities).

However, 'the very fact that . . . [national] identities depend constitutively on difference means that nations are forever haunted by their various definitional others' (Parker *et al.* 1992). As Gilroy (1990) points out, these differences can be internal as well as external (see, for example, the discussion of whiteness and the English rural idyll in Chapter 8). As a consequence, nations often engage in acts of violence, segregation, censorship, economic coercion and political oppression against those (both other nations and their own citizens) who are imagined as threatening or dangerous 'Others'. (See also the discussion of how different groups are granted differential access to rights and resources in section 9.4 below.)

It does not have to be like this, however. Chambers (1993) suggests that there can be different imaginings of British national identity:

> One is Anglo-centric, frequently conservative, backward-looking, and increasingly located in a frozen and largely stereotyped idea of national culture [the Britain of empire and imperialism, Victorian values, cricket playing, claret drinking, and so on]. The other is ex-centric, open-ended, and multi-ethnic. The first is based on a homogeneous 'unity' in which history, tradition, and individual biographies and roles, including ethnic and sexual ones, are fundamentally fixed and embalmed in the national epic, in the mere fact of being 'British'. The other perspective suggests an overlapping network of histories and traditions, a heterogeneous complexity in which positions and identities, including that of the 'national', cannot be taken for granted, are not interminably fixed but are in flux (Chambers 1993: 153–4).

This second vision resonates with Massey's (1993) notion of a progressive sense of place in which she argues that difference and the crossing of boundaries leads to a complexity of vision. Here places (or in this case nations) are no longer understood as homogeneous, bounded (defined as here or there, us or them) insular locations, certain in, and defensive of, their own identities (like Chambers' first imagining of British national identity) but rather are spaces of connection that offer new possibilities of sharing in difference through interaction (see also Chapter 7).

Box 9.1: Imagining 'the other'

Some British children's imaginative geographies of New Zealand

Place

New Zealand is in the southern part of the world, by Australia.

Your countryside is basically palm trees, mountains, beaches and deserts.

I know for a fact that New Zealand is hotter than England. I think one of your main industries would be farming e.g. sheep farming because when I buy lamb in the shop it normally says 'a New Zealand lamb'.

People

We think that you look like us but with a sun tan.

Many people consider New Zealand as being an offspring of Australia and are not considered as being a country of its own.

The Maoris are the natives of the country and when the white people went over they nearly wiped them and their country out.

They do a war dance before every rugby game.

Your national costume is that you either dress up as a kiwi or wear shorts and T shirts, wear sandals on your feet and wear hats with corks hanging down.

Some New Zealand children's imaginative geographies of Britain

Place

It belongs to Europe and it's on the other side of the world.

We think that your climate is cold, wet and misty with a little bit of sun.

We imagine that your country has lots of big green hills, lots of stone walls and castle-like buildings, and thatched crofts with whitewashed walls.

The houses are old and of an older style than seen in New Zealand. They often have two storeys and no front or back gardens like *Coronation Street* [a UK soap opera set in Manchester, which is screened in New Zealand].

People

We think that some are fair-skinned, some are dark-skinned (from Africa), a lot of people are below the poverty line. We think most of them speak English but Great Britain consists of many languages, including Irish and Scottish.

We think that people look very posh and proper and that they act like that as well.

I think that you would do the same sports as us and you might play croquet too.

We think that you eat mashed peas on top of pies, hot dogs, fish and chips and you like to have a cup of tea and a bikkie (like posh people) and also like to eat stuff like us like sweets and chocolate and you drink fizzy (fizzy is coca cola and lemonade).

Holloway and Valentine 2000

■ **Summary**

- Nations are imagined communities. They are often assumed to be the primary axes of identification for their populations.

- The cultural traditions and symbols through which national identities are produced can produce the appearance that the nation is a 'natural' or timeless entity.

- National identities are produced through processes which foster a sense of sameness while also emphasizing a differentiation from others.

- Imagings of national identity can be backward-looking, insular and defensive or progressive, open-ended and heterogeneous.

9.3 Nationalism

Nationalism can be defined as a 'belief that the world's peoples are divided into nations and that each of these nations has the right to self-determination, either as self-governing units within existing nation states or as nation states of their own' (Ignatieff 1994: 3). (Here nation state refers to governance over a spatially bound territory.)

Ignatieff (1994) distinguishes between two forms of nationalism: civic nationalism and ethnic nationalism. He defines civic nationalism as a belief that 'the nation should be composed of all those, regardless of race, colour, creed, gender, language or ethnicity – who subscribe to the nation's political creed. This nationalism is called civic because it envisages the nation as a community of equal, right-bearing citizens, united in patriotic attachment to a sacred set of political practices and virtues. This nationalism is necessarily democratic since it vests sovereignty in all people' (Ignatieff 1994: 3–4). In other words, civic nationalism is a chosen or rational sense of belonging and attachment in which the state precedes the nation. Ignatieff (1994) suggests that some elements of this were first apparent in Great Britain. By the eighteenth century Britain was a nation state formed from four nations: the English, Scots, Irish and Welsh, who shared a sense of belonging predicated on common institutions such as the rule of law, Parliament, the monarchy and a common citizenship, rather than a sense of attachment based on a common ethnicity. The emergence of civic nationalism was furthered by the French and American revolutions, which established their respective republics. Today, most Western nation states are held together by a belief in common citizenship rather than common ethnicity.

In contrast, ethnic nationalism is a belief that attachment to a nation is inherited rather than chosen: in other words it is about blood loyalty. It is predicated not on shared rights and institutions but on what are understood as pre-existing ethnic characteristics and shared experiences such as culture, religion, and language. Here '[i]t is the national community which defines the individual, not the individuals who define the national community' (Ignatieff 1994: 5). Ignatieff (1994: 4) cites the example of Germany, explaining that 'Napoleon's invasion and occupation of the German principalities in 1806 unleashed a wave of German patriotic anger and Romantic polemic against the French ideal of the nation state. The German Romantics argued that it was not the state which created the nation, as the Enlightenment believed, but the nation, its people which created the state.' He goes on to explain how German unification in 1871 provided an example of the success of ethnic nationalism to other nations who at the time were under imperial rule. Although contemporary Germany now thinks of itself as a civic democracy, Ignatieff points out that its citizenship laws are still defined by ethnicity. Contemporary examples of ethnic nationalism also include places such as Serbia and Croatia, where ethnic majority domination is now institutionalized in the form of the nation state, and cases where ethnic nationalists are struggling against what they perceive as cultural or economic subordination, such as Quebecois nationalism in Canada (Box 9.2), and conflicts between Catholics and Loyalist Protestants in Northern Ireland.

Box 9.2: Quebecois nationalism

Since the Canadian Confederation was established in 1867, French-speakers, *les Quebecois*, have sought self-determination and self-government for Quebec. For the Quebecois, Quebec is their nation and Canada their state, whereas for English-speaking Canadians, Canada represents both nation and state. Quebecois nationalism erupted in 1963 when a bomb was planted in an English-speaking district of Montreal. This was followed by other acts of terrorism, including the kidnap and murder of a Canadian politician. However, since the 1960s Quebec has been able to use its freedom within the Canadian federal system to develop its own economy, nationalizing its hydro-electric resources, establishing its own economic institutions independent of Canadian ownership and becoming a player in the global market-place. In effect, Ignatieff (1994) suggests that Quebec has become a state within a state. For a long time English Canadians dominated the Quebec economy and nationalists blamed them for its relative backwardness. Now that Quebec is more economically successful, Ignatieff (1994) suggests, the Quebecois nationalist movement's sense of grievance and motivation for pursuing sovereignty should have waned. Yet he observes that, on the contrary, nationalist sentiment has, if anything, grown, with the emphasis now particularly on language and the desire to be a unilingual nation. To this end Quebec already has laws prohibiting the public use of English and restricting the rights of residents to send their children to English-speaking schools; and considerable control over immigration, which enables it to maximize the selection of French-speaking immigrants. Perhaps somewhat ironically, while Quebec pursues nationhood, it in turn includes an indigenous people, the Cree, who also consider themselves to be a nation and are fighting for their own self-determination over large areas of land.

Nationalism is based on a number of assumptions, namely that nations are 'natural units' of society, that nations have a right to territory or homeland, and that every nation needs its own sovereign state to express its culture. In other words, nationalism presupposes that nation and state are destined for each other, that neither is complete without the other and that each nation has the right to autonomy, freedom and security (Gellner 1983). Yet, this is an ideal rather than reality; there are few places where state boundaries actually match those of a national community in which all citizens share one culture (Johnston *et al.* 2000). Indeed, 'historically speaking most nations have always been culturally and ethnically diverse, problematic, protean and artificial constructs that take shape very quickly and come apart just as fast' (Colley 1992: 5).

Where political units and national units are mismatched, minority groups often experience various forms of social, economic, cultural or political exclusion or injustice. As a result, all round the world nationalist groups are drawing on imaginings of national identity to pursue the aim of achieving self-determination or to lay claims to particular territories in which they perceive their history to be embodied. Anderson (1995) observes that, while there are over 5000 different languages in the world, there are currently fewer than 200 separate states. The means through which separatist aims are pursued vary. While in many cases the use of violence (annexation, terrorism, etc) is often regarded as justified, nationalist movements have also sought independence through the ballot box and around the negotiating table. Spanish Basque nationalism (Box 9.3) represents one example of a nationalist movement which is divided over what is meant by self-determination and how this should be achieved (Raento 1997).

Uneven economic development is often the impetus for nationalism (Nairn 1977). Relatively undeveloped regions may have a sense of economic grievance and of being on the periphery politically. For example, in the 1980s and 1990s there was a surge in support for Scottish nationalism. Scots were resentful that the profits from North Sea oil (located off the coast of Scotland) – which was imagined as Scottish – were not being ploughed back into Scotland by the British government and that, in effect, the English were stealing Scottish oil. These feelings were compounded by the fact that, while Scotland consistently voted Labour (i.e. the political consensus was towards the left), Conservative (i.e. right-wing) governments were returned to power by the British electorate as a whole. In other words, it appeared that English voters, particularly in the south-east of the country, were determining who governed Scotland and that the Conservative government did not therefore have an electoral mandate in Scotland for the controversial policies it pursued. These saw a reduction in public expenditure, privatization, and the reform of institutions such as the health service, legal profession and local government that had contributed to producing and fostering Scotland's quasi-autonomous sense of Scottish identity (Paddison 1993). Following the election of a new Labour government in Britain in 1997, the Scottish people voted in a referendum for devolution: the establishment of a Scottish Parliament with tax-raising powers. Indeed, smaller-scale regions are

---

Box 9.3: Basque nationalism

The Basque nationalist movement aims to establish an independent Basque-speaking state created out of seven historical provinces in Spain and France. Spanish Basque nationalists are divided, however, over what is meant by self-determination and how this should be achieved (Raento 1997). The moderates, who are represented by various political parties, oppose the Spanish state and are attempting to achieve independence through democratic means. In 1980 the Spanish government granted the establishment of what is known as the Basque Autonomous Community (this is made up of three provinces: Alava, Guipuzcoa and Vizcaya). The moderates accept this granting of limited autonomy as the best means through which they can purse their goal of self-determination. In contrast, the radicals understand their country to be occupied by the Spanish. They believe that, by accepting autonomy and co-operating with their oppressors, the moderates have betrayed the Basque people. The radicals are prepared to seek independence, or what they regard as liberation, through violence – most notably acts of terrorism (including kidnapping, murders and bombings) – and public confrontations which challenge the legitimacy of the state. In particular, they have a history of using the street to promote their form of Basque nationalism. Graffiti campaigns, including replacing the words Francia and España with Lapurdi and Gipuzkoa (the province names in Basque) on traffic signs in the border zone between Spain and France, are common. So too are demonstrations – often involving violence – in the provincial capitals and industrial centres at the heart of the radical nationalist area. While such acts of resistance visibly articulate the claims of the Basque Nationalist movement and remind the state of the acuteness of their demands, the moderates argue that the actions of the radicals weaken the bargaining power of the Basque Autonomous Community with the central government.

---

now beginning to get in on the act too. De-industrialization and the marginalization of the North-east within the UK political economy have fostered the rise of regionalism within the North-east and the formation of a Campaign for a Northern Assembly (Tomaney 1995).

While Scottish nationalism and North-east regionalism have been fuelled by a sense of economic injustice and political marginalization, prosperous and politically dominant regions can also seek separation from their less well-developed compatriots. For example, the Italian northern leagues are resentful of the need to subsidize the less prosperous south of Italy.

The emergence of supranational organizations such as the European Union has unwittingly played a part in fostering nationalist movements by appearing to offer a framework that could potentially safeguard the sovereignty of small nation states. The slogan of the Scottish National Party is 'Scotland in Europe – Make it Happen'.

Ignatieff (1994) comments on the paradox that the globalization of capital markets and free trade (see section 9.5) has not eroded but strengthened the desire of small nations to have sovereignty.

Both state nationalism and nationalist movements make much of historical continuity, marshalling the past to make a case for the 'nation' in the present (Daniels 1993). In particular, Lowenthal (1994: 43) argues that 'heritage distils the past into icons of identity, bonding us with precursors and progenitors, with our own earlier selves, and with promised successors'. In doing so the heritage industry is selective, focusing on particular national heroes, battles, monuments, buildings or other emblems of national identity (for example, the English rural landscape, see Chapter 8), cleaning up complex processes of the past to tell a clear story of the nation. As a consequence, Smoult (1994: 108) explains, '[n]ational identities are constructed out of reference to history, or more exactly, to received popular ideas about history that achieve mythic status, irrespective of what modern academic historians perceive to be their actual truth or importance'.

These narratives of the nation can, however, also be rewritten and retold by different generations, as well as by those who hold different political experiences and positions. In a study of Ukrainians living in Bradford, UK, Smith and Jackson (1999) explain that, when the Ukraine was part of the USSR, it was impossible for those in exile to return 'home' or to visit their relatives. The coming of independence in 1991 and the possibilities this offered for people to visit their 'homeland' therefore 'unsettled a stable (though imaginary) sense of "Ukrainianess", forged in exile and with little reference to the changes actually taking place "on the ground" in the Ukraine' (Smith and Jackson 1999: 384). Smith and Jackson (1999: 384) argue that, as a result, the '[r]ecent historical change has led to a greater complexity in the way that the Ukrainian "community" in Bradford is imagined and in the way that the Ukrainian nation can be narrated'.

Nationalism often relies heavily on sentiment. The seduction of nationalism is its promise of a powerful sense of communion and belonging (Schulman 1995). Referring to cultural products of nationalism such as fiction, poetry, music, and so on, Anderson (1983) observes that nations inspire love, a love for which people are prepared to lay down their lives. More cynically, however, Ignatieff (1994: 6) suggests that such sentiments are manipulated to induce people to fight for their nation. He writes, 'The latent purpose of such sentimentality is to imply that one is in the grip of a love greater than reason, stronger than the will, a love akin to fate and destiny. Such love assists the belief that it is fate, however tragic, which obliges you to kill.'

In particular, the nation is often imagined – particularly at times of war – as a woman, for example Britannia or Mother Russia. In this sense, the nation is conceptualized as a mother, as the nurturer of national culture or a chaste daughter in need of protection (McClintock 1993). As Sharp (1996: 101) notes, the rape of Bosnian women by Serb soldiers during the Bosnia-Herzegovina war 'solidifies all too clearly the links between individual female bodies and the nation in both nationalist rhetoric and *realpolitik*'. Nationalism is also used to produce homosocial male bonding or a sense of fraternal comradeship between soldiers. Mosse (1985) observes, for example, how the nationalism of Nazi Germany drew on a nineteenth-century notion of *Mannerbund* – passionate brotherhood – emphasizing physical strength and spiritual virtues such as leadership and heroism.

Nairn (1977) describes nationalism as the modern Janus (a Roman god depicted as looking both forwards and backwards) in that it can be both a positive force and a negative force. For example, democratic movements which pursue social and political liberation from domination by oppressive, authoritarian regimes, or who challenge extreme racial or religious intolerance represent a progressive form of nationalism. Nationalism can also provide some form of social cohesion in periods of rapid social change, unifying rather than dividing or destabilizing communities. The nationalist movements that emerged as a result of the break-up of the Soviet Union represent an example of this (Smith 1995). In contrast, nationalism which has its roots in fear and hatred of the 'Other' can foster insularity, bigotry, racism, aggression and the cultural oppression or denial of human rights for minority groups. The fascism of Hitler's Nazi Germany is perhaps the most obvious example of this.

Of course, nationalism is not only the preserve of popular movements. State nationalism 'reinforces or even exalts the idea of the nation-state . . . state actions can be legitimised in both the domestic and international arena by appealing to "national unity" and "national interests"'' (Johnston *et al.* 2000: 533). Billig (1995) uses the term 'banal nationalism' (Plate 9.1) to describe everyday articulations of national

Plate 9.1   Banal nationalism (© Becky Kennison)

identity and interest such as the flying of national flags outside homes, schools and other public buildings, media reporting of international events, coverage of international sporting competitions, and so on.

■ **Summary**

- Nationalism is a belief that the world's peoples are divided into nations and that each nation needs its own sovereign state to express its culture.

- In reality, there are few places where state boundaries actually match those of a national community in which all citizens share one culture. All round the world nationalist groups are pursuing independence.

- Uneven economic development often provides the impetus for nationalist and regionalist movements. They have also been fostered by the emergence of supranational organizations.

- Nationalism relies heavily on history and sentiment. It can be either a positive or negative force.

- Nationalism is not only the preserve of popular movements but is also mobilized by the state and articulated in banal ways in everyday life.

■ 9.4 Citizenship

'The idea of citizenship refers to relationships between individuals and the community (or State) which impinges on their lives because of who they are and where they live' (Smith 1989: 147). While members of a state have certain duties and obligations towards it, in return they can expect certain rights and benefits. In particular, understandings of citizenship have been developed from the work of Marshall (1950), who defined citizenship in terms of *civil or legal rights* (freedom of speech, assembly, movement, equality in law, etc); *political rights* (right to vote, hold office, engage in political activity, etc) and *social rights* (rights to social security, welfare, basic standard of living) (Richardson 1998, Muir 1997). However, as Muir (1997) points out, members of ethnic minority groups may enjoy these legal rights and benefits but still not feel as if they are citizens because they experience racial discrimination and harassment and regard themselves as less able to exercise their rights before the law.

Despite the fact that the language of citizenship implies inclusion and universality, it is also an exclusionary practice (Lister 1997). Historically, only select groups – notably white property-owning men – have been entitled to citizenship. Other groups, such as women, ethnic, cultural and religious minority groups, those with mental ill-health and indigenous people, have only seen the gradual extension of civil rights (such as the right to vote, to own property or to equal treatment before the law) to include them. McClintock (1995: 260) goes so far as to claim that 'no nation in the world grants women and men the same access to the rights and resources of the nation state'. Indeed, some social groups remain effectively only partial citizens because they are excluded from particular civil, political or social rights (Smith 1989). For example, deaf people are not allowed to sit on a jury in the UK, and children are not entitled to vote.

A number of studies have highlighted how in the eyes of the modern state the citizen is a heterosexual man or woman and consequently, lesbian and gay claims to citizenship rights and political legitimacy are not fully established (Corviono 1997). Indeed, in some states homosexuality is illegal. For example, under Italian law same-sex acts, which are defined as against the common sense of decency in the Criminal Code, may be punished with a prison sentence of between three months and three years. In many states of the USA, sodomy, oral sex and 'unnatural sex acts' are criminalized (Isin and Wood 1999), while in the UK the prosecution of men on assault charges for engaging in consensual same-sex SM activities in the privacy of their homes demonstrated the limits of sexual citizenship (Bell 1995a). Even where same-sex relationships are not explicitly outlawed, lesbians and gay men lack basic civil rights, including rights to anti-discrimination protection in relation to employment, housing, education and so on (Andermahr 1992, Betten 1993, Valentine 1996d, Waaldijk 1993).

The AIDS crisis which emerged at the end of the twentieth century sparked a significant right-wing backlash against lesbians and gay men, further demonstrating the extent to which sexual dissidents are denied full citizenship rights. For example, Isin and Wood (1999: 88) claim that: 'The state's [true of both USA and UK] failure to appropriately respond and assist them [Persons with Aids] is directly related to the state's identification of PWAs [Persons with Aids] as gays, as criminal drug users, as non citizens who are not part of the "general population".' Indeed Yingling (1997) claims that 'the American feeling for the body inscribes disease as foreign and allows AIDS to be read therefore as anti-American' (Yingling 1997: 25) and a subversion of the nation by gay men and drug users.

Some Western countries, such as Australia, the Netherlands, Denmark, Norway, Sweden and Iceland, do recognize lesbian and gay relationships as families or *de facto* marriages. For example, anyone can obtain Dutch nationality if they have been living in a permanent non-marital relationship with a Dutch national for at least three years and have been resident in the country for at least three years. Lesbians and gay men can also gain refugee status in the Netherlands on the grounds of persecution because of their sexuality (Binnie 1994). However, most countries are not so tolerant. As a consequence, '[i]nsofar as marriage between persons of the same sex is not allowed by most legislation, acquisition of citizenship by way of marriage is impossible for lesbian and gay couples of different nationalities' (Tanca 1993: 280 in Binnie 1994: 6). As a result, campaigns have been held across Europe to draw attention to the predicament faced by lesbian and gay couples of different nationalities because of discriminatory partnership legislation (Valentine 1996d). Activists have also sought to highlight the material consequences and rights which follow from the institution of marriage (such as tax benefits, custody rights, adoption, succession to tenancies, inheritance, etc) from which lesbians and gay men are excluded.

The position is more complex for those who define themselves outside the heterosexual/homosexual and male/female binaries. For transsexuals and transgendered people citizenship hangs on the question of their right to self-determination in the face of state definitions of their identity as male or female as classified at birth.

History is littered with examples of these and other groups such as Australian Aborigines, African slaves, Jews and gypsies who have been excluded from membership of nations altogether. As Sibley (1995a: 108) points out, 'key sites of nationalistic sentiment, including the family, the suburb and the countryside, all . . . implicitly exclude black people, gays and nomadic minorities from the nation' (see also Chapter 8). Indeed, often the exclusion of these groups is justified on the grounds that they are somehow closer to nature or less human and so do not belong in civilized society (Sibley 1995a, see also Chapter 2). For example, imperialist practices which categorized races created a hierarchy of people that was used to justify Europeans claiming land and denying indigenous people equal social, civic or political rights in countries such as Australia, Canada and the United States. In such ways indigenous people have been rendered invisible and have had to embark on a long struggle to protect and reclaim land and citizenship rights (Jacobs 1996).

Until 1992, the Australian state had a policy of *terra nullius* (land unoccupied) which denied Aboriginal people titles to land on the basis that because they had not developed a Western-style property system they did not own or have any connection with it. Despite the passing of the *Native Title Act* in 1993, land has not yet been handed back (Isin and Wood 1999). Many indigenous people therefore continue to press for a redistribution of power and some form of self-governance, aware that the rhetoric of equality being espoused by states like Australia and Canada may only result in the incorporation rather than the accommodation of their voices.

Laws in relation to immigration and naturalization, through which those not born in a state can become citizens of it, also enable governments to be selective in terms of who is allocated citizenship (Isin and Wood 1999). For example, Canada has targeted better-qualified immigrants, operates a Canadians-first employment policy, and places strict limits on family reunification (Isin and Wood 1999). In Germany, on the other hand, citizenship is allocated on the basis of ethnicity (Kofman 1995). This means that those migrant workers who in the 1960s were recruited to work in Germany as part of bilateral recruitment agreements with Greece, Spain, Portugal, Turkey, Morocco and Yugoslavia to meet Germany's labour shortage cannot become citizens. Even their children who have been born and educated in Germany must be naturalized to gain citizenship (to qualify for this they must have lived in the country for eight years and attended four years of school). What this means in practice is that *Gastarbeiter* (temporary guest workers) cannot vote, they cannot work in the civil service, and their position within the state will always be tenuous.

The failure of most nation states to guarantee universal citizenship and equality before the law is now increasingly being challenged by groups, such as women, ethnic minorities, indigenous peoples, the disabled and lesbians and gay men, who are demanding full citizenship and social inclusion while also asserting their difference from the white heterosexual male 'norm'. This has led some to claim that such forms of social identification might threaten the emotional attachment and sense of loyalty people feel to the state and its authority, potentially 'disuniting the nation' (Schlesinger 1992, Friedman 1989, Morley and Robins 1995). Hobsbawm (1990: 11 in Munt 1998) observes, 'we cannot assume that for most people national identification – when it exists – excludes, or is always or ever superior to, the remainder of the set of identifications which constitute the social being. In fact it is always combined with identification of another kind, even when it is felt superior to them.' Littleton (1996: 1 in Isin and Wood 1999) is more forthright, claiming that '[i]nstead of regarding themselves as citizens of sovereign nation-states, much less citizens of the world, many people have come to see themselves primarily as members of a racial, ethnic, linguistic, religious or gender group'. It is claimed that this has resulted in the abandonment of so-called universal or common concerns such as humanity, equality and basic material needs in the pursuit of the rights of particular social groups whose interests are sometimes accused of being dangerously narrow or as merely cultural (in other words, of being struggles over representation rather than economic inequality) (Isin and Wood 1999). The following two subsections explore two examples by considering sexual citizenship and diasporic citizenship.

■ 9.4.1 Sexual citizenship

'An ethic of solidarity and commitment informs many liberation movements, and the drive for "rightful" colonisation of "their" space – the desire to occupy their perceived constituency – fuels the imagination, and movement, of radical struggles' (Munt 1998: 3). This is certainly true of lesbians' and gay men's attempts to assert in a range of very different ways their identity, culture and social belonging in the face of social exclusion and invisibility.

In the early 1970s the notion of a Lesbian Nation emerged as a product of lesbian feminism in the USA (see also Chapter 8). In a publication entitled *Lesbian Nation: The Feminist Solution*, Jill Johnston (1973) argued that all women shared a sense of displacement from masculinist US culture and the state. She aimed to inaugurate an autonomous utopian community that would effectively declare secession from the USA and would form a radical state based on the shared identity of the 'majesty of women' (Munt 1998). Like other nationalist movements (see section 9.3) Johnston's vision of a Lesbian Nation also involved a nostalgia for an imagined past and for heroic mythical figures such as the Amazons, warrior women, who were fêted for their strength and agency. However, in the desire of lesbian separatists to create and fix the meaning of this identity fractures up opened between white women and women of colour, middle-class women and working-class women, and so on (Valentine 1997b). Thus '[l]esbian nationalism became a discourse of exclusion' (Munt 1998: 11).

In the 1990s the politics of Queer Nation sought to redefine sexual citizenship in a number of ways. Started by a group of AIDS activists in New York, USA, angry at a number of incidents of 'gay bashing', Queer Nation sought to challenge heterosexual hegemony and to make the nation a safe space for queers. As Berlant and Freeman (1993: 198) point out, 'being queer is not about the right to privacy: it is about the freedom to be public'. Under slogans such as 'We're here! We're Queer! Get used to it!' activists publicly performed queer sexualities in the streets as a means of highlighting heterosexual hegemony while also parodying it. 'Anti-assimilationist in intent, their exhibitionist agenda was self-consciously to shove the homosexual into America's face' (Munt 1998: 14).

Founded on a principle of inclusiveness, Queer Nation rejected the old separatisms which have split gay and lesbian communities, embracing transvestites, bisexual people, sadomasochists and transsexuals, and celebrated subversion, ambiguity and sexual freedom (Bell 1995b). Yet, despite this, it was unable to shake off an incipient white, middle-class masculinity. As Munt explains (1998: 15), 'In its rush to affirm a new political moment, Queer Nation, like many nationalisms, refused and "forgot" the complex lessons of history, and broke apart over the same social divisions evident in Lesbian Nation and other single-issue projects.'

While AIDS activists and Queer Nation have sought to create radical understandings of citizenship (see Brown 1997b), the economic muscle of the pink pound and the pink dollar have also enabled spaces of sexual citizenship to be constituted through consumption (Binnie 1993, Bell 1995b). For example, the 'Buy Gay' campaign initiated by the gay activist Harvey Milk in the USA in the 1970s proved an effective way of securing political representation for sexual dissidents at a local level. By fostering a lesbian and gay business district and gay gentrification, Milk was able to build a political base from which he was elected city supervisor congressman (see also Chapter 7). More recently, the rise of consumer power has meant that companies whose products have a large lesbian and gay market have to pay more attention to customers. Bell (1995b) cites the example of Levi, the clothing company, which faced a boycott by lesbian and gay consumers angry over the company's financial contribution to the homophobic scout movement in the USA. Gay tourism has also brought together consumerism and citizenship since 'being able to go on holiday . . . is presumed to be a characteristic of modern citizenship which has become embedded into people's thinking about health and well-being' (Urry 1990: 24, cited in Bell 1995b: 142). Global gay tourism guides highlight those places where it is safe to be 'out' and identify the different citizenship rights afforded to sexual dissidents around the world.

A recognition of the uneven contours of citizenship rights between countries (Evans 1993) has also led to the formalization of transnational networks such as the International Lesbian and Gay Association into a co-ordinated Federation of Lesbian and Gay Organisations. For gay activists such as Peter Tatchell (1992: 75, cited in Binnie 1994: 3), 'it is through collective solidarity, overriding national boundaries and sectional interests, that we [lesbians and gay men in the European Union] have our surest hope of eventually winning equality'. In July 1992 London staged the first ever Europride festival, in which over 100 000 lesbians and gay men from across Europe marched in the largest ever exhibition of European lesbian and gay consciousness. Since then similar festivals have also been held in Berlin and Amsterdam (Binnie 1994).

■ 9.4.2 Diasporic citizenship

Modern citizenship is predicated on territory, specifically birth and/or residence in a particular nation state. The assumption is that the nation state both represents home for its people and is the site of the production of national culture. Yet political communities throughout history have always been multi-ethnic (Kymlicka 1995). Migration and the mixing of cultures and religions have gone on for thousands of years, such that nation states have always been produced in and through their connections with other places.

Capitalism and colonialization have displaced millions of people around the globe. Between 1500 and 1800 two million Europeans migrated, mainly voluntarily, from Europe to the New World (North and South America and the Caribbean) in search of new opportunities and a better life (Emmer 1993; see also Chapter 8), while 11 million people were forced to migrate from Africa to America, being sold into slavery or kidnapped. This history of violent deportation and transplantation has had the legacy of connecting African-American, Afro-Caribbean, British and South American cultures (Clifford 1989). Even when slavery was abolished people continued to migrate from India and China to work as labourers in the European colonies. Later, the postwar labour shortages in Europe saw more workers voluntarily migrating from South-east Asia and the Caribbean to the UK, and similar patterns of movement from former colonies to other European countries (Dwyer 1999b).

In effect, the processes of colonialism and capitalism have created a global labour market which has perpetuated migration flows. At the end of the twentieth and the beginning of the twenty-first centuries the restructuring of, and uneven economic development within, the global economy has accelerated patterns of migration still further. While many migrants – especially those with skills – choose to move in search of a better life elsewhere, others are forced to do so by political and economic pressures (Boyle et al. 1998, see also section 9.5). Postcolonial struggles, oppressive regimes in Iran, Iraq and Afghanistan, wars in Southern and South-east Asia as well as in Africa, Eastern European countries and the former Soviet Union have all generated large-scale forced migration, both legal and illegal (in the form of human trafficking – see Box 9.4, Figure 9.1), into the West. The UN estimates that the number of refugees (people forcibly displaced from their home country) in the world had risen from three million in 1978 to 18.2 million in 1993 (Isin and Wood 1999).

Box 9.4: Human trafficking: the road from Iraq to Britain

Police arrested Azad, his father and brother in their photocopying shop in Kirkuk, northern Iraq, on the grounds that they were suspected of printing pamphlets for rebel Kurds. Both [the latter] were hanged. When Azad tried to re-open the shop he was warned that he would suffer the same fate. He fled the country paying an agent 1000 dollars (630 pounds) to smuggle him alongside others into south-eastern Turkey and then Istanbul. Here, he, along with a group of about 30 other refugees, paid 850 dollars for a guide to take them on foot across Greece's eastern border. During the journey the group walked by night and slept rough by day. When it snowed they suffered frostbite and deaths by exposure. Once in Salonika Azad and his group were taken by car to Athens and then took the train to Patras. From there they were to wait for a lorry that would cross on the ferry to Ancona in Italy. From Ancona the journey was by rail to Turin where migrants were kept in safe houses and assembled into groups wanting to go to the same countries. From Turin Azad was allocated a truck to Calais and then another to London.

Carroll (2000) *The Guardian* Monday 27 March

Figure 9.1   The road from Iraq to the UK

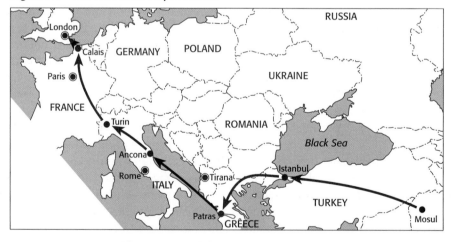

Migration is often assumed to result in a permanent change in residence or home. However, contemporary migration for work is often temporary, though it may involve spending long periods of time, even years, overseas. For example, many women from Mexico, Central America, the Caribbean and the Philippines are employed (legally and illegally) as domestic workers or even prostitutes in Europe, Canada and the USA, earning money which they send back home to support parents, children and husbands (Enloe 1989, Pratt 1997, England 1997). These remittances can be an important component of earnings from export in Third World countries.

Even those who never return hold on to a sense of their homeland, which Rogers (1992: 520) describes as 'a place distant in time and space, and partly lost, but also ever-present in daily life', and can maintain links with people there through letters, telephone conversations, e-mails, photographs and videos. Thanks to developments in transport and communications technologies, interconnections between people and places around the world are increasingly dense (see section 9.5 below). People can therefore have **transnational** or diasporic identities in that they do not identify with or have a sense of belonging fixed to one nation state.

Specifically, 'diaspora' refers to the dispersal or scattering of a population from an original homeland. It was first used to describe the dispersal of the Jews following the Roman conquest of Palestine, but it is now used more loosely to capture the complex sense of belonging that people can have to several different places, all of which they may think of as home (Clifford 1994, see also Chan 1999). In this way, diasporic citizenship questions the assumption of a relationship between group identity and territory by recognizing that people are part of webs of connections and flows between places that cut across national boundaries or borders, and that, as a consequence, their identities and culture are a product of movement and fusion rather than rootedness in one place (Massey and Jess 1995). For example, in a book entitled *The Black Atlantic* Gilroy (1994) examines how the historical pattern of movement, transformation and relocation of black people from Africa across the Americas (described above) has created a hybrid or disaporic culture linking black British, African-American, Caribbean and African peoples. He illustrates this through the example of black music, examining how complex musical traditions have travelled across the diaspora, constantly being reworked and transformed to produce new diaspora musics, from New Orleans jazz, to British Northern soul.

Likewise, studies of borderlands – areas adjoining national boundaries where economic, social and cultural exchanges are common – have focused on these places as sites of crossing or hybridity rather than separation. Work, particularly on the Mexican–US border, clearly shows how fluid everyday cultures defy the political divide between nation states. Anzaldua (1987) describes her sense of being neither Mexican nor American, but somewhere in between, as 'being on both shores at once'. In this way those who speak from in between different cultures are 'always unsettling the assumptions of one culture from the perspective of another, and thus finding ways of being both *the same as* and at the same time *different from* the others among whom they live' (Hall 1995: 206).

Figure 9.2   The number of people seeking asylum in Europe (1999)

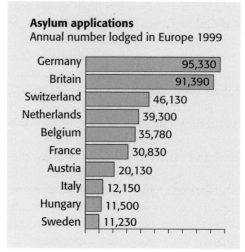

Asylum applications
Annual number lodged in Europe 1999

Germany	95,330
Britain	91,390
Switzerland	46,130
Netherlands	39,300
Belgium	35,780
France	30,830
Austria	20,130
Italy	12,150
Hungary	11,500
Sweden	11,230

Given the emphasis nationalists place on homogeneity and a common culture, it is not surprising that transnational migration and diasporic cultures and identities have provoked fears that the boundedness and distinctiveness of individual nations' cultures are under threat, and that, as a consequence, so too is the nation state (Kymlicka 1995). Across Europe legal and illegal migrants (particularly those who are non-white) are being identified as a threat to the economic and cultural well-being of nation states because they are regarded as a drain on the welfare state and as polluting national culture. For example, the rise in the number of people seeking asylum (see Figure 9.2) in Europe at the end of the twentieth and beginning of the twenty-first centuries has been met with hostility in the UK. Here, parliamentary debates and popular media scares about bogus asylum-seekers living on the welfare state represent little more than thinly veiled racism, with the shadow Home Office minister, Ann Widdicombe, accusing asylum seekers of 'swamping' the UK.

In March 2000 opposition ministers urged the British Labour government to detain all those seeking asylum on entry into the UK and to hold them in camps until their applications could be processed. While resisting these measures, the government nonetheless moved to adopt a system of vouchers for those seeking asylum, in place of welfare benefits. Likewise, in France the arrival of significant numbers of Algerians and Africans has met with a toughening of the government's stance. In 1997 the National Assembly tightened legislation relating to illegal immigration, enabling customs officers 'to fingerprint non-European residence seekers, confiscate passports of suspected illegals, search vehicles and workplaces, and expedite the expulsion process' (Isin and Wood 1999: 53). This fortress mentality is shared across Europe. Despite reducing internal borders through the Schengen Agreement, the European Union has moved to strengthen its borders with the outside world. In doing so, Lister

(1997: 46) claims, 'the dividing line drawn between citizens and foreigners is creating stronger symbolic as well as actual borders between Europeans and non Europeans; between "us" and "them", for which read White and Non White'. In this political context it is perhaps not surprising that across Europe there has been a rise in hate crimes and support for extreme right-wing parties. In 1996, more than 2500 hate crimes were recorded in Germany alone (Isin and Wood 1999).

Such fears about migrants swamping national cultures, or sexual minorities posing a threat to nations are, of course, dangerous nonsense: '[t]here is no common culture that "may cease to be" ' (Isin and Wood 1999: 63). Nations, as the discussion outlined, have always been created out of movements and flows connecting them with other places; and, despite imaginings of homogeneity, have always contained diverse populations. Instead, as Isin and Wood explain (1999: 63), 'What exists is the attempts of the dominant class to homogenise all groups, to impose its own political and social values on them and to ignore or forcefully surpress those who disagree. There never was a common culture of which citizenship was an expression; there are dominated and dominant groups between which citizenship is a mediating institution and a contested field.' Isin and Wood (1999) conclude by arguing that advocates of 'imagined unity' are not fighting for the preservation of national cultures and identities in the face of other forms of social identification, but rather are attempting to suppress diversity and to perpetuate racial, ethnic, gender and sexual divisions and hierarchies.

■ Summary

- Citizenship is defined in terms of rights and responsibilities.

- The language of citizenship implies inclusion and universality but it is also an exclusionary practice.

- Those social groups who are excluded from full citizenship are increasingly fighting to assert their identities, cultures and social belonging.

- Fears that these forms of social identification might threaten people's sense of loyalty or attachment to the state have been criticized as attempts to perpetuate social divisions and hierarchies.

■ 9.5 Globalization

In the nineteenth and twentieth centuries revolutions in transport (particularly the development of rail, motor vehicles and air travel) have speeded up the time it takes to travel between places. A process of *time-space convergence* has thus occurred. In

Figure 9.3 Time-space compression. Redrawn from Harvey (1989).

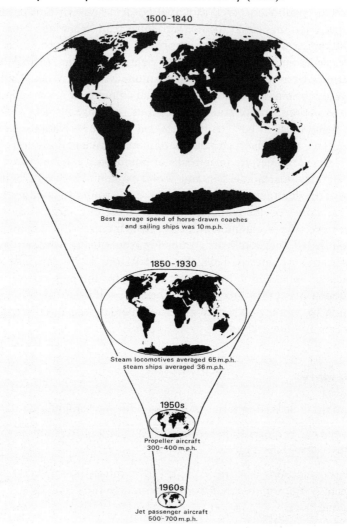

other words, places have effectively moved together or 'converged' on one another. For example, in the 1700s it took four days to travel between Edinburgh and London, now by plane the journey takes about one and a half hours (see Figure 9.3). Marx allegedly referred to this process as the 'annihilation of space by time'. In the same vein, revolutions in information and communications technologies (ICT) from the telegraph to the Internet have enabled instantaneous communication around the globe (Thrift 1990). Whereas in the past places may have existed in a relative state of isolation, ICT has brought more places into contact with one another – and generated a sense of simultaneity. One of the consequences of this is that economic actors who are absent in time and space may have as much influence on local processes as those who are physically present. Another is that social relations can now be extended or stretched over space so that those with access to networked PCs can potentially participate in disembodied global social networks (see Chapter 4). Finally, as the world has become a smaller time-space, so capitalism's urge to exploit global markets has accelerated the pace of life. Harvey (1989) terms this sense that life is speeding up *time-space compression*.

These processes have played an important part in the rapid development of Western economies. Notably, there has been a global shift in the structure and organization of the economy (Dicken 1992), with transnational corporations replacing nation states as the main players shaping investment and economic activities. The advances in technology and communication outlined above mean that these corporations are able to split their functions (e.g. marketing, research and development, production, and so on) and therefore relocate their activities around the globe in pursuit of profit, with no regard for national boundaries or interests. In this way, investment in a place is increasingly dependent on what is going on elsewhere, which undermines the ability of nation states to manage and control their own economies and so threatens their sovereignty.

The activities of transnational corporations have been facilitated by a decline in regulatory barriers, and, more importantly, by the adoption of ICT by banks and financial institutions. This has accelerated the speed at which money can circulate round the global economy, increasing the ability of corporations to move their money in pursuit of new investment opportunities. In turn, this has produced a global financial trading system in which the major stock markets of the world are converging and economies are homogenizing (in other words, states are pursuing similar macroeconomic policies). Global cities (see also Chapter 7) such as New York, London and Tokyo have emerged at the hub of this global economy, co-ordinating and controlling flows of information and investment (Sassen 1991). The behaviour of transnationals and global financial systems therefore means that places round the world are becoming more interdependent (Thrift and Leyshon 1988) in an uneven way, producing structured inequalities. Notably, the gap has widened between the affluent economies of the West and those of some Third World nations which are plagued by debt and poverty. Likewise, within the domestic economies of industrialized nations there have also been uneven patterns of economic growth which have caused regional problems, hitting some social groups – the working class and ethnic minorities – harder than others.

Globalization is not just about increased flows of money, goods and services, however, but also of labour. As section 9.4.2 outlined, spatial patterns of uneven economic development have produced large-scale migration from rural areas to major cities and from low-wage, less developed economies to higher-wage, more developed economies. Many of these flows reflect past economic or historic connections, such as those forged by colonialism. These patterns of migration are also differentiated, reflecting the polarization of labour markets. First, the highly qualified professional classes are moving between global cities such as London, New York and Paris, where major financial institutions, law firms, media and advertising companies, etc are located, in pursuit of specialist jobs and to further their careers. International firms also move staff between their offices to gain experience or to implement new policies or procedures. Second, there are flows of migrants (many from less developed countries) chasing poorly paid, insecure employment. In particular, de-industrialization has resulted in the loss of relatively well-paid skilled manual jobs and the growth of low-paid service sector jobs such as cleaning, personal services, catering industry and 'sweat shop'-style, low-grade manufacturing employment. Global cities are centres not only of production but also of very conspicuous consumption too, and in this sense a marked social polarization is evident within them between the professional classes and those in poorly paid employment or who have no work or homes (see Chapter 7). These inequalities often produce conflicts over the right of different groups to use and occupy so-called 'public' spaces (see Chapter 6).

For some, migration represents hope, excitement and the possibility of forging new homes, identities and lifestyles, but for others, especially those who are forced into involuntary exile or displaced by war, violence and ethnic cleansing, migration can be experienced as dislocation and alienation. Clifford (1989) contrasts such experiences with those of the international traveller who has the security and privilege to move at will in relatively unrestricted ways. The loss of a sense of home or roots (see Chapter 2) and the feeling of being adrift can be profound. Writers have described the sense of being a stranger in a new country, the stigma of difference and the feeling of being constructed as 'Other'. This experience is captured in Eva Hoffman's three-volume autobiography *Lost in Translation* in which she recalls her childhood in Poland, her sense of alienation and dislocation as she struggled to learn English when her family migrated to Canada, and her gradual sense of developing a hybrid identity and feeling at home in her new country (Sarup 1994). For others this feeling of being 'at home' is never achieved, as Sarup (1994) describes in Box 9.5. The experience can be particularly acute for those exiles forced to leave their homeland and become stateless wanderers. Nixon (1994) recounts the life of Bessie Head, a black South African writer who was exiled by apartheid to Botswana, which refused to accept her as a national. In the village where she eventually settled, her identity as a 'mixed-race' South African meant that she was defined as a 'half-caste' or 'low breed' and suffered discrimination. In this way, she was ostracized both nationally and locally.

Globalization has also been associated with the spread of Western, specifically US, culture. Some writers have even gone so far as to suggest that global communication and flows of commodities, people and information will spell the end of geography by eroding geographical distinctiveness such that places will no longer be a source of diversity. This process is often dubbed the McDonaldization of society (Ritzer 1993). Reflecting specifically on the consequences of communications technologies in the twentieth century, McLuhan (1962: 31) coined the term 'global village' (which has now become something of a cliché) to describe what he saw as the increased

---

Box 9.5: The exile

'Born in India, I was brought over here to be educated at the age of 9. The war
started and my father returned to India. My mother had died when I was 5 and
the only memory I have of her is that I was not allowed to go into the room where
she was dying. My father died in the partition of India in 1947, and I believe that
when I was told that he had died I did not understand. To me he just continued
to be absent. I will not tell the full story here, but I do often think of my fifty years
here in England. Am I British? Yes, I have, as a friend pointed out, a 'white man's
house', and I've forgotten my mother tongue, but I do not feel British. I think of
myself as an exile and it's painful here, and *there* in India when I return for short
visits. I don't have the confidence to become, as some have suggested, cosmopolitan. But like so many others, I am pre-occupied by ideas of home, displacement, memory and loss.'

Sarup 1994: 93

---

interconnection of different places. He argued that electronic media such as television would lead to detraditionalization (in other words, the loss of cultures, languages, etc) – a point also made by Harvey (1989) in his reference to local social
practices being overrun by capitalist relations of production.

Associated with this is the devaluation of difference. There is a danger that, because
world cities such as New York and London dominate the global economy, non-
American or non-European places and local cultures may become regarded as backward or inferior rather than different. In a study of British Asian youth, Gillespie
(1995) found that most of those who took part in her research ate Indian food at
home because that was what was prepared by their families, but chose to eat
English or American food at school. In their eyes, Western food had more cultural
capital than Indian food, which they described as 'village food'.

The concept of tradition is not, however, straightforward. Tradition is often romanticized in a static or fixed way (particularly in nationalist narrations of the nation),
but Massey and Jess (1995: 232) point out that 'there is an alternative version of
the past which sees it as a process of constant mixing and accumulating: a tradition
which is not "lost" as you move further from some (anyway mythical) origin, but
rather is built as new developments, contacts and innovations are introduced'.
Using the example of London, where American burger chains are juxtaposed with
Chinese, Italian and Indian restaurants and 'local' butchers and bakers, they point
out that the process of global mixing (albeit one which occurs in unequal and uneven
ways) has been multidirectional rather than a unidirectional process of westernization. They further note that it has not produced homogeneity but rather permeability,
hybridity and local distinctiveness.

Elsewhere, Massey (1998b) reflects on youth culture in Yucatec Maya in Mexico. She observes that here young people play computer games, drink cola and wear T-shirts covered in English slogans and American-style baseball caps and that they might therefore be characterized as part of a global youth culture. However, she goes on to explain how these cultural products and activities are mixed in with locally distinctive cultures that have their own histories. She points out that the so-called global commodities and slogans are made sense of, and actively adopted and adapted, within the interpretative framework of the local culture. In other words, youth culture in Yucatec, Mexico is not a closed local culture, but nor is it an undifferentiated global culture. Rather, it is a complex product of interaction. What is true for Mexican youth culture is also true for other cultures too. All cultures are hybrid, but the mix of global and local is always different so that the distinctiveness of local places is a product of their particular global connections (see also Massey 1991, Miller 1992). In this way, the local is never destroyed by the global but can actually be produced through it. All global flows have local geographies.

Global firms are sensitive to and often exploit these differences, packaging and marketing them and targeting certain groups for particular products. Thus local differences can also be reworked through the global market. Robins (1991: 31) explains: 'Cultural products are assembled from all over the world and turned into commodities for a new "cosmopolitan" market-place: world music and tourism, ethnic arts, fashion and cuisine; Third World writing and cinema. The local and "exotic" are torn out of place and time to be repackaged for the world bazaar.' Likewise, Desforges (1998) observes how, for Western young people, international travel itself can be built upon a representation of the world as a series of differences from home which fix 'other' people and places as knowledges or experiences that can be 'collected' or 'consumed' and then used once back home as a form of cultural capital (see Chapter 2) at work and amongst friends.

■ Summary

- Transport and communications revolutions have produced time-space convergence and time-space compression.

- These processes have helped to cause a global shift in the organization of the economy, with transnational corporations replacing nation states as the main players shaping investment and economic activities.

- People and places are linked into global systems in different and unequal ways.

- Globalization is producing large-scale migration. While some migrants forge positive new ways of life, others are forced into involuntary exile and experience dislocation and alienation.

- Globalization is often accused of being a homogenizing process and of devaluing difference and threatening the distinctiveness of place.

- However, all global flows have local geographies because they are made sense of, and are actively adapted in different ways within the framework of different local cultures.

9.6 Global citizenship

As section 9.4 outlines, citizenship has traditionally been equated with nationality and the nation state. However, the increased pace of the globalization of the economy, politics, culture and society, and the rise of information technologies (see section 9.5), are problematizing this meaning of citizenship. Neither political nor economic systems are now coextensive with boundaries of nation states (Castells 1995, Thrift 1995) and the ability of nation states to regulate economic and social matters is diminishing (Harvey 1989). Likewise, people's lives around the world are being woven together in increasingly complex ways by the global economy, which Appadurai (1990) suggests has five dimensions: people, media, technology, capital and ideology. The borders of nation states are therefore becoming increasingly less relevant. One consequence, claim Isin and Wood (1999), is that new spaces are opening for different forms of politics to develop. While various international regimes of governance are emerging which are challenging the sovereignty of the state in what is known as 'deterritorialization', so too are new notions of citizenship that transcend national boundaries in terms of political and legal rights, responsibilities and practices. Two examples are technological citizenship and ecological citizenship.

9.6.1 Technological citizenship

Information and communications technologies (ICT) are creating new spaces for the development of fluid and ephemeral networks or 'communities' (see Chapter 4) which cross the political boundaries of nation states. In doing so, they are claimed to be creating a forum for people to pursue their rights and exercise their responsibilities on a global scale. Whereas the broadcast media (such as television and radio), as one-to-many forms of communication, can be readily controlled by gatekeepers, states are finding it difficult to control the content of the Internet, which offers the possibility of many-to-many forms of communication. As a consequence, ICT enables people to access a wider spread of materials and a more eclectic range of views than are usually found on national television or radio programmes. In this sense, the global freedom of expression and the rights of electronic assembly offered by the Internet represent an obvious threat to authoritarian regimes that want to restrict public information, criticism and political activism. The Chinese, for example, have tried to restrict access to modems, while Saudi Arabia has adopted a complex filter system in an attempt to block access to material which is perceived to threaten Muslim values, traditions and culture (Whitaker 2000).

The Internet also poses a challenge, however, to democratic regimes and the legitimacy of the state. Most notably, the 'transcending qualities of ICT' are said to potentially 'facilitate the demise of modernist forms of governance based upon territory, hierarchical managerial control of populations and policing' (Loader 1997: 1). For example, ICT is imagined to offer a future where representative government might be substituted by more direct forms of government (at a national or even international scale), enabling us to vote or take part in debates on-line rather than electing members of parliament in what is currently a pyramid form of authority and participation (Kitchin 1998). The USA has already experimented with public electronic networks (PENS) to enable citizens to access information, e-mail local leaders and take part in conferences and debates about local issues using their own networked PCs or those in public places such as libraries (Kitchin 1998).

Indeed, it is claimed that traditional models of political organization, based, for example, around social class, narrow political parties or place, will be displaced by the ability of people to mobilize globally. In this ICT is imagined as contributing to a process of democratization that can extend notions of representation and accountability on a global scale. The Internet allows groups such as children and young people, who are restricted in terms of their mobility, financial independence and political voice, to create representations of their own lives and to connect with other young people around the world (e.g. Leonard 1998, Valentine and Holloway 2000). Already, advocacy and liberal rights groups are using web pages, e-mail campaigns and list servers to disseminate information (for example, web sites offer information to refugees about how to get round restrictions on entry to the UK and about their legal and welfare entitlements), to raise issues and to make politicians more accountable to both national and international constituencies. When the USA and Britain launched airstrikes against the Serb government in the late 1990s, Serbian citizens used e-mail and bulletin boards to articulate their experiences and views to a global audience, while the success of the Zapatistas' fight against the Mexican government has been attributed to the effective way they have used the Internet (Riberio 1998). In such ways, the ICT can provide a space for groups who are disenfranchised within their own nation state or have limited access to national print and broadcast media.

Indeed, the Internet has been used to inform and organize international protests against global capitalism, including large-scale demonstrations to disrupt the World Trade Organisation talks in Seattle (November 1999) and to prevent bankers attending an IMF meeting in Washington (April 2000). May Day anti-capitalist rallies have also become an annual international event, with the Net used as a means to co-ordinate actions in countries around the world (see Box 9.6 and Plate 9.2).

Box 9.6: DIY culture

DIY culture is a 'youth-centred and directed cluster of interests and practices around green radicalism, direct action politics, new musical sounds and experience, it's a kind of 1990s counter culture' (McKay 1998: 2).

One example is the May Day 2000: Festival of Anti-Capitalist Action. This was born out of young people's frustration at orthodox politics; the failure of conventional political parties to deal with global issues such as environmental degradation; the widening of the global gap between the 'haves' and 'have nots', and the reluctance of governments to try to check the power of global corporations. Faced with the narrow vision of politicians and their inability to communicate with young people, this direct action aimed to express its opposition to the global economy by striking at its very heart.

The campaign of action, which took place simultaneously in 33 countries, was organized globally using the Internet. Protesters disrupted city life around the world as much as possible. Actions included Guerrilla Gardening in London when people armed with plants, soil and water tried to create a community garden in the centre of London while chanting 'Resistance is Fertile'. Participants described it as part of an attempt to promote social and environmental harmony and to reclaim the streets from capitalism: 'to bolster Planet Earth and declare war on Planet Inc.'

As Vidal (1997: 277, in McKay 1998: 3) explains: 'There is increasing evidence that globalism and localism . . . may be intimately linked, opposite sides of the same coin. The more that corporations globalise and lose touch with the concerns of ordinary people, the more that the seeds of grass-roots revolt are sown; equally the more that governments hand responsibility to remote supranational powers the more they lose their democratic legitimacy and alienate.'

Anarchist and Far Right groups in particular have promoted the use of the Internet as a means of subverting the nation state (see Figure 9.4). The Cyberspace Minuteman, a bulletin board service, serves as a means for linking much of the Far Right in the USA and Europe. These types of bulletin board services, which include digitized pictures of David Duke (an American Ku Klux Klan leader) and Adolf Hitler, allow users to meet on-line, share information, disseminate racist and homophobic materials and plan events (Whine 1997). Both the US and German governments' attempts to regulate hate groups on the Internet have met with little success. Whine (1997: 223) observes, 'The Internet's global reach presents a seemingly insurmountable challenge to would-be regulators.'

In May 2000 a computer virus that was attached to an e-mail headed 'I LOVE YOU', initiated in Malaysia, hit computers worldwide, infecting an estimated 80 per

Plate 9.2 May Day anti-capitalist rally, London (© Jonathan Brookes)

Figure 9.4 Activists are using the Internet to subvert the state. (© Dave Chisholm)

Source: *The Sunday Times*, 20 February 2000

cent of German and 70 per cent of British PCs and causing damage estimated at several billion dollars (Clark 2000: 8). The virus fuelled fears that cyberterrorists might be able to take advantage of the World Wide Web to wreak global chaos. Already the Internet is facilitating the globalization of crime (such as drug trafficking and money laundering). In contrast, policing is still largely organized on the basis of the boundaries of nation states, albeit with a recognition of the need to foster links and co-operation between national forces. Lenk (1997: 129) explains: 'The mindset related to territorial boundaries of action and influence is still very strong. Letting foreign police tread a square metre of their soil in hard pursuit of criminals is still an anathema to European national governments. And senior police officials complain that police co-operation is advancing only slowly while national borders are becoming insignificant for crime.' Not surprisingly, all these trends are raising fears that conflicts will increasingly arise between the emerging cybersociety and modernist democratic institutions.

ICT, however, also presents national governments and police forces with potential new means and opportunities for the technological surveillance of citizens. The monitoring of everyday electronic transactions potentially enables corporations and governments to build highly sophisticated profiles of individuals. Interactive polling could be used to undermine the institution of the secret ballot, while biometrics – technologies based on mapping and digitizing unique bodily characteristics (such as fingerprints, voice patterns and retinas) – offer a spectre in which such information could be used to track individuals' movements and even be stored in central databases and traded (Lyon 1994, Graham 1998, see also Chapter 6). As such, new technologies might also pose a potential threat to individuals' civil and political rights, creating in the worst-case scenario an Orwellian world in which global panopticism is possible (Lyon 1994, see also Chapter 2 and Chapter 6).

In the midst of such speculations about the possible benefits and risks of technological citizenship it is important to remember that, while, on the one hand, the Internet is lauded for 'creating a world that all may enter without privilege or prejudice accorded by race, economic power, military force or station of birth' (Barlow 1996, see also Chapter 2), on the other hand, globalization has not been happening equally, and at the same speed, all over the world (Massey and Jess 1995). Some parts of the world are literally being missed out of the technological revolution; in other places people feel constrained by these developments. In other words, different groups of people are linked into global systems in different and unequal ways. The reality is that there are very real barriers to people's ability to participate in the emerging cybersociety and to exercise their rights and responsibilities as technological citizens in a networked world. Not least, individuals must have public access to ICT, or the economic capital to their own networked PC, and the social and cultural capital to know how to use one. Even though the figures for those connected to the information superhighway are growing exponentially, at the moment it is still only a small percentage of the world's population who are on-line and able to operate in a global frame of reference. They are, of course, largely the affluent, educated, professional and managerial classes of the Western world – though even within the West patterns of access to and use of ICT are highly uneven (Valentine and Holloway 2000).

■ ### 9.6.2 Ecological citizenship

The earth is facing a range of major environmental problems: the loss of habitats, which threatens biodiversity, non-renewable resource depletion, acid rain, global warming. For example, rising levels of greenhouse gases in the atmosphere are predicted to be causing a rise in global temperatures which will eventually melt the ice-caps, causing global flooding and a change in the world's climatic systems. In 1992 the United Nations Earth Summit in Rio de Janeiro focused attention on the fact that such global problems cannot be tackled by nation states acting individually; rather, an international response is needed because these problems are global in terms of their nature and significance. In turn, this requires not only nations to recognize their interdependence but also citizens too. In other words, we are not just citizens of a particular country, we do not just share citizenship with particular groups (such as diasporic citizenship or sexual citizenship), we are all also citizens of the earth. As such, we need to be internationalist in our outlook and, in doing so, to recognize some form of universal concerns and rights and responsibilities that transcend national boundaries. In this sense, the discourse of ecological citizenship represents an inclusionary model of citizenship in contrast to the more exclusionary nature of citizenship as defined by individual nation states.

Isin and Wood (1999: 117) observe that ' "[c]itizenship" refers not only to legal and political rights but also to various practices in which humans act as political and moral agents'. They suggest that one dimension of ecological citizenship is that we share an *intergenerational responsibility*: that we have obligations not only to the present generation of global citizens but also to future generations. This involves present-day citizens making tough choices, for example between industrial developments which might provide jobs and a higher standard of living for this generation and recognizing ecological justice for future generations.

Ecological citizenship also requires us to acknowledge the interdependency between human activities and the environment and therefore our obligations and responsibilities, not just towards each other but also towards the environment. In this sense we need to have an *ethic of care* for nature, both because we are dependent upon it and because we have a moral responsibility towards it in its own right. In other words, we need to recognize the anthropocentrism of modern societies and the place of humans within nature (see also Chapter 8). However, Isin and Wood (1999) warn of the dangers of harking back to an imagined golden age when humans and nature were in harmony, in which nature is essentialized as pristine and pure and humans as dominant and exploitative. They write, 'There are no universal or essential grounds on which an ecological citizen can defend nature but contingent and unstable grounds on which she weighs her responsibilities towards the environment as well as her source of identification and loyalty' (Isin and Wood 1999: 116).

The transnational nature of physical and biological systems, the transboundary impact of pollution, hazardous waste, and the loss of biodiversity and renewable minerals mean that most environmental problems lie beyond the jurisdiction of individual nation states. Rather, *international governance* is required, in which each state must be willing to surrender some degree of sovereignty, for example by being bound by global agreements or accepting the need to change domestic policies on transport, energy and industry, in the name of the common good. Here supranational organizations such as the UN and the EU have been instrumental in proposing and negotiating transnational treaties. The effectiveness of these agreements depends ultimately, however, on international laws which can prescribe and enforce these rights and duties on a transnational scale. In particular, within the global system there are no effective legal measures to monitor nation states' performances, enforce compliance or penalize those that break international agreements.

The willingness of nation states to co-operate in the global interest is hampered by the fact that global problems are not experienced in a universal way; rather, they

have an uneven impact. Different countries therefore have different levels of investment in supporting transnational initiatives. For example, ozone depletion is occurring most rapidly nearest the poles and thus is a more immediate problem for countries
in higher latitudes than for those around the equator; likewise low-lying countries
such as the Netherlands have more reason to be concerned about rising sea levels than
landlocked states that are high above sea level, such as Austria and Switzerland (Yearley
1995). Nation states also have differential technical skills and resources to take pre-
ventive measures and to manage the impact of global problems. Notably, affluent
Western nations such as the Netherlands are better placed to respond to rising sea
levels than low-lying countries in the South such as Bangladesh, which is also already
a victim of flooding.

The South has also criticized countries from the affluent West for promoting what
is effectively their own agenda under the guise of international interest (Middleton
et al. 1993). In particular, by promoting the need for transnational reductions in
energy levels, the industrialized nations of the West effectively shift the blame for
global warming from their own historic and present high levels of pollution onto
the future emissions of the South. Not surprisingly, the countries of the South resent
the fact that, if they are expected to limit their energy levels, their economic development and ability to raise their living standards to the level of the West will be
harmed. They consider that, if cuts are to be made, it is the industrialized countries
of the West that should bear the brunt of them (Yearley 1995). Global problems
are also sometimes manipulated by countries which use claims about the common
good to advance what are actually narrower national interests. Such tactics can undermine global initiatives. As Yearley (1995: 229) explains, 'some governments favour
the widespread imposition of high environmental standards not so much for the environment's sake but to ensure that costs of industry are similar in competing countries'. The willingness of some nation states to act in the common good can also be
eroded by private corporations – many of which are transnational – which lobby
governments to oppose tighter regulations in order to protect their own interests
(Yearley 1995), and by the difficulties of producing transnational consensus amongst
scientists about the nature of problems and the most appropriate responses to them.
In this way, the failings of transnational agreements implicitly question the legitimacy
of the nation state as an appropriate framework within which to exercise international governance.

It is often transnational activists and pressure groups campaigning around issues
such as animal rights, biodiversity, renewable energy, eco-feminism, and various
environmental management campaigns, who are taking the lead in publicizing the
need for nation states to act individually and collectively. For example, Greenpeace,
which has members all round the world, plays an active role in monitoring and drawing attention to governmental abuses of the environment, using the green credentials of some nation states to berate the governments of others. These groups
represent the beginnings of a global civil society and challenge the legitimacy of the

nation state by highlighting the lack of congruence between ecological issues and state boundaries and by questioning the way modern societies exploit nature in processes of production, consumption and exchange.

At a different scale, community groups across the world are also thinking global but acting local (Rucht 1993). In Brisbane, Australia, local environmentalists and city planners have joined forces to develop recycling and energy efficiency schemes with the aim of cutting emissions and pollution (see also Chapter 7). Such initiatives give local authorities the moral high ground over national governments and have even been used by regional authorities seeking more autonomy to challenge national government's power (Yearley 1995). In such ways local social and political relationships may help to shape and influence the global context.

Thomashow (1995: 139, cited in Isin and Wood 1999: 117) sums up ecological citizenship when he writes: 'The ecologically aware citizen takes responsibility for the place where he or she lives, understands the importance of making collective decisions regarding the commons, seeks to contribute to the common good, identifies with bioregions and ecosystems rather than obsolete nation-states or transnational corporations, considers the wider impact of his or her actions, is committed to mutual and collaborative community building, observes the flow of power in controversial issues, attends to the quality of interpersonal relationships in political discourse, and acts according to his or her convictions.'

Ecological citizens challenge conventional political boundaries and expose the need for a new, more progressive, politics which acknowledges emerging global solidarities. For such transnational forms of citizenship to be effective and not to be infringed, Newby (1996: 214) suggests that it will be necessary to develop 'new institutions of democratic accountability'. In a similar vein, Falk (1994: 128, cited in Lister 1997: 61) argues, 'Citizenship is tied to democracy, and global citizenship should in some way be tied to global democracy, at least to a process of democratisation that extends some notions of rights, representation and accountability to the operation of international institutions, and gives some opportunity to the peoples whose lives are being regulated to participate in the selection of leaders.'

Drawing on the work of Held (1995), Isin and Wood (1999) argue for a new model of global democracy, termed 'cosmopolitanism'. This vision imagines a transnational structure of political action to replace the present structure of international laws mediated between sovereign nation states. This would, in turn, provide a framework for political practices at other scales from the national to the local, so that 'individuals who composed the states and societies whose constitutions were formed in accordance with cosmopolitan law might be regarded as citizens, not just of their national communities or regions, but of a universal system of "cosmopolitical governance" ' (Held 1995: 233, cited in Isin and Wood 1999: 120). In this way, the notion of sovereignty which is currently embedded in nation states with fixed borders and territories would become redundant. Rather, sovereignty would emerge in diverse self-governing, overlapping socio-spatial centres of power (Held 1995).

Despite the superficial attractions of this vision of cosmopolitanism, it is important to remember that implicit in such a model is always a danger of Western cultural imperialism. Globalization is effectively the westernization of the planet, in that global flows of trade, finance and so on originate from the West and are controlled by the West (Massey and Jess 1995). The formation of a global civil society may only be another form of this. Those who become power brokers in this global civil society would probably be Western-born or educated individuals (Zolo 1997). As Zolo (1997: 137) comments, 'Nor moreover, does it [global civil society] go much beyond the optimistic expectation of affluent Westerners to be able to feel and be universally recognised as citizens of the world – citizens of a welcoming, peaceful, ordered and democratic "global village" – without for a moment or in any way ceasing to be "themselves" i.e. Western citizens.'

Irrespective of these reservations about the possibilities of a truly global civil society, what this section on global citizenship and subsections 9.4.1 and 9.4.2 demonstrate is that globalization and postmodernism have brought a shift away from the notion of a homogeneous and universal concept of citizenship towards one which is 'multi-dimensional and plural' (Isin and Wood 1999: 22). The identity of the postmodern citizen is an ensemble of different overlapping and intersecting forms of citizenship (diasporic, sexual, ecological, etc) rather than being merely the result of having membership of a particular nation state. In these terms, citizenship is not conceptualized as a fixed right or privilege but rather is the product of an 'on-going negotiation of identity and difference' (Isin and Wood 1999: 22).

■ **Summary**

- Globalization and information technologies are problematizing the meaning of citizenship.

- Various international regimes of governance are emerging to challenge the sovereignty of the state, as are new notions of citizenship which transcend national boundaries, e.g. technological and ecological citizenship.

- Cosmopolitanism imagines a transnational structure of political action to replace the structure of international laws mediated between sovereign states.

- Globalization and postmodernism are causing a shift away from a homogeneous and universal concept of citizenship based on membership of a nation state. Rather, the identity of the postmodern citizen is an ensemble of overlapping and interconnected forms of citizenship.

■ Further Reading

- The classic place to begin is with Anderson, B. (1983) *Imagined Communities: Reflections on the Origin and Spread of Nationalism*, Verso, London. Other key books on nation and nationalism include: Billig, M. (1995) *Banal Nationalism*, Sage, London; Ignatieff, M. (1994) *Blood and Belonging: Journeys into the New Nationalism*, Vintage, London, and Smith, A. (1995) *Nations and Nationalism in a Global Era*, Polity Press, Cambridge. A gender perspective is offered by Yuval-Davis, N. (1997) *Gender and Nation*, Sage, London.

- Isin, E.F. and Wood P.K. (1999) *Citizenship and Identity*, Sage, London, provides an excellent and very comprehensive overview of the debates about citizenship. Another useful source is Lister, R. (1997) *Citizenship*, Macmillan, Basingstoke.

- There are many complex debates about globalization. Waters, M. (1995) *Globalization*, Routledge, London, offers an introduction to the main arguments. Important geographical contributions include: Harvey, D. (1989) *The Condition of Postmodernity: An Inquiry into the Origins of Cultural Change*, Butterworth, Cambridge; Massey, D. and Jess, P. (eds) (1995) *A Place in the World? Places, Cultures and Globalization*, Open University Press, Milton Keynes, and Sassen, S. (1991) *The Global City: New York, London, Tokyo*, Princeton University Press, Princeton, NJ.

- Boyle, P., Halfacree, K. and Robinson V. (1998) *Exploring Contemporary Migration*, Longman, Harlow is a good starting point to think about migration, while Robertson, G., Mash, M., Tickner, L., Bird, J., Curtis, B. and Putnam, T. (eds) (1994) *Travellers' Tales: Narratives of Home and Displacement*, Routledge, London includes several essays on the experience of displacement and exile. A non-academic perspective on this issue is offered by Eva Hoffman's biography (1991), *Lost in Translation*, Minerva, London, in which she recalls her childhood in Poland, her sense of alienation and dislocation as she struggles to learn English when her family migrate to Canada and her gradual sense of developing a hybrid identity and feeling at home in Canada.

- Contemporary speculation about the emergence of new forms of global citizenship is comprehensively explored in Held, D. (1995) *Democracy and the Global Order: From the Modern State to Cosmopolitan Governance*, Stanford University Press, Stanford, CA, and Isin, E.F. and Wood P.K. (1999) *Citizenship and Identity*, Sage, London.

■   Exercises

1. Think about which nation(s) you identify with and why you do so. How do you define your 'national' identity?

2. Make a list of all the examples of banal nationalism that you encounter in your everyday life. Do these forms of nationalism matter? If so, why?

3. Use the Internet to look up the record of British parliamentary debates, look in *Hansard* or search back copies of your national newspapers to locate political and popular debates about immigration, refugees, asylum seekers and human trafficking. How are migrants represented in these sources, and by whom? What discourses are employed by different protagonists in these debates? What do these debates say about nationalism and citizenship?

4. In a small group discuss how you might construct a new model of global democracy to replace the current system of nation states. What do you envisage would be the problems in trying to establish such a model? How realistic do you consider this model to be?

■   Essay Titles

1. Using examples, examine some of the processes through which the nation might be understood to be gendered.

2. Critically evaluate the characterization of nationalism as Janus-headed.

3. Critically assess Littleton's (1996) claim that many people, instead of regarding themselves as citizens of sovereign nation states, see themselves as primarily members of racial, ethnic, religious or gender groups.

4. To what extent do you agree with the claim that globalization is effectively the westernization of the planet?

Appendix A: A guide to doing a project or dissertation

1. Choosing a topic
2. Preliminary research
3. Research design
4. Writing

▪ 1. Choosing a topic

If you are able to select your own research project you have only yourself to blame if you are bored with what you choose. The best research topics are often those that are based on students' own academic or personal interests. The ideal topic is something specific and highly focused, but it is often easiest to begin by brainstorming general areas of interest to identify a theme or topic. This can then be narrowed down to produce a set of specific research questions. To brainstorm possible topics:

- Follow up ideas from your undergraduate modules and reading lists. This book alone contains hundreds of potential topics from body modification and fear of crime, to animal rights and human trafficking.

- Think about current debates in geography by reading recent articles in *Progress in Human Geography*.

- Identify gaps in the geographical literature from your lectures, critical reading and from your personal experiences of topics which you have been taught or read about.

- Do you have a hobby (e.g. hill walking, see Chapter 8), an interest (e.g. the Internet, see Chapter 4, or science fiction, see Chapter 2), social activity (e.g. clubbing, see Chapter 7) or do you take part in voluntary work (e.g. caring for the elderly, see Chapter 2) which could form the basis of a project?

- Do you already have plans for a future career? If so, can you base a project on it which might give you useful work experience, or be of use or interest to a potential employer (e.g. if you are planning a career in teaching, can you base your project on working in a school? See Chapter 5).

• Is there an issue in the national or local news that has caught your attention (e.g. the siting of a motorway or waste dump which is causing community conflict, see Chapter 4, or the global carnival against capitalism, see Chapter 9).

■ 2. Preliminary research

All projects should begin with a review of the literature in your chosen field (the guides to further reading at the end of each chapter might help you to get started). This will enable you to set your own work in the context of what has gone before and to evaluate the conclusions from your own empirical work in the light of existing understandings of the topic (for example, by outlining to what extent your findings confirm or challenge previous studies).

Start with your module reading lists. Search library catalogues and on-line databases using key words and the names of key authors. When you find useful articles and books look at the references the author(s) have cited and track them down. Ask your tutors for their advice. Search the World Wide Web for relevant information (do not forget to keep details of the web page address, title of the page/author and the date/time it was accessed). Remember, however, that there is no system for checking the validity of what is posted on the Web, because although a lot of the information available on-line is relevant and useful, much of it is also spurious or unsubstantiated. Identify the key authors in your field and check *Geoabstracts* or their web pages to obtain a list of their relevant publications. If you have a *specific* question you would like to ask, for example about a reference or a method they have used, try e-mailing them (but do not send a vague e-mail just telling them your topic title and asking them to tell you everything they know about it).

If there is not much literature on your specific chosen subject this can be a good sign because it suggests that this is an original topic which has been underresearched. This does not mean, however, that you can stop looking for previous work. Rather, it is important to identify literatures on other related topics or work from other disciplines that you might usefully draw upon. Although the ideal project is specific and highly focused, you must be able to relate your work to broader areas of the discipline and to understand the wider implications of your research.

Critically evaluate the literature which you have collected. This means that, rather than just listing, summarizing or describing previous studies, you need to ask questions such as: What are the aims of the studies? What theories have the authors drawn upon? To what extent have the authors fulfilled their aims and objectives? What evidence did the authors present to make their arguments? How was this material collected? Were the research methods used appropriate and rigorous? How convincing are the findings, arguments and the conclusions of the authors? Are there competing approaches to, or understandings of, the same topic by different authors? If so, how and why do they differ, whose arguments are most convincing and why? What weaknesses, gaps or new issues can you identify in the existing literature?

Explore other available data sets. These might include census data, historical archives, newspaper cuttings and secondary data such as government reports or documentary/statistical material held by commercial organizations and public institutions such as the police. If this sort of information is crucial to your study make sure that this data is available before embarking on your research. Large organizations may not keep their data in a form you require (for example, students working on projects about crime often require statistics at the scale of the neighbourhood, whereas the police may only record crime statistics on a citywide or regional basis); they may be reluctant to reveal this information to an 'outsider', especially if it is commercially sensitive, or they may charge for supplying data.

■ 3. Research design

Once you have read around your topic and have a sense of what you are trying to achieve, you will need to identify your research questions and to think about choosing an appropriate methodology. The philosophy and methodology you adopt will shape the sort of questions you can ask; the type of data you collect; and the sort of findings you will generate.

Depending on the topic you have chosen, your project may include quantitative (questionnaire surveys) or qualitative methods (interviews, focus groups, participant observation, self-directed photography, textual analysis, etc), or a combination of them both. How you plan to analyse the data you collect should always be considered as part of the research design. For example, if you use a questionnaire it is important that an appropriate number are completed and that the questions are worded in a way that produces data which is in a suitable form for analysis. A pilot study is a very helpful way of identifying any problems with the research design and fine tuning it accordingly. When designing your fieldwork you need to ensure that you have access to the necessary equipment and/or technical support (tape recorders, transcribers, cameras, computer packages, etc).

Common problems arise in relation to:

- *Where to do the research*: Choosing where to work is important. Locally based projects which can be carried out in your university or home town are usually logistically easiest. If you have more ambitious plans, particularly to carry out work abroad, remember to take into account potential language and cultural barriers, problems of access to materials, the cost of travel and accommodation, and the fact that it is more important to be well prepared because, unlike those who do their research in the local area, there is usually no chance to go back to your field area if you forget to collect important material. However, with determination and foresight, all these problems usually can be overcome. A limited amount of financial assistance may be available from various funding bodies. Libraries usually contain directories of grant-awarding bodies; if not, librarians are often quite knowledgeable about where such information might be found.

- *Scale/time constraints*: Be realistic in the scope of the project. Bear in mind that you must not only allow enough time to research the background to the project and collect primary data, but you must also allocate sufficient time to analyse the information and write up the findings. It often takes a lot longer to analyse and write up research material than you expect!

- *Access*: It is important to check out the availability of information and your ability to access key informants at an early stage. For example, if you spend time researching and designing a project to be based in an institution such as a school or prison, only to find out as you try to embark on your fieldwork that none of these local institutions will grant you access to them, then all your preparation time and effort (literature review, research design, etc) will have been wasted.

- *Ethics*: If your research involves the collection of confidential information about individuals, commercially sensitive data, or working with vulnerable groups (such as children, the elderly, individuals with mental health problems, and so on), you must be aware of the ethical dimensions of your work.

Figure A1   I used to hate writing...

- *Safety considerations*: As part of the planning of your fieldwork you should consider any possible risks associated with what you propose to do. Always remember to inform someone of your departure, route, activity, location and estimated time of your return (and, if possible, take a friend or relative with you). Carry a personal attack alarm. Never enter a home or office to conduct an interview if you feel unsafe, wary or in any way uneasy about the person or the place. Make sure you have permission before entering private property or private land.

■ 4. Writing

It is never too soon to begin writing up your dissertation or project. It always takes longer than you think! First, find a comfortable place where you find it easy to concentrate. For some people this might be a 'private' space such as a study bedroom, whereas other people work more effectively in communal spaces such as the library or a computer room.

Always draw up a plan of how you intend to structure your material and your argument, and which information will be included where, before you start writing. This will help you to develop your analysis of your findings and will also help you to avoid becoming waffly or repetitive when you start writing. Brainstorming several possible different structures on a large sheet of paper, or putting different points/data onto index cards which can then be shuffled and rearranged, are both useful techniques for helping you to look at the same material from different perspectives before settling on the most appropriate plan (see Figure A1).

Then just start writing. Some people like to work in a linear way, beginning with the introduction and following on with each section/chapter in numerical order. However, it is often easiest to begin by writing the part that you feel most familiar with or happy about. Indeed, the introduction is often best left until last because this must set up everything which follows, and it is often easier, therefore, to make the introduction fit the main body of the text than vice versa. If you have writer's block, do not do nothing. Get on with something simple that does not require too much thought, even if this is only writing the acknowledgements, typing up references or sorting out illustrations. This is never time wasted because it is something that will otherwise need to be done in the future. If all else fails, clean the house or do the ironing. At least this way, although you may not have made much academic progress, you can at least feel virtuous about something.

When you are writing always tell yourself you are only aiming to produce a draft; this helps to take the pressure off and sometimes you can surprise yourself with the first effort. If you cannot find the right words, do not succumb to the temptation to stare out of the window or to put the television on; instead hack out a rough outline of what you are trying to say, and then move on to the next section. This way you do not get demoralized because you have a sense that you are making progress and, if the next section flows more readily, it is often easier then to come back to the part which needs polishing into proper sentences. If you are really stuck, try explaining to someone else what you are trying to write about. Verbalizing an argument often makes it easier to see things clearly, and even if you still cannot make sense of your ideas, with any luck your friend might utter the magic words: 'So what you are trying to say is . . .'. Failing this, they can hopefully at least provide tea and sympathy.

Setting yourself targets and rewards can be an effective way of motivating yourself: if I write 500 words by the end of the afternoon then I can go to the gym/watch the movie/go out for a drink with friends.

As you are writing, remember to save your file regularly and to make backup copies on disk(s). Try to print out your project before the final deadline because there can be long print queues and university networked computers may crash if a lot of students are trying to print out their work simultaneously at the last minute.

Finally, do not forget the presentation: spell-check the text, number the pages, check that the references are included and correctly laid out, proofread the finished product. A neat and well-illustrated project can put the marker in a good frame of mind, whereas silly typographical and spelling errors can undermine the impression created by the work itself. Where appropriate, maps, graphs, diagrams and photographs should be used to illustrate the text and provide detailed information. Illustrations should be clear and simple and should be numbered so that they can be clearly referred to in the body of the text. Each illustration should have a caption title and the source should also be given.

■ Further Reading

- At the stage of choosing and planning your topic it would be useful to read Flowerdew, R. (1997) 'Finding previous work on the topic', in Flowerdew, R. and Martin, D. (eds) *Methods in Human Geography: a Guide for Students Doing a Research Project*, Longman, Harlow, and also to consult Kneale, P. (1999) *Study Skills for Geographers*, Arnold, London. The former book also contains a useful chapter on the importance of philosophy and methodology in research design: Graham, E. (1997) 'Philosophies underlying human geography research'.

- Helpful sources in relation to **qualitative methods** include Cook, I. and Crang, M. (1995) *Doing Ethnographies*, CATMOG Series No. 58: School of Environmental Sciences, UEA, Norwich; chapters on interviewing, participant observation, and analysis of texts in Flowerdew, R. and Martin, D. (eds) (1997) *Methods in Human Geography*, Longman: Harlow; the special issue of the journal *Area*, 1996, Vol. 28 on focus groups; May, T. (1993) *Social Research: Issues, Methods and Process*, Open University Press, Buckingham; Yin, R.K. (1994) *Case Study Research: Design and Methods*, Sage, Thousand Oaks; Feldman, M. (1995) *Strategies for Interpreting Qualitative Data*, Sage, Beverly Hills.

- For guidance on **quantitative techniques** see chapters on questionnaire design and sampling and analysing numerical and categorical data in Flowerdew, R. and Martin, D. (eds) (1997) *Methods in Human Geography*, Longman, Harlow; Bryman, A. and Cramer, D. (1996) *Quantitative Data Analysis with Minitab: A Guide for Social Scientists*, Routledge, London.

- An invaluable guide to writing is Becker, H.S. (1986) *Writing for Social Scientists: How to Start and Finish Your Thesis, Book or Article*, University of Chicago Press, Chicago. A useful general source for any project work is Parsons, T. and Knight, P.G. (1995) *How to Do Your Dissertation in Geography and Related Disciplines*, Chapman & Hall, London.

Appendix B: Glossary

- This contains brief/simplified definitions of the key terms used in this book.
- The first time that the key terms are referred to in this text they are highlighted in bold.
- The definitions are cross-referenced using capitals.

Agency: Literally, the ability to act. Commonly used to refer to the ability of people to make choices or decisions which shape their own lives. This notion of self-determination has formed an important part of the HUMANISTIC critique of STRUCTURALIST approaches to geography.

Capital accumulation: The use or investment of capital to reproduce more capital. This is the aim of, or driving force in, a CAPITALIST society. It results in patterns of uneven development.

Capitalism: A specific economic and social system in which there is a fundamental division between those who own the means of production (factories, machinery, etc) and those who have to sell their labour power through their ability to do paid work. This separation of society into two classes – capital and labour – is based on exploitative relationships in which capital gains more from the process of production, by appropriating surplus value, than labour. It is also characterized by competition between organizations in which profit (CAPITAL ACCUMULATION) is the main goal.

Citizenship: The relationship between individuals and a political body (i.e. the NATION STATE). Citizenship is usually conceptualized in terms of the rights/privileges individuals can expect in return for fulfilling certain obligations to the state.

Class: A system of social stratification based on people's economic position (specifically the social relations of property and work). Understandings and definitions of class are highly contested (see also UNDERCLASS).

Commodification: The processes through which people, ideas or things are converted into commodities that can be bought and sold. As such, it is a manifestation of capitalism. Such is the extensiveness of commodification that it is alleged by some writers to be contributing to the creation of a global culture (see GLOBALIZATION).

Corporeal/corporeality: A recognition of the importance of the body as crucial to understanding our lived experiences, in which the body is understood to be socially constructed rather than a biological given.

Cultural capital: The way people gain social status through cultural practices (particularly different forms of consumption or lifestyles) which enable them to demonstrate their taste and judgement. It was first used by the French sociologist Pierre Bourdieu (see Chapter 2).

Cultural turn: A trend in the late twentieth and early twenty-first centuries which has seen the social sciences and humanities increasingly focus on culture (and specifically the construction, negotiation and contestation of meanings). Linked to POSTMODERNIST philosophies.

Cyberspace: A term that comes from the science fiction writer William Gibson's novel *Neuromancer*. Gibson coined it to describe an alternative computer-generated world. 'Cyberspace' is now used more broadly by academics and in popular culture to refer to diverse forms of on-line space produced and experienced through information and communication technologies.

De-institutionalization: A process through which institutional spaces such as asylums are closed and replaced with various forms of alternative care in the home or 'community' (see Chapter 5).

Diaspora/diasporic: The scattering or dispersal of a population. Because TRANSNATIONAL linkages develop across diasporic communities, diaspora is also used as a theoretical concept to challenge fixed understandings of identity and place.

Discourse: Sets of connected ideas, meanings and practices through which we talk about or represent the world.

Dualism/istic: Where two factors (e.g. home/work, body/mind, nature/culture, private/public) are assumed to be distinct and mutually exclusive and to have incompatible characteristics.

Environmental determinism: An approach which suggests that people's behaviour and activities are controlled (or determined) by their environment.

Essentialism/essentialist: The belief that social differences (such as gender or race) are determined by biology and that bodies therefore have fixed properties or 'essences'.

Ethnicity: The social categorization of people on the basis of learnt cultural differences/lifestyle. Often confused with RACE.

Flâneur: Someone who strolls around the streets in the role of an onlooker, voyeuristically observing rather than participating in city life. Although this is often assumed to be a man, feminists have drawn attention to examples of the flâneuse. The concept of the flâneur is associated with the writings of the poet Charles Baudelaire.

Gender: The socially constructed characteristics attributed to each sex: masculinity and femininity. It is important to remember, however, that this term is contested and that the sex/gender distinction has been challenged (see Chapter 2).

Gentrification: A process through which traditional working-class areas have been taken over and 'done up' by middle-class people (mainly single people and dual-income households). Gentrification changes both the social and physical landscape. Property values are reordered, low-income residents are displaced by middle-class residents, and the built environment is remade as dilapidated buildings are transformed into luxury apartments and new shops and services move in to meet the needs and lifestyle of the gentrifiers.

Global city: A term used by Saskia Sassen to describe those cities (such as New York and Tokyo) which are at the hub of the world economy, co-ordinating and controlling flows of information, investment, etc.

Globalization: The way in which economic, political, social and cultural processes operate at a global scale, connecting people and places into a global system, albeit in different and unequal ways (see also TIME-SPACE COMPRESSION, TIME-SPACE CONVERGENCE). It is a highly contested term: some writers argue it is a product of contemporary technological change, while others claim it is nothing new; some argue that globalization is a homogenizing process which is threatening the distinctiveness of places and the significance of the nation state; others point out that all global flows have local geographies because they are actively adapted in different ways within local cultures.

Habitus: A term employed by the French sociologist Pierre Bourdieu to describe the class-oriented, unintentional ways we have of behaving which give away our class origins. For example, he suggests that everything from how we use our cutlery to the way we walk betrays our social location.

Hegemony/hegemonic: The power of a dominant group to persuade a subordinate group – through ideology and everyday practices/institutions rather than by force – to accept its moral, political and cultural values as the 'natural order'.

Homeland: The geographical space to which a particular nation or ethnic group feels attached. It is particularly associated with DIASPORIC communities, who often retain emotional links with their country of origin.

Humanistic: Relating to a theoretical approach to geography which is characterized by an emphasis on human AGENCY, consciousness and meanings and values. It developed in the 1970s partially as a critique of the spatial science of POSITIVISM.

Hybridity: The new forms produced by the positive combining or mixing of cultures.

Identity: A highly contested term. Early understandings of identities conceptualized them primarily in terms of social categories such as class and gender, in which identity was assumed to reflect a core or fixed sense of self. More contemporary theorizations understand identities as a REFLEXIVE project, emphasizing their multiple, fluid and unstable nature.

Ideology: A set of meanings, ideas or values which (re)produce relations of domination and subordination.

Imagined community: A term by employed by Benedict Anderson to describe the way individuals feel part of a nation and have an image of communion with other citizens, even though they will never know or meet all their fellow members of the nation. It has subsequently been used to describe collective identities at a range of other scales.

Individualization: According to Beck (1992), this can be understood as a shift in ways of thinking about how individuals relate to society. He argues that contemporary changes in the labour market, familial relations and class cultures are creating new life situations and biographical development patterns. The lifecourse is no longer organized around employment history, with the consequence that the possible pathways people can follow after school are becoming more diversified. Traditional agencies such as the nuclear family, school and church are no longer key agencies of social reproduction, channelling individuals into set roles, and social change is eroding traditional forms of knowledge and communication (e.g. expert knowledge). This destructuring of people's life situations is placing us all in a state of ambivalence. Faced with a proliferation of choices, our individual biographies are increasingly reflexive in that we can choose between different lifestyles, subcultures and identities.

Institutional racism: This includes processes, attitudes and behaviour which intentionally, or unintentionally (e.g. through collective failure or unwitting prejudice), disadvantage people from ethnic minority groups.

Marxism: A set of theories developed from the writing of Karl Marx, a nineteenth-century German philosopher. Marxist approaches to geography use these insights to examine geographies of CAPITALISM, challenging the processes that produce patterns of uneven development.

Moral panic: A term coined by the sociologist Stanley Cohen (1972: 9) to describe the way in which a 'condition, episode, person or group of persons emerges to become defined as a threat to societal values and interests'.

Nation state: A territorial unit, defined by political boundaries and governed by a range of institutions.

Neighbourhood community: Residential area in which households participate in common activities, offer each other varying degrees of support or aid, and have some sense of shared identity or belonging.

NIMBYISM: An acronym (Not-In-My-Back-Yard) for community groups that oppose various forms of change or development in their neighbourhood.

Objectivity: The assumption that knowledge is produced by individuals who can detach themselves from their own experiences, positions and values and therefore approach the object being researched in a neutral or disinterested way (see also SUBJECT/SUBJECTIVE).

Ontological security: 'Confidence or trust that the natural and social worlds are as they appear to be' (Giddens 1984: 375).

Other/othering: The 'Other' refers to the person that is different or opposite to the self. Othering is the process through which the other is often defined in relation to the self in negative ways; for example, woman is often constructed as other to man, black as the other of white, and so on.

Panopticon: A model prison designed in the nineteenth century by Jeremy Bentham. This had a circular design to ensure that all the inmates would be kept under constant surveillance from a central watch tower. Crucially, the design intended that the occupants of the cells would not be able to tell if, or when, they were actually being watched and so would exercise self-surveillance. The notion of 'imperfect panopticism' (Hannah 1997) has been used to describe contemporary processes whereby people are disciplined and controlled, for example through closed circuit television.

Patriarchy: Literally, 'rule of the father'. This term is used to describe a social system and practices whereby men dominate or oppress women. These unequal relations occur at a range of scales from the household to society as a whole.

Performance/performativity: The notion that identity is produced through the repetition of particular acts within a regulatory framework.

Positional good: A good that indicates the social status of its consumer.

Positionality: The way in which our own experiences, beliefs and social location affect the way we understand the world and go about researching it.

Positivism: A theoretical approach to human geography, characterized by the adoption of a scientific approach in which theories/models derived from observations are empirically verified through scientific methods to produce spatial laws. Positivism came to the fore in the 1950s and 1960s in what was known as the quantitative revolution.

Postmodernism: A theoretical approach to human geography which rejects the claims of grand theories or metanarratives. Instead it recognizes that all knowledge is partial, fluid and contingent and emphasizes a sensitivity to difference and an openness to a range of voices. Deconstruction is a postmodern method. Postmodernism is also a style, associated with a particular form of architecture and aesthetics.

Psychoanalytic theory: A theory of human subjectivity.

Queer: Originally a term of abuse for lesbians and gay men. However, it has now been reclaimed by sexual dissidents and is used in an ironic way to resist and subvert heterosexuality. Queer theory challenges both hetero-normativity and stable or fixed ways of defining gender and sexuality, instead emphasizing their diverse, fluid and fragmented nature.

Race: The social categorization of people on the basis of perceived physical characteristics such as skin colour (see also ETHNICITY). The notion of race is a social construction which is historically and spatially specific. Race is often thought of in terms of a black–white binary in which black is constructed as inferior to white. Such discourses, assumptions and practices are racist (see also INSTITUTIONAL RACISM).

Reflexivity: A process of reflection about who we are, what we know, and how we come to know it (see also POSITIONALITY).

Revanchist city: A term used by Neil Smith to describe the backlash or 'revenge' being exercised against marginal and disadvantaged groups by the US urban elite through exclusionary policies such as 'zero tolerance' of crime and disorder, anti-homelessness laws, the privatization of 'public spaces', etc.

Sexuality: Sexual desire/behaviour and, more broadly, the expression of sexual identities such as lesbian, gay, bisexual, heterosexual, etc (see also QUEER).

Social exclusion: Process through which some individuals/groups become marginalized from or unable to participate in 'normal' society (see also UNDERCLASS). It involves not only exclusion from the labour market but also financial exclusion by institutions such as banks and insurance companies, the inability to access the legal or political system, etc.

Social justice: The distribution of income and other forms of material benefits within society.

Structuralism: A theoretical approach to human geography which is characterized by a belief that, in order to understand the surface patterns of human behaviour, it is necessary to understand the structures underlying them which produce or shape human actions.

Subject/subjective: This refers to the individual human agent, including both physical embodiment and thought/emotional dimensions. Subjective research is that which acknowledges the personal judgements, experiences, tastes, values and so on of the researcher (see also OBJECTIVE).

Third space: This is produced through processes that challenge, displace or exceed binary divisions/oppositions. It enables new ways of thinking, and new practices and identities to emerge.

Time-space compression: The speeding up of the pace of life.

Time-space convergence: Developments in communications and transport which are effectively shrinking the world, bringing places closer together (e.g. in the 1700s it took four days to travel between London and Edinburgh, whereas today it takes about one and a half hours by plane).

Transnational identity: This refers to people who have social, cultural, political and economic links, and therefore a simultaneous sense of identification, with more than one nation.

Underclass: Those multiply deprived individuals whose experiences are characterized by intergenerational poverty, dependency on welfare, unstable employment/unemployment, low skills, poor access to education, and a high level of health problems.

Urban social movements: Popular protest groups which operate outside the conventional political party framework to challenge the state provision of services or the regulation of the environment.

Bibliography

Adler, S. and Brenner, J. (1992) 'Gender and space: lesbians and gay men in the city', *International Journal of Urban and Regional Research*, **16**, 24–34.

Agyeman, J. (1995) 'Environment, heritage and multi-culturalism', *Interpretation: A Journal of Heritage and Environmental Interpretation*, **1**, 5–6.

Agyeman, J. and Spooner, R. (1997) 'Ethnicity and the rural environment', in Cloke, P. and Little, J. (eds) *Contested Countryside: Otherness, Marginalisation and Rurality*, Routledge, London.

Ainley, R. (1998) 'Watching the detectors: control and Panopticon', in Ainley, R. (ed.) *New Frontiers of Space, Bodies and Gender*, Routledge, London.

Aitken, S. (1994) *Putting Children in Their Place*, Association of American Geographers, Washington, DC.

Aitken, S. and Herman, T. (1997) 'Gender, power and crib geography: transitional spaces and potential places', *Gender, Place and Culture*, **4**, 63–88.

Allan, G. and Crow, G. (eds) (1989) *Home and Family: Creating the Domestic Sphere*, Macmillan, Basingstoke.

Allen, J., Massey, D. and Pryke, M. (eds) (1999) *Unsettling Cities*, Open University Press, Buckingham.

Allison, L. (1986) 'On dirty things', *Political Geography Quarterly*, **5**, 241–51.

Andermahr, S. (1992) 'Subjects or citizens? Lesbians in the new Europe', in Ward, A., Gregory, J. and Yuval-Davis, N. (eds) *Women and Citizenship in Europe: Borders, Rights and Duties: Women's Differing Identities in a Europe of Contested Boundaries*, Trentham Books, London.

Anderson, B. (1983) *Imagined Communities: Reflections on the Origin and Spread of Nationalism*, Verso, London.

Anderson, J. (1995) 'The exaggerated death of the nation-state', in Anderson, J., Brook, C. and Cochrane, A. (eds) *A Global World?* The Open University Press, Milton Keynes.

Anderson, K. (1991) *Vancouver's Chinatown: Racial Discourse in Canada, 1875–1980*, McGill-Queen's University Press, Montreal.

Anderson, K. (1995) 'Culture and nature at the Adelaide zoo: at the frontiers of human geography', *Transactions of the Institute of British Geographers*, **20**, 275–94.

Anderson, K. (1997) 'A walk on the wild side: a critical geography of domestication', *Progress in Human Geography*, **21**, 463–85.

Anderson, K. (1998a) 'Sites of difference: Beyond a cultural politics of race polarity', in Fincher, R. and Jacobs, J.M. (eds) *Cities of Difference*, Guilford Press, London.

Anderson, K. (1998b) 'Animal domestication in geographic perspective', *Society and Animals*, **6**, 119–35.

Anzaldua, G. (1987) *Borderlands/La Frontera: The New Mesitza*, Aunt Lute, San Francisco.

Aoki, D. (1996) 'Sex and muscle: the female bodybuilder meets Lacan', *Body and Society*, **16**, 87–106.

Appadurai, A. (1990) 'Disjuncture and difference in the global cultural economy', *Theory, Culture and Society*, 7, 295–310.

Ashworth, G.J., White, P.E. and Winchester, H.P.M. (1988) 'The red-light district in the West European city: a neglected aspect of the urban landscape', *Geoforum*, 19, 201–12.

Athenaeum (1996) *The Business Lunch*, Athenaeum Club, Pall Mall, London, SW1 UK.

Barlow, J.P. (1990) 'Being in nothingness: virtual reality and the pioneers of cyberspace', *Mondo 2000*, **2**, 34–43.

Barlow, S. (1996) 'Declaration of the independence of cyberspace', *Cyber Rights Electronic List*, 8 February.

Barthes, R. (1981) 'Semiology and the urban', in Gottdiener, M. and Lagopoulos, A.P. (eds) *The City and the Sign: an introduction to urban semistics*, Columbia University Press, New York.

Bassett, C. and Wilbert, C. (1999) 'Where you want to go today (like it or not): leisure practices in cyberspace', in Crouch, D. (ed.) *Leisure/Tourism Geographies: Practices and Geographical Knowledge*, Routledge, London.

Baumgartner, M.P. (1988) *The Moral Order of a Suburb*, Oxford University Press, New York.

Baxter, J. and Western, M. (1998) 'Satisfaction with housework: examining the paradox', *Sociology*, **32**, 101–20.

Baym, M. (1995) 'The emergence of community in computer mediated communication', in Jones, S. (ed.) *Cybersociety: Computer Mediated Communication and Community*, Sage, London.

Beardsworth, A. and Keil, T. (1990) 'Putting the menu on the agenda', *Sociology*, 24, 139–51.

Beauregard, R.A. (1986) 'The chaos and complexity of gentrification', in Smith, N. and Williams, P. (eds) *Gentrification of the City*, Allen & Unwin, London.

Beck, U. (1992) *Risk Society: On the Way to Another Modernity*, Sage, London.

Bell, C. and Newby, H. (1971) *Community Studies: An Introduction to the Sociology of the Local Community London*, Allen & Unwin, London.

Bell, D. (1995a) 'Perverse dynamics, sexual citizenship and the transformation of intimacy', in Bell, D. and Valentine, G. (eds) *Mapping Desire: Geographies of Sexualities*, Routledge, London.

Bell, D. (1995b) 'Pleasure and danger: the paradoxical spaces of sexual citizenship', *Political Geography*, 14, 139–53.

Bell, D. (1997) 'Anti-idyll: rural horror', in Cloke, P. and Little, J. (eds) *Contested Countryside: Otherness, Marginalisation and Rurality*, Routledge, London.

Bell, D. (2000) 'Eroticising the rural', in Phillips, R., Watt, D. and Shuttleton, D. (eds) *De-Centring Sexualities: Politics, and Representation Beyond the Metropolis*, Routledge, London.

Bell, D. and Binnie, J. (1998) 'Theatres of Cruelty, rivers of desire: the erotics of the street', in Fyfe, N. (ed.) *Images of the Street*, Routledge, London.

Bell, D. and Valentine, G. (1995a) *Mapping Desire: Geographies of Sexualities*, Routledge, London.

Bell, D. and Valentine, G. (1995b) 'The sexed self: strategies of performance, sites of resistance', in Pile, S. and Thrift, N. (eds) *Mapping the Subject: Geographies of Cultural Transformation*, Routledge, London.

Bell, D. and Valentine, G. (1995c) 'Queer Country: rural lesbian and gay lives', *Journal of Rural Studies*, **11**, 113–22.

Bell, D. and Valentine, G. (1997) *Consuming Geographies: We Are Where We Eat*, Routledge, London.

Bell, D., Binnie, J., Cream, J. and Valentine, G. (1994) 'All hyped up and no place to go', *Gender, Place and Culture*, **1**, 31–47.

Bell, M. (1992) 'The fruit of difference: the rural–urban continuum as a system of identity', *Rural Sociology*, **57**, 65–82.

Bell, M. (1994) *Childerley: Nature and Morality in a Country Village*, University of Chicago Press, Chicago.

Benedikt, M. (ed.) (1991) *Cyberspace: First Steps*, MIT Press, Cambridge, MA.

Bennett, T. (1989) 'Factors related to participation in neighbourhood watch schemes', *British Journal of Criminology*, **29**, 207–18.

Bennett, T. (1992) 'Themes and variations in neighbourhood watch', in Evans, D.J., Fyfe, N.R. and Herbert, D.T. (eds) *Crime, Policing and Place: Essays in Environmental Criminology*, Routledge, London.

Bennington, G. (1994) 'Postal politics and the institution of nation', in Bhabha, H. (ed.) *The Location of Culture*, Routledge, London.

Bentham, J. (1995) *The Panopticon Writings* (ed. Miran Bozovic), Verso, London.

Bergman, E. (1979) 'The geography of the New York business lunch', *The Geographical Review*, **69**, 235–8.

Berlant, L. and Freeman, E. (1993) 'Queer nationality', in Warner, M. (ed.) *Fear of a Queer Planet: Queer Politics and Social Theory*, University of Minnesota Press, Minneapolis.

Berman, M. (1986) 'Take it to the streets: conflict and community in public space', *Dissent*, **33**, 476–85.

Bernstein, B. (1971) *Class, Codes and Control*, volume 1, Routledge & Kegan Paul, Andover, Hants.

Betten, L. (1993) 'Rights in the workplaces', in Waaldijk, K. and Clapham, A. (eds) *Homosexuality: A European Community Issue: Essays on Lesbian and Gay Rights in European Law and Policy*, Martinus Nijhoff, Dordrecht.

Bhabha, H. (1994) *The Location of Culture*, Routledge, London.

Bianchini, F. (1990) 'The crises of urban public social life in Britain: origins of the problem and possible responses', *Planning, Policy and Research*, **5**, 4–8.

Bianchini, F. (1995) 'Night cultures, night economies', *Planning Practice and Research*, **10**, 121–6.

Binnie, J. (1993) 'Invisible cities/hidden geographies: sexuality and the city', Paper presented at the Social Policy and the City Conference, Liverpool, July.

Binnie, J. (1994) 'Invisible European: sexual citizenship in the new Europe', Paper presented at the International Geographical Union, Prague, Czech Republic. Available from the author.

Binnie, J. (1995) 'Trading places: consumption, sexuality and the production of Queer space', in Bell, D. and Valentine, G. (eds) *Mapping Desire: Geographies of Sexualities*, Routledge, London.

Billig, M. (1995) *Banal Nationalism*, Sage, London.

Bishop, K.D. and Phillips, A.A.C. (1993) 'Seven steps to market – the development of the market-led approach to countryside conservation and recreation', *Journal of Rural Studies*, 9, 315–38.

Bishop, P. (1996) 'Off road: four-wheel drive and the sense of place', *Environment and Planning D: Society and Space*, 14, 257–71.

Blum, V. and Nast, H. (1996) 'Where's the difference? The heterosexualization of alterity in Henri Lefebvre and Jacques Lacan', *Environment and Planning D: Society and Space*, 14, 559–80.

Blunt, A. and Wills, J. (2000) *Dissident Geographies*, Prentice Hall, Harlow.

Bocock, R. (1993) *Consumption*, Routledge, London.

Boddy, T. (1992) 'Underground and overhead: building the analogous city', in Sorkin, M. (ed.) *Variations on a Theme Park: The New American City and the End of Public Space*, Hill and Wang, New York.

Boethius, U. (1995) 'Youth, the media and moral panics', in Fornas, J. and Bolin, G. (eds) *Youth Culture in Late Modernity*, Sage, London.

Bondi, L. (1991) 'Gender divisions and gentrification: a critique', *Transactions of the Institute of British Geographers*, 16, 190–8.

Bondi, L. (1997) 'In whose words? On gender identities, knowledge and writing practices', *Transactions of the Institute of British Geographers*, 22, 245–58.

Bonnett, A. (1996) 'The new primitives: identity, landscape and cultural appropriation in the mythopoetic men's movement', *Antipode*, 28, 273–91.

Bonnett, A. (2000) *White Identities: Historical and International Perspectives*, Longman, Harlow.

Bordo, S. (1993) *Unbearable Weight: Feminism, Western Culture and the Body*, University of California Press, Los Angeles.

Bourdieu, P. (1984) *Distinction: A Social Critique of the Judgement of Taste*, Routledge, London.

Bowlby, S.R., Foord, J., McDowell, L. and Momsen, J. (1982) 'Environment planning and feminist theory: a British perspective', *Environment and Planning A*, 14, 711–16.

Bowring, F. (1997) 'Communitarianism and morality: in search of the subject', *New Left Review*, 222, 93–113.

Boyer, C. (1993) 'The city of illusion: New York's public spaces', in Knox, P. (ed.) *The Restless Urban Landscape*, Prentice Hall, Englewood Cliffs, NJ.

Boyle, P., Halfacree, K. and Robinson, V. (1998) *Exploring Contemporary Migration*, Longman, Harlow.

Boys, J. (1984) 'Is there a feminist analysis of architecture?', *Built Environment*, 10, 25–34.

Boys, J. (1998) 'Beyond maps and metaphors? Rethinking the relationships between architecture and gender', in Ainley, R. (ed.) *New Frontiers of Space, Bodies and Gender*, Routledge, London.

Bridge, L. (1982) Review of Peach, C. *et al.* (eds) *Ethnic Segregation in Cities* and Jackson, P. and Smith, S. (eds) *Social Interaction and Ethnic Segregation*, *Race and Class*, **24**, 83–6.

Bridges, L. (1983) 'Policing the urban wasteland', *Race and Class*, **25**, 31–48.

Brown, M. (1995) 'Ironies of distance: an ongoing critique of the geographies of AIDS', *Environment and Planning D: Society and Space*, **13**, 159–83.

Brown, M. (1997a) 'Radical politics out of place? The curious case of ACT UP Vancouver', in Pile, S. and Keith, M. (eds) *Geographies of Resistance*, Routledge, London.

Brown, M. (1997b) *RePlacing Citizenship*, Guilford Press, London.

Brown, R.M. (1975) *Strain of Violence*, Oxford University Press, New York.

Buckingham, D. (1997) 'Electronic child abuse? Rethinking the media's effects on children', in Barker, M. and Petley, J. (eds) *Ill Effects: The Media/Violence Debate*, Routledge, London.

Bunce, M. (1994) *The Countryside Ideal: Anglo-American Images of Landscape*, Routledge, London.

Burgess, J. (1998) 'But is it worth taking the risk? How women negotiate access to urban woodland: a case study', in Ainley, R. (ed.) *New Frontiers of Space, Bodies and Gender*, Routledge, London.

Burgess, J., Harrison, C.M. and Limb, M. (1988) 'People, parks and the urban green: a study of popular meanings and values for open spaces in the city', *Urban Studies*, **25**, 455–73.

Butler, J. (1990) *Gender Trouble: Feminism and the Subversion of Identity*, Routledge, London.

Butler, J. (1993a) *Bodies That Matter: On the Discursive Limits of 'Sex'*, Routledge, London.

Butler, J. (1993b) 'Endangered/endangering: schematic racism and white paranoia', in Gooding-Williams, R. (ed.) *Reading Rodney King/Reading Urban Uprising*, Routledge, London.

Butler, R. (1994) 'Geography and vision-impaired and blind populations', *Transactions of the Institute of British Geographers*, **19**, 366–8.

Butler, R. (1998) 'Rehabilitating the images of disabled youths', in Skelton, T. and Valentine, G. (eds) *Cool Places: Geographies of Youth Cultures*, Routledge, London.

Butler, R. (1999) 'Double the trouble or twice the fun? Disabled bodies in the gay community', in Butler, R. and Parr, H. (eds) (1999) *Mind and Body Spaces: Geographies of Illness, Impairment and Disability*, Routledge, London.

Butler, R. and Bowlby, S. (1997) 'Bodies and spaces: an exploration of disabled people's experiences of public space', *Environment and Planning D: Society and Space*, **15**, 411–33.

Butler, R. and Parr, H. (eds) (1999) *Mind and Body Spaces: Geographies of Illness, Impairment and Disability*, Routledge, London.

Buttel, F. (1998) 'Nature's place in the technological transformation of agriculture. Some reflections on the recombinant BST controversy in the USA', *Environment and Planning A*, **30**, 1151–63.

Buttimer, A. (1971) 'Sociology and planning', *Town Planning Review*, **42**, 145–80.

Campbell, B. (1993) *Goliath*, Methuen, London.

Campbell, D. (1993) 'Big brother is here', *The Guardian*, 13 May, pt 2, pp. 2–3.

Carroll, R. (2000) 'On the long road to Britain', *The Guardian*, 27 March, p. 4.

Casey, C. (1995) *Work, Self and Society: After Industrialisation*, Routledge, London.

Castells, M. (1983) *The City and the Grassroots*, University of California Press, Berkeley, CA.

Castells, M. (1995) *The Network Society*, Blackwell, Oxford.

Cater, J. and Jones, T. (1989) *Social Geography: An Introduction to Contemporary Issues*, Edward Arnold, London.

Chamberlain, J. (1988) [1977] *On Our Own: Patient-Controlled Alternatives to the Mental Health Care System*, MIND, London.

Chambers, I. (1993) 'Narratives of nationalism: being "British" ', in Carter, E., Donald, J., Squires, J. (eds) *Space and Place: Theories of Identity and Location*, Lawrence & Wishart, London.

Chan, W.W.Y. (1999) 'Identity and home in the migratory experience of recent Hong Kong Chinese Canadian migrants', in Teather, E.K. (ed.) *Embodied Geographies: Spaces, Bodies and Rites of Passage*, Routledge, London.

Chaudhary, V. (1995) 'Property boom town where home dreams founder', *The Guardian*, 10 June, p. 11.

Chauncey, G. (1995) *Gay New York: The Making of the Gay World 1890–1940*, Flamingo, London.

Cheney, J. (1985) *Lesbian Land*, Word Weavers Press, Minneapolis, MN.

Chouniard, V. (1994) 'Geography, law and legal struggles: which ways ahead?', *Progress in Human Geography*, **18**, 415–40.

Chouniard, V. (1997) 'Making space for disabling differences: challenges to ableist geographies', *Environment and Planning D: Society and Space*, **15**, 379–87.

Chouniard, V. (1999) 'Body politics: disabled women's activism in Canada and beyond', in Butler, R. and Parr, H. (eds) (1999) *Mind and Body Spaces: Geographies of Illness, Impairment and Disability*, Routledge, London.

Clark, M. (2000) 'When love comes in . . .', *The Guardian*, 11 May, p. 8.

Clifford, J. (1988) *The Predicament of Culture: Twentieth Century Ethnography, Literature and Art*, Harvard University Press, Cambridge, MA.

Clifford, J. (1992) 'Traveling cultures', in Grossberg, L., Nelson, C. and Treichter, P. (eds) *Cultural Studies*, Routledge, London.

Clifford, J. (1994) 'Diasporas', *Cultural Anthropology*, **9**, 302–28.

Cline, S. (1990) *Just Desserts*, Andre Deutsch, London.

Cloke, P. (1993) 'The countryside as commodity: new spaces for rural leisure', in Glyptis, S. (ed.) *Leisure and the Environment*, Belhaven, London.

Cloke, P. (1997) 'Poor country: marginalisation, poverty and rurality', in Cloke, P. and Little, J. (eds) *Contested Countryside: Otherness, Marginalisation and Rurality*, Routledge, London.

Cloke, P. (1999) 'The country', in Cloke, P., Crang, P. and Goodwin, M. (eds) *Introducing Human Geography*, Arnold, London.

Cloke, P. and Little, J. (1997) *Contested Countryside: Otherness, Marginalisation and Rurality*, Routledge, London.

Cloke, P., Milbourne, P. and Thomas, C. (1994) *Lifestyles in Rural England*, Rural Development Commission, London.

Cloke, P., Philo, C. and Sadler, D. (1991) *Approaching Human Geography*, Paul Chapman, London.

Cloke, P. and Milbourne, P. (1992) 'Deprivation and lifestyles in rural Wales', *Journal of Rural Studies*, 8, 321–33.

Cloke, P., Milbourne, P. and Thomas, C. (1996) 'The English National Forest: local reactions to plans for renegotiated nature–society relations in the countryside', *Transactions of the Institute of British Geographers*, 21, 552–71.

Cloke, P. and Perkins, H. (1998) 'Cracking the canyon with the awesome foursome: representations of adventure tourism in New Zealand', *Environment and Planning D: Society and Space*, 16, 185–218.

Cloke, P. and Thrift, N. (1987) 'Intra-class conflict in rural areas', *Journal of Rural Studies*, 3, 321–33.

Cloke, P. and Thrift, N. (1990) 'Class and change in rural Britain', in Marsden, T., Lowe, P. and Whatmore, S. (eds) *Rural Restructuring: Global Processes and Their Responses*, Fulton, London.

Coakley, J. and White, A. (1992) 'Making decisions: gender and sport participation among British adolescents', *Sociology of Sport Journal*, 9, 20–35.

Cockburn, C. (1983) *Brothers: Male Dominance and Technological Change*, Pluto Press, London.

Cockburn, C. (1985) *Machinery of Dominance*, Pluto Press, London.

Cohen, S. (1972) *Folk Devils and Moral Panics: The Creation of Mods and Rockers*, MacGibbon and Kee, London.

Coleman, A. (1990) *Utopia on Trial*, Hilary Shipman, London.

Coles, J. (1993) 'Private security versus the short arm of the law', *The Guardian*, 6 August, p. 18.

Colley, L. (1992) *Britons: Forging the Nation, 1707–1837*, Yale University Press, New Haven, CT.

Connell, R.W. (1983) *Which Way is Up? Essays on Sex, Class and Culture*, Allen & Unwin, Sydney.

Connell, R.W. (1995) *Masculinities*, Polity Press, Cambridge.

Cook, I. and Crang, P. (1996) 'The world on a plate: culinary culture, displacement and geographical knowledges', *Journal of Material Culture*, 1, 131–54.

Cordell, H. and Hendee, J. (1982) *Renewable Resources Recreation in the United States*, American Forestry Association, Washington, DC.

Cornwall, J. (1984) *Hard-Earned Lives: Accounts of Health and Illness from East London*, Tavistock, London.

Corrigan, P. (1979) *Schooling the Smash Street Kids*, Macmillan, London.

Cortese, A. (1995) 'The rise, hegemony and decline of the Chicago School of Sociology in the 1920s and 30s', *Sociological Review*, **44**, 474–94.

Corviono, J. (ed.) (1997) *Same Sex: Debating the Ethics, Science and Culture of Homosexuality*, Rowman and Littlefield, Lanham, MD.

Cottingham, C. (1982) *Race, Poverty and the Urban Underclass*, Heath, Lexington, DC.

Crang, M. and Thrift, N. (eds) (2000) *Thinking Space*, Routledge, London.

Crang, P. (1994) 'It's showtime: on the workplace geographies of display in a restaurant in Southeast England', *Environment and Planning D: Society and Space*, **12**, 675–704.

Cream, J. (1995) 'Re-solving riddles: the sexed body', in Bell, D. and Valentine, G. (eds) *Mapping Desire: Geographies of Sexualities*, Routledge, London.

Creed, G. and Chang, B. (1997) 'Recognising rusticity: identity and the power of place', in Ching, B. and Creed, G. (eds) *Knowing Your Place: Rural Identity and Cultural Hierarchy*, Routledge, New York.

Cresswell, T. (1992) 'The crucial "where of graffiti": a geographical analysis of reactions to graffiti in New York', *Environment and Planning D: Society and Space*, **10**, 329–44.

Cresswell, T. (1996) *In Place/Out of Place: Geography, Ideology and Transgression*, University of Minnesota Press, Minneapolis, MN.

Crouch, D. (ed.) (1999) *Leisure/Tourism Geographies: Practices and Geographical Knowledge*, Routledge, London.

Crow, L. (1996) 'Including all of our lives: renewing the social model of disability', in Morris, J. (ed.) *Encounters with Strangers: Feminism and Disability*, The Women's Press, London.

Curry, D. (1993) 'Decorating the body politic', *New Formations*, **19**, 69–82.

Dalby, S. and McKenzie, F. (1997) 'Reconceptualising local community: environment, identity and threat', *Area*, **29**, 99–108.

Dalla Costa, M. and James, S. (1975) *The Power of Women and the Subversion of the Community*, Falling Wall Press, Bristol.

Daniels, S. (1993) *Fields of Vision: Landscape, Imagery and National Identity in England and the United States*, Polity Press, Cambridge.

D'Augelli, A. and Hart, M. (1987) 'Gay women, men and families in rural settings: towards the development of helping communities', *American Journal and Community Psychology*, **15**, 79–93.

Davidoff, L. and Hall, C. (1987) *Family Fortunes: Men and Women of the English Middle Class 1750–1850*, Hutchinson, London.

Davidoff, L., L'Esperance, J. and Newby, H. (1979) 'Landscapes with figures: home and community in English society', in Mitchell, J. and Oakley, A. (eds) *The Rights and Wrongs of Women*, Penguin, Harmondsworth.

Davidson, R. (1982) *Tracks*, Granada, St Albans.

Davie, M.R. (1937) 'The pattern of urban growth', in Murdock, G.P. (ed.) *Studies in the Science of Society*, New Haven Press, New Haven, CT.

Davies, W.K.D. and Herbert, D. (1993) *Communities Within Cities: An Urban Social Geography*, Belhaven Press, London.

Davis, J. and Ridge, T. (1997) *Same Scenery, Different Lifestyle: Rural Children on Low Income*, The Children's Society, London.

Davis, M. (1990) *City of Quartz: Excavating the Future in Los Angeles*, Verso, London.

Davis, T. (1995) 'The diversity of Queer politics and the redefinition of sexual identity and community in urban spaces', in Bell, D. and Valentine, G. (eds) *Mapping Desire: Geographies of Youth Cultures*, Routledge, London.

Dear, M. (1992) 'Understanding and overcoming the NIMBY syndrome', *Journal of the American Planning Association*, **58**, 286–300.

Dear, M. and Gleeson, B. (1991) 'Community attitudes towards the homeless', *Urban Geography*, **12**, 155–76.

Dear, M. and Taylor, S.M. (1982) *Not On Our Street: Community Attitudes to Mental Health Care*, Pion, London.

Dear, M. and Wolch, J. (1987) *Landscapes of Despair from Deinstitutionalisation to Homelessness*, Princeton University, Princeton, NJ.

Dear, M., Wilton, R., Gaber, S.L. and Takahashi, L. (1997) 'Seeing people differently: the sociospatial construction of disability', *Environment and Planning D: Society and Space*, **15**, 455–80.

De Jong, G. (1983) 'Defining and implementing the independent living concept', in Crewe, N.M. and Zola, I.K. (eds) *Independent Living for Physically Disabled People*, Jossey-Bass, San Francisco, CA, pp. 4–27.

Delph-Janiurek, T. (1999) 'Sounding gender(ed): vocal performances in English university teaching spaces', *Gender, Place and Culture*, **6**, 137–54.

D'Emilio, J. (1992) *Making Trouble: Essays on Gay History, Politics and University*, Routledge, New York.

Dennis, N., Henriques, F. and Slaughter, C. (1956) *Coal is Our Life*, Tavistock, London.

Desforges, L. (1998) ' "Checking out the planet": global representations/local identities and youth travel', in Skelton, T. and Valentine, G. (eds) *Cool Places: Geographies of Youth Cultures*, Routledge, London.

Dicken, P. (1992) *Global Shift: The Internationalisation of Economic Activity*, Paul Chapman, London.

Dobash, R.E. and Dobash, R.R. (1980) *Violence Against Wives*, Open Books, Shepton Mallet.

Dobash, R.E. and Dobash, R.R. (1992) *Women, Violence and Social Change*, Routledge, London.

Donald, J. (1993) 'How English is it? Popular literature and national culture', in Carter, E., Donald, J. and Squires, J. (eds) *Space and Place: Theories of Identity and Location*, Lawrence & Wishart, London.

Dorn, M. (1998) 'Beyond nomadism: the travel narratives of a "cripple" ', in Nast, H. and Pile, S. (eds) *Places Through the Body*, Routledge, London.

Dorn, M. and Laws, G. (1994) 'Social theory, body politics and medical geography', *Professional Geographer*, **46**, 106–10.

Douglas, M. (1978) [1966] *Purity and Danger*, Routledge & Kegan Paul, London.

Downes, D. (1988) *Contrasts in Tolerance: Post-war Penal Policy in the Netherlands and England and Wales*, Clarendon Press, Oxford.

Driver, F. (1985) 'Power, space and the body: a critical assessment of Foucault's *Discipline and Punish*', *Environment and Planning D: Society and Space*, **3**, 301–24.

Driver, F. (1993) *Power and Pauperism: The Workhouse System 1834–1884*, Cambridge University Press, Cambridge.

Driver, F. (1995) 'Submerged identities: familiar and unfamiliar histories', *Transactions of the Institute of British Geographers*, **20**, 410–13.

Drumm, T.L. (1994) 'The new enclosures: racism in the normalised community', in Gooding-Williams, R. (ed.) *Reading Rodney King/Reading Urban Uprising*, Routledge, London.

Du Gay, P. (1996) *Consumption and Identity at Work*, Sage, London.

Dumm, T.L. (1994) 'The new enclosures: racism in the normalised community', in Gooding-Williams, R. (ed.) *Reading Rodney King/Reading Urban Uprising*, Routledge, London.

Duncan, N. (ed.) (1996) *Bodyspace: Destabilizing Geographies of Gender and Sexuality*, Routledge, London.

Dwyer, C. (1997) 'Negotiating differences: questions of identity for young British Muslim women', Paper presented at *Women in the Asia-Pacific region: persons, powers and politics*, Singapore, 11–13 August. Copy available from the author.

Dwyer, C. (1999a) 'Contradictions of community: questions of identity for young British Muslim women', *Environment and Planning A*, **31**, 53–68.

Dwyer, C. (1999b) 'Migrations and diasporas', in Cloke, P., Crang, P. and Goodwin, M. (eds) *Introducing Human Geography*, Arnold, London.

Dyck, I. (1990) 'Space, time, and renegotiating motherhood: an exploration of the domestic workplace', *Environment and Planning D: Society and Space*, **8**, 459–83.

Dyck, I. (1995) 'Hidden geographies: the changing lifeworlds of women with multiple sclerosis', *Social Science and Medicine*, **40**, 307–20.

Dyck, I. (1999) 'Body troubles: women, the workplace and negotiations of a disabled identity', in Butler, R. and Parr, H. (eds) *Mind and Body Spaces: Geographies of Illness, Impairment and Disability*, Routledge, London.

Edington, J. and Edington, M. (1986) *Ecology, Recreation and Tourism*, Cambridge University Press, Cambridge.

Edwards, D. (1987) 'Provoking her own demise: from common assault to homicide', in Hanmer, J. and Maynard, M. (eds) *Women, Violence and Social Control*, Macmillan, Basingstoke.

Egerton, J. (1990) 'Out but not down: lesbians' experiences of housing', *Feminist Review*, **36**, 75–88.

Ekins, R. and King, D. (1999) 'Towards a sociology of transgendered bodies', *The Sociological Review*, **47**, 580–602.

Elias, N. (1978/1982) [orig. 1939] *The Civilising Process, Vol I, The History of Manners* and *Vol II, State Formation and Civilisation*, Basil Blackwell, Oxford.

Elkin, T., McLaren, D. and Hillman, M. (1991) *Reviving the City*, Friends of the Earth, London.

Elwood, S. (2000) 'Lesbian living spaces: multiple meanings of home', *Journal of Lesbian Studies*, 4, 11–28.

Emmer, P. (1993) 'Intercontinental migration as a world historical process', *European Review*, 1, 67–74.

England, K. (1991) 'Gender relations and the spatial structure of the city', *Geoforum*, 22, 135–47.

England, K. (1997) *Who Will Mind the Baby?* Routledge, London.

Enloe, C. (1989) *Bananas, Beaches and Bases: Making Feminist Sense of International Politics*, Pandora, London.

Epstein, D. and Johnston, R. (1994) 'On the straight and narrow: the heterosexual presumption, homophobia and schools', in Epstein, D. (ed.) *Challenging Lesbian and Gay Inequalities in Education*, Open University Press, Buckingham.

Epstein, S. (1987) 'Gay politics, ethnic identity: the limits of social constructionism', *Socialist Review*, August, 9–54.

Erikson, T. (1997) 'The nation as a human being – a metaphor in a mid-life crisis? Notes on the imminent collapse of Norwegian national identity', in Olwig, K.F. and Hastrup, K. (eds) *Siting Culture: the Shifting Anthropological Object*, Routledge, London.

Ettore, E. (1978) 'Women, urban social movements and the lesbian ghetto', *International Journal of Urban and Regional Research*, 2, 499–520.

Etzioni, A. (1991) *Community: rights, responsibilities and the Communitarian Agenda*.

Etzioni, A. (1993) *The Spirit of Community: The Reinvention of American Society*, Simon & Schuster, New York.

Etzioni, A. (1995) *The Spirit of Community: Rights, Responsibilities and the Communitarian Agenda*, Fontana, London.

Evans, D.T. (1993) *Sexual Citizenship: The Material Construction of Sexualities*, Routledge, London.

Faderman, L. (1991) *Odd Girls and Twilight Lovers: a History of Lesbian Life in Twentieth Century America*, Penguin, Harmondsworth.

Falk, R. (1994) 'The making of global citizenship', in van Steenbergen, B. (ed.) *The Condition of Citizenship*, Sage, London.

Falk, P. and Campbell, C. (1997) 'Introduction', in Falk, P. and Campbell, C. (eds) *The Shopping Experience*, Sage, London.

Featherstone, M. (1991) *Consumer Culture and Postmodernism*, Sage, London.

Featherstone, M. (1998) 'The *Flâneur*, the city and virtual public life', *Urban Studies*, 35, 909–25.

Featherstone, M. (1999) 'Body modification: an introduction', *Body and Society*, 5, 1–14.

Featherstone, M. and Burrows, R. (eds) (1995) *Cyberspace, Cyberbodies and Cyberpunk: Cultures of Technological Embodiment*, Sage, London.

Featherstone, M., Hepworth, M. and Turner, B.S. (eds) (1991) *The Body*, Sage, London.

Featherstone, M. and Wernick, A. (1995) *Images of Ageing: Cultural Representations of Later Life*, Routledge, London.

Fielding, S. (2000) 'Walk on the left! Children's geographies and the primary school', in Holloway, S. and Valentine, G. (eds) *Children's Geographies: Playing, Living, Learning*, Routledge, London.

Fincher, R. and Jacobs, J.M. (eds) (1998) *Cities of Difference*, Guilford Press, London.

Fine, M., Weis, L., Addleston, J. and Mazuza, J. (1997) '(In)secure times: constructing white working class masculinities in the late twentieth century', *Gender and Society*, **11**, 52–68.

Finkelstein, J. (1989) *Dining Out: a Sociology of Modern Manners*, Polity Press, Cambridge.

Firey, W. (1945) 'Sentiment and symbolism as ecological variables', *American Sociological Review*, **10**, 140–8.

Firey, W. (1947) *Land Use in Central Boston*, MIT Press, Cambridge, MA.

Flowerdew, R. (ed.) (1982) *Institutions and Geographical Patterns*, Croom Helm, London.

Foucault, M. (1977) *Discipline and Punish: The Birth of the Prison*, Allen Lane, London.

Foucault, M. (1978) *The History of Sexuality, Vol. 1: An Introduction*, Random House, New York (trans. Robert Hurley).

Foucault, M. (1980) 'Body/Power', in Gordon, C. (ed.) *Michel Foucault: Power/Knowledge*, Harvester, Brighton.

Frankenberg, R. (1966) *Communities in Britain: Social Life in Town and Country*, Penguin, Harmondsworth.

Frankenberg, R. (1993a) 'Growing up white: feminism, racism and the social geography of childhood', *Feminist Review*, **45**, 51–4.

Frankenberg, R. (1993b) *The Social Construction of Whiteness: White Women, Race Matters*, Routledge, London.

Friedan, B. (1963) *The Feminine Mystique*, Pelican, Harmondsworth.

Fuss, D. (1990) *Essentially Speaking: Feminism, Nature and Difference*, Routledge, London.

Fyfe, N. (1995) 'Controlling the local spaces of democracy and liberty? 1994 criminal justice legislation', *Urban Geography*, **16**, 192–7.

Fyfe, N. and Bannister, J. (1996) 'City watching: closed-circuit television surveillance in public spaces', *Area*, **28**, 37–47.

Fyfe, N. and Bannister, J. (1998) ' "The eyes upon the street": closed-circuit television, surveillance and the city', in Fyfe, N. (ed.) *Images of the Street*, Routledge, London.

Gagen, E. (2000) 'Playing the part: performing gender in America's playgrounds', in Holloway, S. and Valentine, G. (eds) *Children's Geographies: Playing, Living, Learning*, Routledge, London.

Gans, H. (1962) *The Urban Villagers*, Free Press, New York.

Garner, R. (1993) *Animals, Politics and Morality*, Manchester University Press, Manchester.

Gathorne-Hardy, F. (1999) 'Accommodating difference: social justice, disability and the design of affordable housing', in Butler, R. and Parr, H. (eds) *Mind and Body Spaces: Geographies of Illness, Impairment and Disability*, Routledge, London.

Gellner, E. (1983) *Nations and Nationalism*, Blackwell, Oxford.

Genovese, R.G. (1981) 'A women's self-help network as a response to service needs in the suburbs', in Stimpson, C.R., Dixler, E., Nelson, M. and Yatrakis, K.B. (eds) *Women and the American City*, University of Chicago Press, Chicago.

Gibson, K. (1991) 'Company towns and class processes', *Environment and Planning D: Society and Space*, **10**, 609–50.

Gilbert, M. (1996) 'On space, sex and being stalked', *Women and Performance: a Journal of Feminist Theory*, **9**, 125–50.

Gillespie, M. (1995) *Television, Ethnicity and Cultural Change*, Routledge, London.

Gilligan, C. (1982) *In a Different Voice*, Harvard University Press, Cambridge, MA.

Gilroy, P. (1987) *There Ain't No Black in the Union Jack: The Cultural Politics of Race and Nation*, Hutchinson, London.

Gilroy, P. (1990) 'One nation under a groove: the cultural politics of "race" and racism in Britain', in Goldberg, D.T. *Anatomy of Racism*, University of Minnesota Press, Minneapolis.

Gilroy, P. (1994) *Black Atlantic*, Verso, London.

Gleeson, B. (1996) 'A geography for disabled people?', *Transactions of the Institute of British Geographers*, **59**, 55–76.

Gleeson, B. (1998) 'Justice and the disabling city', in Fincher, R. and Jacobs, J. (eds) *Cities of Difference*, Guilford Press, New York.

Gleeson, B. (1999) *Geographies of Disability*, Routledge, London.

Glenn, J. (1995) 'Americana', *The Observer*, preview section, 31 September, p. 46.

Goffman, E. (1961) *Asylums: Essays on the Social Situation of Mental Patients and Other Inmates*, Penguin, Harmondsworth.

Goheen, P.G. (1998) 'Public space and the geography of the modern city', *Progress in Human Geography*, **22**, 479–96.

Gold, M. (1999) *Animal Century*, Jon Carpenter Publishing, London.

Goodman, D. and Redclift, M. (1991) *Refashioning Nature*, Routledge, London.

Goodman, D. and Watts, M. (1994) 'Reconfiguring the rural or fording the divide: capitalist restructuring and the global agrofood system', *Journal of Peasant Studies*, **22**, 1–49.

Goodman, D. and Wilkinson, J. (1987) *From Farming to Biotechnology*, Basil Blackwell, Oxford.

Gordon, T., Holland, J. and Lahelma, E. (1998) 'Moving bodies/still bodies: embodiment and agency in schools', Paper presented at the British Sociological Association Annual Conference, University of Edinburgh, 6–9 April.

Goss, J. (1992) 'Modernity and post-modernity in the retail landscape', in Anderson, K. and Gale, F. (1992) *Inventing Places: Studies in Cultural Geography*, Longman, Melbourne, Australia.

Goss, J. (1993) 'The "magic of the mall": an analysis of form, function and meaning in the contemporary retail built environment', *Annals of the Association of American Geographers*, **83**, 18–47.

Graham, S. (1998) 'Spaces of surveillant simulation: new technologies, digital representa-
    tions and material geographies', *Environment and Planning D: Society and Space*,
    16, 483–504.
Green, S. (1991) 'Making transgressions: the use of style in a women-only community
    in London', *Cambridge Anthropology*, **15**, 71–87.
Greene, K. (1997) 'Fear and loathing in Mississippi: the attack on Camp Sister Spirit',
    Women and Politics, **17**, 17–39.
Gregory, D. (1994) *Geographical Imaginations*, Blackwell, Oxford.
Gregson, N. and Lowe, M. (1995) ' "Home"-making: on the spatiality of daily social
    reproduction in contemporary middle class Britain', *Transactions of the Institute of
    British Geographers*, **20**, 224–35.
Gregson, N. and Robinson, F. (1992) 'The "underclass": a class apart?' *Critical Social
    Policy*, 38–51.
Griffin, C. (1985) *Typical Girls?* Routledge and Kegan Paul, London.
Griffin, S. (1978) *Women and Nature: The Roaring Inside Her*, Harper and Row, New
    York.
Griffiths, H., Poulter, I. and Sibley, D. (2000) 'Feral cats in the city', in Philo, C. and
    Wilbert, C. (eds) *Animal Spaces, Beastly Places*, Routledge, London.
Grosz, E. (1989) *Sexual Subversions: Three French Feminists*, Allen & Unwin, Sydney.
Grosz, E. (1993) 'Bodies-cities', in Colomina, B. (ed.) *Sexuality and Space*, Princeton
    Architectural Press, New York.
Grosz, E. (1994) *Volatile Bodies: Towards a Corporeal Feminism*, Indiana University
    Press, Bloomington and Indianapolis.

Hague, G. (1998) 'Domestic violence policy in the 1990s', in Watson, S. and Doyal, L.
    (eds) *Engendering Social Policy*, Open University Press, Buckingham.
Hahn, H. (1985) 'Experiencing the environment in a wheelchair', unpublished paper,
    Dept. of Political Science, University of Southern California, Los Angeles, CA.
Hahn, H. (1986) 'Disability and the urban environment: a perspective on Los Angeles',
    Environment and Planning D: Society and Space, **4**, 272–88.
Halfacree, K. (1996) 'Out of place in the country: travellers and the "rural idyll" ', *Antipode*,
    28, 42–72.
Halford, S. and Savage, M. (1998) 'Rethinking restructuring: embodiment, agency and
    identity in organisational change', in Lee, R. and Wills, J. (eds) *Geographies of Eco-
    nomies*, Arnold, London.
Hall, E. (1999) 'Workspaces: refiguring the disability–employment debate', in Butler, R.
    and Parr, H. (eds) *Mind and Body Spaces: Geographies of Illness, Impairment and
    Disability*, Routledge, London.
Hall, S. (1995) 'New cultures for old', in Jess, P. and Massey, D. (eds) *A Place in the
    World? Places, Cultures and Globalisation*, Open University Press, London.
Hamnett, C. (1984) 'Gentrification and urban location theory: a review and assessment',
    in Herbert, D.T. and Johnston, R.J. (eds) *Geography and the Urban Environment,
    volume 6: Progress in Research and Applications*, John Wiley, New York.

Hamnett, C. (1991) 'The blind men and the elephant: the explanation of gentrification', *Transactions of the Institute of British Geographers*, **16**, 173–89.

Hamnett, C. (1999) 'The city', in Cloke, P., Crang, P. and Goodwin, M. (eds) *Introducing Human Geography*, Arnold, London.

Hannah, M. (1997) 'Imperfect Panopticism: envisioning the construction of normal lives', in Benko, G. and Strohmayer, U. (eds) *Space and Social Theory: Interpreting Modernity and Postmodernity*, Blackwell, Oxford.

Hanson, S. and Pratt, G. (1988) 'Reconceptualising the links between home and work in urban geography', *Economic Geography*, **64**, 299–321.

Haraway, D. (1990) [1985] 'A manifesto for cyborgs: science, technology, and socialist feminism in the 1980s', in Nicholson, L. (ed.) *Feminism/Postmodernism*, Routledge, London. First published in *Socialist Review*, 80, 1985.

Haraway, D. (1991) *Simians, Cyborgs and Women: The Reinvention of Nature*, Free Association Books, London.

Haraway, D. (1997) *Modest_Witness@Second_Millennium.FemaleMan^(C)_Meets_OncoMouse^(TM)*. Routledge, London.

Harrison, C. and Burgess, J. (1994) 'Social constructions of nature: a case study of conflicts over the development of Rainham Marshes'. *Transactions of the Institute of British Geographers*, **19**, 291–310.

Harrison, C., Limb, M. and Burgess, J. (1987) 'Nature in the city: popular values for a living world', *Journal of Environmental Management*, **25**, 347–62.

Harper, S. (1987) 'The British rural community: an overview of perspectives', *Journal of Rural Studies*, **3**, 309–19.

Harper, S. and Laws, G. (1995) 'Rethinking the geography of ageing', *Progress in Human Geography*, **19**, 2: 199–221.

Hart, A. (1995) '(Re)constructing a Spanish red-light district: prostitution, space and power', in Bell, D. and Valentine, G. (eds) *Mapping Desire: Geographies of Sexualities*, Routledge, London.

Hart, A. (1998) *Buying and Selling Power: Anthropological Reflections on Prostitution in Spain*, Westview Press, Oxford.

Hart, R. (1979) *Children's Experience of Place*, Irvington, New York.

Hartley, J. (1992) *The Politics of Pictures: the creation of the public in the age of the popular media*, Routledge, London.

Harvey, D. (1973) *Social Justice and the City*, Johns Hopkins University Press, Baltimore, MD.

Harvey, D. (1989) *The Condition of Postmodernity: An Inquiry into the Origins of Cultural Change*, Butterworth, Cambridge.

Hayden, D. (1980) 'What would a non-sexist city be like? Speculations on housing, urban design and human work', *SIGNS: Journal of Women in Culture and Society*, **5**, 170–87.

Hayden, D. (1984) *Redesigning the American Dream: The Future of Housing, Work and Family Life*, Norton, New York.

Haywood, C. and Mac An Ghaill, M. (1995) 'The sexual politics of the curriculum: contesting values', *International Studies in Sociology of Education*, **5**, 221–36.

Hearn, J., Sheppard, D.L., Tancred-Sheriff, P. and Burrell, G. (1989) *The Sexuality of Organization*, Sage, London.

Heelas, P. (1996) *The New Age Movement*, Blackwell, Oxford.

Held, D. (1995) *Democracy and the Global Order: From the Modern State to Cosmopolitan Governance*, Stanford University Press, Stanford, CA.

Hengst, H. (1997) 'Negotiating "us" and "them": children's constructions of collective identity', *Childhood*, 4, 43–62.

Herbert, S. (1998) 'Policing contested space: on patrol at Smiley and Hauser', in Fyfe, N. (ed.) *Images of the Street*, Routledge, London.

Herek, G. and Berrill, K. (1992) *Hate Crimes: Confronting Violence Against Lesbians and Gay Men*, Sage, London.

Herrnstein, R. and Murray, C. (1994) *The Bell Curve: Intelligence and Class Structure in American Life*, Free Press, New York.

Hetherington, K. (1992) 'Stonehenge and its festival: spaces of consumption', in Shields, R. (ed.) *Lifestyle Shopping*, Routledge, London.

Hetherington, K. (1996) 'The utopics of social ordering – Stonehenge as a museum without walls', in MacDonald, S. and Fyfe, G. (eds) *Theorising Museums*. Sociological Review Monograph, Blackwell, Oxford.

Hey, V. (1997) *The Company She Keeps: an Ethnography of Girls' Friendships*, Open University Press, Buckingham.

Hill Collins, P. (1990) *Black Feminist Thought: Knowledge, Consciousness and the Politics of Empowerment*, Harper Collins, London.

Hillery, G. (1955) 'Definitions of community: areas of agreement', *Rural Sociology*, 20, 111–23.

Hillier, L., Harrison, L. and Bowditch, K. (1999) 'Neverending Love' and 'Blowing your load: the meanings of sex to rural youth', *Sexualities*, 2, 69–88.

Hinchcliffe, S. (1999) 'Cities and natures: intimate strangers', in Allen, J., Massey, D. and Pryke, M. (eds) *Unsettling Cities*, Routledge, London.

Hobsbawm, E.J. (1990) *Nations and Nationalism since 1780*, Cambridge University Press, Cambridge.

Hoggart, R. (1957) *The Uses of Literacy: Aspects of Working Class Life*, Chatto & Windus, London.

Holcomb, B. (1986) 'Geography and urban women', *Urban Geography*, 7, 448–56.

Holland, J., Ramazanoglu, C., Sharpe, S. and Thomson, R. (1998) *The Male in the Head: Young People, Heterosexuality and Power*, Tufnell Press, London.

Holloway, S.L. (1998) 'Local childcare cultures: moral geographies of mothering and the social organisation of pre-school education', *Gender, Place and Culture*, 5, 29–53.

Holloway, S.L. (1999) 'Mother and worker? The negotiation of motherhood and paid employment in two urban neighbourhoods', *Urban Geography*, 20, 438–60.

Holloway, S.L. and Valentine, G. (eds) (2000) *Children's Geographies: Living, Playing, Learning*, Routledge, London.

Holloway, S.L. and Valentine, G. (2000) 'Corked hats and Coronation Street: British and New Zealand children's imaginative geographies of the Other', *Childhood*, 7, 335–58.

Holloway, S.L., Valentine, G. and Bingham, N. (2000) ' "They're gorgeous": masculinities, femininities and information technologies', *Environment and Planning A*, **32**, 617–33.

hooks, b. (1991) *Yearning: Race, Gender and Cultural Politics*, Turnaround, London.

hooks, b. (1992) *Black Looks: Race and Representation*, Turnaround, London.

Howkins, A. (1986) 'The discovery of rural England', in Colls, R. and Dodd, P. (eds) *Englishness, Politics and Culture 1880–1920*, Croom Helm, London.

Hoyt, H. (1939) *The Structure and Growth of Residential Neighbourhoods in American Cities*, Federal Housing Administration, Washington, DC.

Hubbard, P. (1999) *Sex and the City: Geographies of Prostitution in the Urban West*, Ashgate, Aldershot.

Hugman, R. (1994) *Ageing and the Care of Older People in Europe*, Macmillan, Basingstoke.

Hugman, R. (1999) 'Embodying old age', in Teather, E.K. (ed.) *Embodied Geographies: Spaces, Bodies and Rites of Passage*, Routledge, London.

Hunt, G., Riegel, S., Morales, T., and Waldorf, D. (1993) 'Changes in prison culture: prison gangs and the case of the "Pepsi generation" ', *Social Problems*, **40**, 398–409.

Husain, S. (1988) *Neighbourhood Watch in England and Wales: A Locational Analysis*, Home Office Crime Prevention Unit Paper 12, HMSO, London.

Hyder, K. (1995) 'Ecstasy's deadly cocktails', *Observer*, 13 August, p. 10.

Hyams, M. (2000) ' "Pay attention in class . . . [and] don't get pregnant": a discourse of academic success amongst adolescent Latinas', *Environment and Planning A*, **32**, 571–760.

Ignatieff, M. (1994) *Blood and Belonging: Journeys into the New Nationalism*, Vintage, London.

Imrie, R. (1996) *Disability and the City: International Perspectives*, Paul Chapman, London.

Ingram, G.B. (2000) 'Mapping decolonization of male homoerotic space in Pacific Canada', in Phillips, R., Watt, D. and Shuttleton, D. (eds) *De-centring Sexualities: Politics and Representations Beyond the Metropolis*, Routledge, London.

Irwin, J. (1980) *Prisons in Turmoil*, Little Brown, Chicago.

Isin, E.F. and Wood, P.K. (1999) *Citizenship and Identity*, Sage, London.

Jaakson, R. (1986) 'Second-home domestic tourism', *Annals of Tourism Research*, **13**, 367–91.

Jackson, E.L. (1986) 'Outdoor recreation and environmental attitudes', *Leisure Studies*, **5**, 1–24.

Jackson, P. (1984) 'Social disorganisation and moral order in the city', *Transactions of the Institute of British Geographers*, **9**, 168–80.

Jackson, P. (1987) 'The idea of "race" and the geography of racism', in Jackson, P. (ed.) *Race and Racism: Essays in Social Geography*, Allen & Unwin, London.

Jackson, P. (1988) 'Street life: the politics of carnival', *Environment and Planning D: Society and Space*, **6**, 213–27.

Jackson, P. (1989) *Maps of Meaning*, Routledge, London.

Jackson, P. (1992) 'The racialization of labour in post-war Bradford', *Journal of Historical Geography*, **18**, 190–209.

Jackson, P. (1994) 'Constructions of criminality: police–community relations in Toronto', *Antipode*, **26**, 216–35.

Jackson, P. (1998a) 'Domesticating the street: the contested spaces of the high street and the mall', in Fyfe, N. (ed.) *Images of the Street: Planning, Identity and Control in Public Space*, Routledge, London.

Jackson, P. (1998b) 'Constructions of "whiteness" in the geographical imagination', *Area*, 30, 99–106.

Jackson, P. and Penrose, J. (eds) (1993) *Constructions of Race, Place and Nation*, UCL Press, London.

Jacobs, J. (1961) *The Death and Life of Great American Cities*, Vintage, New York.

Jacobs, J. (1996) *Edge of Empire: Postcolonialism and the City*, Routledge, London.

Jacobs, J. and Fincher, R. (1998) 'Introduction', in Fincher, R. and Jacobs, J. (eds) *Cities of Difference*, Guilford Press, London.

James, A. (1993) *Childhood Identities: Self and Social Relationships in the Experience of the Child*, Edinburgh University Press, Edinburgh.

James, A., Jenks, C. and Prout, A. (eds) (1998) *Theorizing Childhood*, Polity Press, Cambridge.

James, A. (1993) *Childhood Identities*, Edinburgh University Press, Edinburgh.

James, S. (1990) 'Is there a "place" for children in geography?', *Area*, **22**, 278–83.

Jamieson, L. and Toynbee, C. (1989) 'Shifting patterns of parental authority, 1900–1980', in Corr, H. and Jamieson, L. (eds) *The Politics of Everyday Life*, Macmillan, London.

Jarlöv, L. (1999) 'Leisure lots and summer cottages as places for people's own creative work', in Crouch, D. (ed.) *Leisure/Tourism Geographies: Practices and Geographical Knowledge*, Routledge, London.

Jay, E. (1992) *Keep Them in Birmingham*, Commission for Racial Equality, London.

Jenks, C. (1996) *Childhood*, Routledge, London.

Johnston, J. (1973) *The Lesbian Nation: The Feminist Solution*, Touchstone Books/ Simon and Schuster, New York.

Johnson, L. (1990) 'New patriarchal economies in the Australian textile industry', *Antipode*, **22**, 1–32.

Johnson, L. (1989) 'Embodying geography – some implications of considering the sexed body in space', *Proceedings of the 15th New Zealand Geographers' Conference*, Dunedin.

Johnson, L. (1997) 'Queen(s') street or Ponsonby poofters? The embodied HERO Parade site', *New Zealand Geographer*, **53**, 29–33.

Johnson, L. and Valentine, G. (1995) 'Wherever I lay my girlfriend that's my home: the performance and surveillance of lesbian identities in domestic environments', in Bell, D. and Valentine, G. (eds) *Mapping Desire: Geographies of Sexualities*, Routledge, London.

Johnston, L. (1998) 'Reading the sexed bodies and spaces of the gym', in Nast, H. and Pile, S. (eds) *Places Through the Body*, Routledge, London.

Johnston, R.J. (1991) *Geography and Geographers*, Edward Arnold, London.

Johnston, R.J., Gregory, D., Pratt, G. and Watts, M. (eds) (2000) *The Dictionary of Human Geography*, Blackwell, Oxford.

Jones, A. (1993) 'Defending the border: men's bodies and vulnerability', *Cultural Studies from Birmingham*, **2**, 77–123.

Jones, O. (1997) 'Little figures, big shadows: country childhood stories', in Cloke, P. and Little, J. (eds) *Contested Countryside Cultures: Otherness, Marginalisation and Rurality*, Routledge, London.

Jones, O. (2000) 'Melting geography: purity, disorder, childhood and space', in Holloway, S. and Valentine, G. (eds) *Children's Geographies: Playing, Living, Learning*, Routledge, London.

Jordan, B. and Jones, M. (1988) 'Poverty, the underclass and probation practice', *Probation Journal*, **35**(4), 123–7.

Jordanova, L. (1989) *Sexual Visions: Images of Gender in Science and Medicine Between the Eighteenth and Twentieth Centuries*, Harvester Wheatsheaf, New York.

Katz, C. and Monk, J. (eds) *Full Circles: Geographies of Women over the Life Course*, Routledge, London.

Kearns, R. and Smith, C. (1993) 'Housing stressors and mental health among marginalized urban populations', *Area*, **25**, 267–78.

Keith, M. (1995) 'Making the street visible: placing racial violence in context', *New Community*, **21**, 551–65.

Kelling, G. (1987) 'Acquiring a taste for order: the community and the police', *Crime and Delinquency*, **1**, 90–102.

Kelly, M.P.F. (1994) 'Towanda's triumph: social and cultural capital in the transition to adulthood in the urban ghetto', *International Journal of Urban and Regional Research*, **18**, 88–111.

Kennedy, E.L. and Davis, M.D. (1993) *Boots of Leather, Slippers of Gold: The History of a Lesbian Community*, Routledge, New York.

Kimmel, M. and Kaufman, M. (1994) 'Weekend warriors: the new men's movement', in Brod, H. and Kaufman, M. (eds) *Theorising Masculinities*, Sage, Thousand Oaks, CA.

King, R. and McDermott, K. (1995) *The State of Our Prisons*, Clarendon Press, Oxford.

Kirby, K. (1996) 'Cartographic vision and the limits of politics', in Duncan, N. (ed.) *Bodyspace: Destabilising Geographies of Gender and Sexuality*, Routledge, London.

Kirby, V. (1995) *Indifferent Boundaries: Spatial Concepts of Human Subjectivity*, Guilford Press, London.

Kitchin, R. (1998) *Cyberspace*, John Wiley & Sons, Chichester.

Kitchin, R. (forthcoming) ' "Out of place", "knowing one's place": space, power and the exclusion of disabled people', *Disability and Society*.

Klesse, C. (1999) ' "Modern primitivism": non-mainstream body modification and racialised representation', *Body and Society*, **5**, 39–50.

Knopp, L. (1992) 'Sexuality and the spatial dynamics of capitalism', *Environment and Planning D: Society and Space*, **10**, 651–69.

Knopp, L. (1998) 'Sexuality and urban space: gay male identity politics in the United States, the United Kingdom and Australia', in Fincher, R. and Jacobs, J. (eds) *Cities of Difference*, Guilford Press, London.

Kofman, E. (1995) 'Citizenship for some but not for others: spaces of citizenship in contemporary Europe', *Political Geography*, **14**, 121–37.

Kramer, J.L. (1995) 'Bachelor farmers and spinsters: lesbian and gay identity and community in rural North Dakota', in Bell, D. and Valentine, G. (eds) *Mapping Desire: Geographies of Sexualities*, Routledge, London.

Krenichyn, K. (1999) 'Messages about adolescent identity: coded and contested spaces in a New York City high school', in Teather, E.K. (ed.) *Embodied Geographies: Spaces, Bodies and Rites of Passage*, Routledge, London.

Kymlicka, W. (1995) *Multicultural Citizenship*, Oxford University Press, Oxford.

Langer, M. (1989) *Merleau-Ponty's Phenomenology of Perception: a Guide and Commentary*, Macmillan, Basingstoke.

Lash, S. and Urry, J. (1994) *Economies of Signs and Spaces*, Sage, London.

Lauria, M. and Knopp, L. (1985) 'Towards an analysis of the role of gay communities in the urban renaissance', *Urban Geography*, **6**, 152–69.

Laurie, N., Dwyer, C., Holloway, S. and Smith, F. (1999) *Geographies of New Femininities*, Longman, Harlow.

Laurier, E. (1999) 'That sinking feeling: elitism, working leisure and yachting', in Crouch, D. (ed.) *Leisure/Tourism Geographies: Practices and Geographical Knowledge*, Routledge, London.

Laws, G. (1994) 'Oppression, knowledge and the built environment', *Political Geography*, **13**, 7–32.

Laws, G. (1995) 'Embodiment and emplacement: identities, representation and landscape in Sun City retirement communities', *International Journal of Ageing and Human Development*, **40**, 253–80.

Laws, G. (1997a) 'Women's life courses, spatial mobility, and state policies', in Jones III, J.P., Nast, H. and Roberts, S.M. (eds) *Thresholds in Feminist Geography: Difference, Methodology and Representation*, Rowman & Littlefield, Oxford.

Laws, G. (1997b) 'Spatiality and age relations', in Jamieson, A., Harper, S. and Victor, C. (eds) *Critical Approaches to Ageing and Later Life*, Open University Press, Buckingham.

Le Doeuff, M. (1991) *Hipparchia's Choice: An Essay Concerning Women, Philosophy, etc.*, Blackwell, Oxford.

Lee, R. and Wills, J. (eds) (1998) *Geographies of Economies*, Arnold, London.

Lees, L. (1998) 'Urban renaissance and the street: spaces of control and contestation', in Fyfe, N. (ed.) *Images of the Street*, Routledge, London.

Leidner, R. (1991) 'Serving hamburgers and selling insurance', *Gender and Society*, **5**, 154–77.

Leidner, R. (1993) *Fast Food, Fast Talk: The Routinisation of Everyday Life*, University of California Press, Berkeley, CA.

Lenk, K. (1997) 'The challenge of cyberspatial forms of human interaction to territorial governance and policing', in Loader, B. (ed.) *The Governance of Cyberspace*, Routledge, London.

Leonard, M. (1998) 'Paper planes: travelling the new grrrl geographies', in Skelton, T. and Valentine, G. (eds) *Cool Places: Geographies of Youth Cultures*, Routledge, London.

Lewis, C. and Pile, S. (1996) 'Woman, body, space: Rio carnival and the politics of performance', *Gender, Place and Culture*, **3**, 23–41.

Ley, D. (1980) 'Liberal ideology and the post industrial city', *Annals of the Association of American Geographers*, **70**, 238–58.

Ley, D. (1996) *The New Middle Class and the Remaking of the Central City*, Oxford University Press, Oxford.

Ley, D. and Cybriwsky, R. (1974) 'Urban graffiti as territorial markers', *Annals of the Association of American Geographers*, **64**, 491–505.

Liepins, R. (1998) ' "Women of broad vision": nature and gender in the environmental activism of Australia's "Women in Agriculture" movement', *Environment and Planning A*, **30**, 1179–96.

Lister, R. (1997) *Citizenship: Feminist Perspectives*, New York University Press, New York.

Little, J. (1984) 'Gender relations and the rural labour process', in Whatmore, S., Marsden, T. and Lowe, P. (eds) *Gender and Rurality*, David Fulton, London.

Little, J. (1986) 'Feminist perspectives in rural geography: an introduction', *Journal of Rural Studies*, **2**, 1–8.

Little, J. (1987) 'Gentrification and the influence of local level planning', in Cloke, P. (ed.) *Rural Planning: Policy into Action?*, Harper & Row, London.

Little, J. and Austin, P. (1996) 'Women and the rural idyll', *Journal of Rural Studies*, **12**, 101–11.

Littleton, J. (ed.) (1996) *Clash of Identities: Essays on Media, Manipulation, and Politics of the Self*, Prentice-Hall, Englewood Cliffs, NJ.

Livingstone, D. (1992) *The Geographical Tradition*, Basil Blackwell, Oxford.

Livingstone, S. (1992) 'The meanings of domestic technologies: a personal construct analysis of familial gender relations', in Silverstone, R. and Hirsch, E. (eds) *Consuming Technologies: Media and Information in Domestic Spaces*, Routledge, London.

Loader, B. (1997) 'The governance of cyberspace: politics, technology and global restructuring', in Loader, B. (ed.) *The Governance of Cyberspace*, Routledge, London.

Lofland, L. (1973) *A World of Strangers: Order and Action in Urban Public Space*, Basic Books, New York.

Longhurst, R. (1996) 'Refocusing groups: pregnant women's geographical experiences of Hamilton, New Zealand', *Area*, **28**, 143–9.

Longhurst, R. (1997) '(Dis)embodied geographies', *Progress in Human Geography*, **21**, 486–501.

Lovatt, A. and O'Connor, J. (1995) 'Cities and the night-time economy', *Planning Practice and Research*, **10**, 127–33.

Low, M. (1999) 'Communitarianism, civic order and political representation', *Environment and Planning A*, **31**, 87–111.

Lowenthal, D. (1994) 'Identity, heritage and history', in Gillis, J.R. (ed.) *Commemorations: The Politics of National Identity*, Princeton University Press, Princeton, NJ.

Lucas, T. (1998) 'Youth gangs and moral panics in Santa Cruz, California', in Skelton, T. and Valentine, G. (eds) *Cool Places: Geographies of Youth Cultures*, Routledge, London.

Luke, T. (1997) 'At the end of Nature: cyborgs, "humachines", and the environments in postmodernity', *Environment and Planning A*, **29**, 1367–80.

Lupton, D. (1996) *Food, the Body and the Self*, Sage, London.

Lupton, D. (1995) 'The embodied computer/user', in Featherstone, M. and Burrows, R. (eds) *Cyberspace, Cyberbodies and Cyberpunk: Cultures of Technological Embodiment*, Sage, London.

Luxton, M. (1980) *More Than a Labour of Love*, Women's Press, Toronto.

Lyon, D. (1994) *The Electronic Eye: The Rise of the Surveillance Society*, Polity Press, Cambridge.

Mac an Ghaill, M. (ed.) (1996) *Understanding Masculinities*, Open University Press, Buckingham.

MacKenzie, S. (1984) 'Editorial introduction', *Antipode*, **16**, 3–10.

MacKenzie, S. and Rose, D. (1983) 'Industrial change, the domestic economy and home life', in Anderson, J., Duncan, S. and Hudson, R. (eds) *Redundant Spaces in Cities and Regions*, Academic Press, New York.

Mackintosh, H. (1952) *Housing and Family Life*, Cassell & Co., London.

Mairs, N. (1995) *Remembering the Bone House*, Beacon Press, Boston, MA.

Malbon, B. (1998) 'Clubbing: consumption, identity and the spatial practices of everynight life', in Skelton, T. and Valentine, G. (eds) *Cool Places: Geographies of Youth Cultures*, Routledge, London.

Malbon, B. (1999) *Clubbing: Dancing, Ecstasy and Vitality*, Routledge, London.

Malik, S. (1992) 'Colours of the countryside – a whiter shade of pale', *Ecos*, **13**, 33–40.

Marsden, T., Lowe, P. and Whatmore, S. (eds) (1990) *Rural Restructuring*, David Fulton, London.

Marshall, T.H. (1950) *Citizenship and Social Class*, Cambridge University Press, Cambridge.

Marston, S. (1990) 'Who are the people? Gender, citizenship and the making of the American nation', *Environment and Planning D: Society and Space*, **8**, 449–58.

Marston, S. (2000) 'The social construction of scale', *Progress in Human Geography*, **24**, 219–42.

Mascia-Lees, F.E. and Sharpe, P. (eds) (1992) *Tattoo, Torture, Mutilation and Adornment: The Denaturalisation of the Body in Culture and Text*, University of New York Press, New York.

Massey, D. (1991) 'A global sense of place', *Marxism Today*, June, 24–9.

Massey, D. (1993) 'Power-geometry and a progressive sense of place', in Bird, J., Curtis, B., Putnam, T., Robertson, G. and Tickner, L. (eds) *Mapping the Futures: Local Cultures, Global Changes*, Routledge, London.

Massey, D. (1995) 'Rethinking radical democracy spatially', *Environment and Planning D: Society and Space*, **13**, 283–8.

Massey, D. (1998a) 'Blurring the binaries? High tech in Cambridge', in Ainley, R. (ed.) *New Froniters of Space, Bodies and Gender*, Routledge, London.

Massey, D. (1998b) 'The spatial constructions of youth cultures', in Skelton, T. and Valentine, G. (eds) *Cool Places: Geographies of Youth Cultures*, Routledge, London.

Massey, D. (1999) 'Spaces of politics', in Massey, D., Allen, J. and Sarre, P. (eds) *Human Geography Today*, Polity Press, Cambridge.

Massey, D. and Jess, P. (1995) 'Places and cultures in an uneven world', in Massey, D. and Jess, P. (eds) *A Place in the World? Places, Cultures and Globalization*, Open University Press, Milton Keynes.

Massey, D., Allen, J. and Pile, S. (eds) (1999) *City Worlds*, Open University Press, Buckingham.

Massey, D., Allen, J. and Sarre, P. (1999) *Human Geography Today*, Polity Press, Cambridge.

Matthews, H., Taylor, M., Sherwood, K., Tucker, F. and Limb, M. (2000) 'Growing up in the countryside: children and the rural idyll', *Journal of Rural Studies*, **16**, 141–54.

Matrix (1984) *Making Space: Women and the Man Made Environment*, Pluto Press, London.

May, J. (1973) 'Innocence and experience: the evolution of the concept of juvenile delinquency in the mid-19th century', *Victorian Studies*, **17**, 7–29.

May, J. (1996) 'A little taste of something more exotic: the imaginative geography of everyday life', *Geography*, **81**, 57–64.

McCellan, J. (1994) 'Netsurfers', *The Observer*, 13 February, p. 10.

McClintock, A. (1993) 'Family feuds: gender, nationalism and the family', *Feminist Review*, **44**, 61–80.

McClintock, A. (1995) *Imperial Leather: Race, Gender, and Sexuality in the Colonial Contest*, Routledge, London.

McDowell, L. (1983) 'Towards an understanding of the gender division of urban space', *Environment and Planning D: Society and Space*, **1**, 59–72.

McDowell, L. (1995) 'Body work: heterosexual gender performances in City workplaces', in Bell, D. and Valentine, G. (eds) *Mapping Desire: Geographies of Sexualities*, Routledge, London.

McDowell, L. (1997) *Capital Culture: Gender at Work in the City*, Blackwell, Oxford.

McDowell, L. (1997) 'A tale of two cities? Embedded organisations and embodied workers in the City of London', in Lee, R. and Wills, J. (eds) *Geographies of Economies*, Arnold, London.

McDowell, L. (2000a) 'The trouble with men? Young people, gender transformations and the crisis of masculinity', *International Journal of Urban and Regional Research*, **24**, 201–9.

McDowell, L. (2000b) 'Learning to serve? Employment aspirations and attitudes of young working-class men in an era of labour market restructuring.' *Gender, Place and Culture*, **7**, 389–416.

McDowell, L. and Court, G. (1994) 'Performing work: bodily representations in merchant banks', *Environment and Planning D: Society and Space*, **12**, 727–50.

McHenry, H. (1998) 'Wild flowers in the wrong field are weeds! Examining farmers' constructions of conservation', *Environment and Planning A*, **30**, 1039–53.

McInroy, N. and Boyle, M. (1996) 'The refashioning of civic identity: constructing and consuming the "new" Glasgow', *Scotlands*, **3**, 70–87.

McKay, G. (1998) *DIY Culture: Party and Protest in Nineties Britain*, Verso, London.

McKendrick, J.H., Bradford, M. and Fielder, A. (2000) 'Time for a party: making sense of the commercialisation of leisure space for children', in Holloway, S. and Valentine, G. (eds) *Children's Geographies: Playing, Living, Learning*, Routledge, London.

McLaren, D. (1992) 'London as ecosystem', in Thornley, A. (ed.) *The Crisis of London*, Routledge, London.

McLaughlin, M.L., Osbourne, K.K. and Smith, C.B. (1995) 'Standards of conduct on Usenet', in Jones, S. (ed.) *Cybersociety: Computer Mediated Communication and Community*, Sage, London.

McLuhan, M. (1962) *The Gutenburg Galaxy: The Making of Typographic Man*, Routledge, London.

McNay, L. (1994) *Foucault: a Critical Introduction*, Blackwell, Oxford.

McRobbie, A. (ed.) (1989) *Zoot Suits and Second Hand Dresses*, Macmillan, Basingstoke.

McRobbie, A. (1994) *Postmodernism and Popular Culture*, Routledge, London.

Mehta, A. and Bondi, L. (1999) 'Embodied discourse: on gender and fear of violence', *Gender, Place and Culture*, **6**, 67–84.

Mennell, S., Murcott, A. and van Otterloo, A. (1992) *The Sociology of Food: Eating, Diet and Culture*, Sage, London.

Mercer, K. and Race, I. (1988) 'Sexual politics and black masculinity: a dossier', in Chapman, R. and Rutherford, J. (eds) *Male Order, Unwrapping Masculinity*, Lawrence & Wishart, London.

Merchant, C. (1990) *The Death of Nature: Women, Ecology and the Scientific Revolution*, Harper & Row, San Francisco, CA.

Merleau-Ponty, M. (1962) *Phenomenology of Perception*, Humanities Press, New York (trans. Colin Smith).

Middleton, N., O'Keefe, P. and Moyo, S. (1993) *The Tears of a Crocodile: from Rio to Reality in the Developing World*, Pluto, London.

Milbourne, P. (1997) *Revealing Rural Others: Representation, Power and Identity in the British Countryside*, Pinder, London.

Miles, R. (1982) *Racism and Migrant Labour*, Routledge and Kegan Paul, London.

Miller, D. (1987) *Material Culture and Mass Consumption*, Blackwell, Oxford.

Miller, D. (1988) 'Appropriating the state on the council estate', *Man*, **23**, 353–72.

Miller, D. (1992) 'The young and the restless in Trinidad. A case study of the local and the global in mass consumption', in Silverstone, R. and Hirsch, E. (eds) *Consuming Technologies*, Routledge, London.

Miller, J. (1997) 'Country life is a killer', *Sunday Times*, 27 April, p. 3.

Milligan, C. (1999) 'Without these walls: a geography of mental ill-health in a rural environment', in Butler, R. and Parr, H. (eds) *Mind and Body Spaces: Geographies of Illness, Impairment and Disability*, Routledge, London.

Mills, C. (1988) ' "Life on the upslope": the postmodern landscape of gentrification', *Environment and Planning D: Society and Space*, 6, 169–89.

Mingay, G. (ed.) (1989) *The Rural Idyll*, Routledge, London.

Mitchell, D. (1995) 'The end of public space? People's Park, definitions of the public and democracy', *Annals of the Association of American Geographers*, 85, 108–33.

Mitchell, D. (1996) 'Political violence, order, and the legal construction of public space: power and the public forum doctrine', *Urban Geography*, 17, 158–78.

Mitchell, D. (1997) 'The annihilation of space by law: the roots and implications of anti-homeless laws in the United States', *Antipode*, 29, 303–35.

Mitchell, D. (2000) *Cultural Geography: A Critical Introduction*, Blackwell, Oxford.

Monaghan, L. (1999) 'Creating "the perfect body": a variable project', *Body and Society*, 5, 267–90.

Montgomery, J. (1995a) 'The story of Temple Bar: Creating Dublin's cultural quarter', *Planning Practice and Research*, 10, 135–69.

Montgomery, J. (1995b) 'Urban vitality and the culture of cities', *Planning Practice and Research*, 10, 101–10.

Moore Milroy, B. and Wismer, S. (1994) 'Communities, work and public/private sphere models', *Gender, Place and Culture*, 1, 71–90.

Moore, J. (1997) *A Cultural Geography of the New Age Movement*, Unpublished dissertation, University of Coventry.

Mordue, T. (1999) 'Heartbeat Country: conflicting values, coinciding visions', *Environment and Planning A: Society and Space*, 31, 629–46.

Morgan, D. (1996) *Family Connections*, Polity Press, Cambridge.

Morley, D. (1986) *Family Television: Cultural Power and Domestic Leisure*, Methuen, London.

Morley, D. and Robins, K. (1995) *Spaces of Identity: Global Media, Electronic Landscapes and Cultural Boundaries*, Routledge, London.

Mormont, M. (1987a) 'Rural nature and urban natures', *Sociologia Ruralis*, 27, 3–20.

Mormont, M. (1987b) 'The emergence of rural struggles and their ideological effects', *International Journal of Urban and Regional Research*, 7, 559–75.

Mormont, M. (1990) 'Who is rural? Or how to be rural: towards a sociology of the rural', in Marsden, T., Lowe, P. and Whatmore, S. (eds) *Rural Restructuring*, David Fulton, London.

Morrell, J. (1990) *The Employment of People with Disabilities: Research into the Policies and Practices of Employers*, HMSO, London.

Morris, J. (1991) *Pride Against Prejudice*, Women's Press, London.

Mort (1989) 'The politics of consumption', in Hall, S. and Jacques, M. (eds) *New Times: the Changing Face of Politics in the 1990s*, Lawrence & Wishart, London.

Morse, M. (1994) 'What do cyborgs eat? Oral logic in an information society', *Discourse*, 16, 86–123.

Moss, P. (1999) 'Autobiographical notes on chronic illness', in Butler, R. and Parr, H. (eds) *Mind and Body Spaces: Geographies of Illness, Impairment and Disability*, Routledge, London.

Moss, P. and Dyck, I. (1996) 'Inquiry into environment and body: women, work and chronic illness', *Environment and Planning D: Society and Space*, 14, 737–53.

Moss, P. and Dyck, I. (1999) 'Journeying through M.E.: identity, the body and women with chronic illness', in Teather, E.K. (ed.) *Embodied Geographies: Spaces, Bodies and Rites of Passage*, Routledge, London.

Mosse, G.L. (1985) *Nationalism and Sexuality*, University of Wisconsin, Madison, WI.

Muir, R. (1997) *Political Geography*, Macmillan, Basingstoke.

Mumford, L. (1945) 'On the future of London', *Architectural Review*, 97, 3–10.

Munt, S. (1995) 'The lesbian flâneur', in Bell, D. and Valentine, G. (eds) *Mapping Desire: Geographies of Sexualities*, Routledge, London.

Munt, S. (1998) 'Sisters in exile: the Lesbian nation', in Ainley, R. (ed.) *New Frontiers of Space, Bodies and Gender*, Routledge, London.

Murray, C. (1984) *Losing Ground: American Social Policy, 1950–1980*, Basic Books, New York.

Murdoch, J. and Marsden, T. (1994) *Reconstituting Rurality*, UCL Press, London.

Murdoch, J. and Pratt, A. (1993) 'Rural studies: modernism, postmodernism and the "post-rural" ', *Journal of Rural Studies*, 9, 411–28.

Murphy, P. and Watson, S. (1997) *Surface City: Sydney at the Millennium*, Pluto Press, Sydney.

Myers, J. (1992) 'Nonmainstream body modification: genital piercing, branding, burning and cutting', *Journal of Contemporary Ethnography*, 21, 276–306.

Myrdal, G. (1962) *Challenge to Affluence*, Pantheon, New York.

Myslik, W. (1996) 'Renegotiating the social/sexual identities of place: gay communities as safe havens or sites of resistance', in Duncan, N. (ed.) *BodySpace: Destabilising Geographies of Gender and Sexuality*, Routledge, London.

Nairn, T. (1977) *The Break Up of Britain: Crisis and Neo-Nationalism*, New Left Books, London.

Nast, H. and Pile, S. (eds) (1998) *Places Through the Body*, Routledge, London.

National Law Center on Homelessness and Poverty (1999) *Out of Sight, Out of Mind? A Report on Anti-Homeless Laws, Litigation and Alternatives in 50 United States Cities*, available from the National Law Center on Homelessness and Poverty, 918 F Street, NW, Suite 412, Washington, DC, USA.

Nayak, A. (1999a) ' "White English ethnicities": racism, anti-racism and student perspectives', *Race Ethnicity and Education*, 2, 177–202.

Nayak, A. (1999b) ' "Pale warriors": skinhead culture and the embodiment of white masculinities', in Braj, A., Hickman, M.J. and Mac an Ghaill, M. (eds) *Thinking Identities: Ethnicity, Racism and Culture*, Macmillan, Basingstoke.

Nelson, L. (1999) 'Bodies (and spaces) *do* matter: the limits of performativity', *Gender, Place and Culture*, 6, 331–54.

Newby, H. (1979) *Green and Pleasant Land? Social Change in Rural England*, Hutchinson, London.

Newby, H. (1986) 'Locality and rurality: the restructuring of rural social relations', *Regional Studies*, 20, 209–15.

Newby, H. (1996) 'Citizenship in a green world: global commons and human steward-ship', in Bulmer, M. and Rees, A.M. (eds) *Citizenship Today*, UCL Press, London.

Newby, H., Bell, C., Rose, D. and Sanders, P. (1978) *Property, Power and Paternalism*, Hutchinson, London.

Newman, O. (1973) *Defensible Space*, Collier, New York.

Newton, T. (1995) *Managing Stress: Emotion and Power at Work*, Sage, London.

Nieva, V. and Gutek, B. (1981) *Women and Work: A Psychological Perspective*, Praeger, New York.

Nixon, R. (1994) 'Refugees and homecomings: Bessie Head and the end of exile', in Robertson, G., Mash, M., Tickner, L., Bird, J., Curtis, B. and Putnam, T. (eds) *Travellers Tales: Narratives of Home and Displacement*, Routledge, London.

Noin, D. and White, P. (1997) *Paris*, John Wiley & Sons, London.

Nordin, U. (1993) 'Second homes', in Aldskogrus, H. (ed.) *Cultural Life, Recreation and Tourism* (National Atlas of Sweden), Royal Swedish Academy of Science, Stockholm.

Norris, C. and Armstrong, G. (1997) *Categories of Control: the Social Construction of Suspicion and Intervention in CCTV Systems*, Report to the ESRC. Copy available from the Dept. of Social Policy, University of Hull, Hull, UK.

Oberhauser, A. (1995) 'Gender and household economic strategies in rural Appalachia', *Gender, Place and Culture*, **2**, 51–70.

O'Brien, M. (1989) *Reproducing the World*, Westview Press, Boulder, CO.

Orbach, S. (1988) *Fat is a Feminist Issue*, Arrow Books, London.

Ostwald, M. (1997) 'Virtual urban futures', in Holmes, D (ed.) *Virtual Politics: Identity and Community in Cyberspace*, Sage, London.

Oswell, D. (1998) 'The place of "childhood" in Internet content regulation', *International Journal of Cultural Studies*, **1**, 131–51.

Pacione, M. (1983) 'Neighbourhood communities in the modern city: some evidence from Glasgow', *Scottish Geographical Magazine*, **99**, 169–81.

Paddison, R. (1993) 'New nationalism in an old state: Scotland and the UK in the 1980s', in Williams, C.H. (ed.) *The Political Geography of the New World Order*, Belhaven Press, London.

Pahl, R. (1965) 'Class and community in English commuter villages', *Sociologia Ruralis*, **5**, 5–23.

Pahl, R. (1970) *Readings in Urban Sociology*, Pergamon, Oxford.

Pain, R. (1999) 'Women's experiences of violence over the life course', in Teather, E.K. (ed.) *Embodied Geographies: Spaces, Bodies and Rites of Passage*, Routledge, London.

Park, R., Burgess, E. and McKenzie, R. (1967) [1925] *The City* (4th edn), University of Chicago Press, Chicago.

Parker, A., Russo, M., Sommer, D. and Yaeger, P. (eds) (1992) *Nationalisms and Sexualities*, Routledge, London.

Parr, H. (1997) 'Mental health, public space and the city: questions of individual and collective access', *Environment and Planning D: Society and Space*, **15**, 435–54.

Parr, H. (1999) 'Bodies and psychiatric medicine: interpreting different geographies of mental health', in Butler, R. and Parr, H. (eds) *Mind and Body Spaces: Geographies of Illness, Impairment and Disability*, Routledge, London.

Parr, H. and Butler, R. (1999) 'New geographies of illness, impairment and disability', in Butler, R. and Parr, H. (eds) *Mind and Body Spaces: Geographies of Illness, Impairment and Disability*, Routledge, London.

Parr, H. and Philo, C. (1995) 'Mapping "mad" identities', in Pile, S. and Thrift, N. (eds) *Mapping the Subject: Geographies of Cultural Transformation*, Routledge, London.

Parr, H. and Philo, C. (1996) *A Forbidding Fortress of Locks, Bars and Padded Cells? The Locational History of Nottingham's Mental Health Care*, Historical Geography Research Series, No. 32.

Paterson, K. (1998) 'Disability studies and phenomenology: finding a space for both the carnal and the political', Paper presented at the British Sociological Association, Edinburgh, April.

Peach, C. (ed.) (1975) *Urban Social Segregation*, Longman, London.

Peach, C., Robinson, V. and Smith, S. (eds) (1981) *Ethnic Segregation in Cities*, Croom Helm, London.

Pearson, G. (1983) *Hooligan: A History of Respectable Fears*, Macmillan, London.

Pendry, R. (1993) 'Neighbourhood watch', *The Guardian, Weekend*, 20 February, pp. 30–1.

Penley, C. and Ross, A. (1991) 'Cyborgs at large: interview with Donna Haraway', *Social Text*, **25/26**, 8–23.

Pepper, D. (1993) *Eco-socialism: From Deep Ecology to Social Justice*, Routledge, London.

Phillips, M. (1993) 'Rural gentrification and the process of class colonisation', *Journal of Rural Studies*, **9**, 123–40.

Philo, C. (1987) ' "Fit localities for an asylum": the historical geography of "the mad business" in England viewed through the pages of the Asylum Journal', *Journal of Historical Geography*, **13**, 398–415.

Philo, C. (1989) 'Enough to drive one mad: the organisation of space in nineteenth century lunatic asylums', in Wolch, J. and Dear, M. (eds) *The Power of Geography*, Unwin Hyman, London.

Philo, C. (1992) 'Neglected rural geographies: a review', *Journal of Rural Studies*, **8**, 193–207.

Philo, C. and Kearns, G. (1993) 'Culture, history, capital: a critical introduction to selling places', in Kearns, G. and Philo, C. (eds) *Selling Places: The City as Cultural, Capital, Past and Present*, Pergamon, Oxford.

Philo, C. and Parr, H. (2000) 'Institutional geographies: introductory remarks', *Geoforum*, **31**, 513–21.

Philo, C. and Wilbert, C. (2000) 'Animal spaces, beastly places: an introduction', in Philo, C. and Wilbert, C. (eds) *Animal Spaces, Beastly Places*, Routledge, London.

Pierson, E. (1992) 'Creating our own event', in Harding, C. (ed.) *Wingspan*, St Martin's Press, New York.

Pile, S. (1996) *The Body and The City: Psychoanalysis, Space and Subjectivity*, Routledge, London.

Pile, S. and Thrift, N. (1995) (eds) *Mapping the Subject*, Routledge, London.

Pile, S., Brook, C. and Mooney, G. (eds) (1999) *Unruly Cities?*, Open University Press, Buckingham.

Pilger, J. (1993) 'The pit and the pendulum', *Guardian Weekend*, 30 January, pp. 6–9.

Pilkington, E. (1994) 'Killing the age of innocence', *The Guardian*, 30 May, p. 18.

Ploszajska, T. (1994) 'Moral landscapes and manipulated spaces: gender, class and space in Victorian reformatory schools', *Journal of Historical Geography*, 20, 413–29.

Pollard, I. (1989) 'Pastoral interludes third text: Third World perspectives', *Contemporary Art and Culture*, 7, 41–6.

Pratt, G. (1997) 'Stereotypes and ambivalence: the construction of domestic workers in Vancouver, British Columbia', *Gender, Place and Culture*, 4, 159–78.

Pratt, G. (1998) 'Grids of Difference: place and identity formations', in Finder, R. and Jacobs, J. (eds) *Cities of Difference*, Guilford Press, London.

Pratt, M.B. (1992) [1988] 'Identity: skin blood heart', in Crowly, H. and Himmelweit, S. (eds) *Knowing Women: Feminism and Knowledge*, Polity Press, Cambridge.

Prins, B. (1995) 'The ethics of hybrid subjects: feminist constructivism according to Donna Haraway', *Science, Technology and Human Values*, 20, 352–67.

Putnam, T. and Newton, C. (1990) *Household Choices*, Futures, London.

Quilley, S. (1995) 'Manchester's "village in the city": the gay vernacular in a post-industrial landscape of power', *Transgressions: A Journal of Urban Exploration*, 1, 36–50.

Raento, P. (1997) 'Political mobilisation and place-specificity: radical nationalist street campaigning in the Spanish Basque Country', *Space and Polity*, 1, 191–204.

Ravenscroft, N. (1992) *Recreation Planning and Development*, Macmillan, Basingstoke.

Ravenscroft, N. (1998) 'Rights, citizenship and access to the countryside', *Space and Polity*, 2, 33–49.

Ravenscroft, N. (1999) 'Hyper-reality in the official (re)construction of leisure sites: the case of rambling', in Crouch, D. (ed.) *Leisure/Tourism Geographies: Practices and Geographical Knowledge*, Routledge, London.

Ravetz, A. (1989) 'The Home of Women: a view from the interior', in Attfield, J. and Kirkham, P. (eds) *A View from the Interior: Feminism, Women and Design*, Women's Press, London.

Redclift, M. (1992) 'The meaning of sustainable development', *Geoforum*, 23, 395–403.

Reeve, A. (1996) 'The private realm of the managed town centre', *Urban Design International*, 1, 61–80.

Reid, E. (1995) 'Virtual worlds: culture and imagination', in Jones, S. (ed.) *Cybersociety: Computer Mediated Communication and Community*, Sage, London.

Rheingold, H. (1993) *The Virtual Community: Homesteading on the Electronic Frontier*, Addison Wesley, Reading, MA.

Riberio, G.L. (1998) 'Cybercultural politics: political activism at a distance in a transnational world', in Alvarez, S., Dagnino, E. and Escobar, A. (eds) *Cultures of Politics/Politics of Cultures: Re-visioning Latin American Social Movements*, Westview Press, Boulder, CO.

Rich, A. (1986) [1984] *Blood, Bread and Poetry: Selected Prose 1979–1985*, Norton & Co., London.

Richardson, D. (1998) 'Sexuality and citizenship', *Sociology*, **32**, 83–100.

Ritzer, G. (1993) *The McDonaldization of Society*, Pine Forge Press, Newbury Park, CA.

Rivlin, L.G. and Wolfe, M. (1985) *Institutional Settings in Children's Lives*, Wiley, New York.

Roberts, M. (1991) *Living in a Man-Made World: Gender Assumptions in Modern Housing Design*, Routledge, London.

Robertson, G., Mash, M., Tickner, L., Bird, J., Curtis, B. and Putnam, T. (eds) (1994) *Travellers' Tales: Narratives of Home and Displacement*, Routledge, London.

Robins, K. (1991) 'Tradition and translation: national culture in its global context', in Corner, J. and Harvey, S. (eds) *Enterprise and Heritage: Crosscurrents in National Culture*, Routledge, London.

Robins, K. (1995) 'Collective emotion and urban culture', in Healey, P. (ed.) *Managing Cities: New Urban Context*, John Wiley & Sons, Chichester.

Robinson, J. (1990) ' "A perfect system of control?": State power and native locations in South Africa', *Environment and Planning D: Society and Space*, **8**, 135–62.

Robinson, F. and Gregson, N. (1992) 'The "underclass": a class apart?', *Critical Social Policy*, **12**, 38–51.

Rodaway, P. (1994) *Sensuous Geographies: Body, Sense and Place*, Routledge, London.

Rogers, A. (1992) 'The boundaries of reason: the world, the homeland and Edward Said', *Environment and Planning D: Society and Space*, **10**, 511–26.

Rohlen, T. (1974) *For Harmony and Strength: Japanese White-Collar Organisation in Anthropological Perspective*, University of California Press, Berkeley.

Rojek, C. (1988) 'The convoy of pollution', *Leisure Studies*, 7, 21–31.

Rose, D. (1984) 'Rethinking gentrification: beyond the uneven development of Marxist urban theory', *Environment and Planning D: Society and Space*, **2**, 47–74.

Rose, D. (1989) 'A feminist perspective of employment restructuring and gentrification: the case of Montreal', in Wolch, J. and Dear, M. (eds) *The Power of Geography: How Territoriality Shapes Social Life*, Unwin Hyman, London.

Rose, G. (1990) 'Imagining Poplar in the 1920s: contested concepts of community', *Journal of Historical Geography*, **16**, 425–37.

Rose, G. (1993) *Feminism and Geography: The Limits of Geographical Knowledge*, Polity Press, Cambridge.

Rose, G. (1996) 'As if the mirrors had bled: masculine dwelling, masculinist theory and feminist masquerade', in Duncan, N. (ed.) *BodySpace: Destabilising Geographies of Gender and Sexuality*, Routledge, London.

Rose, G. (1997) 'Situating knowledges: positionality, reflexivities and other tactics', *Progress in Human Geography*, **21**, 305–20.

Rose, H. (1978) 'In practice supported, in theory denied: an account of an invisible urban movement', *International Journal of Urban and Regional Research*, **2**, 521–37.

Rosen, M. (1985) 'Breakfast at Spino's: dramaturgy and dominance', *Journal of Management*, **11**, 2, 31–48.

Rosen, M. (1988) 'You asked for it: Christmas at the bosses' expense', *Journal of Management Studies*, **25**, 463–80.

Rosenbaum, D.P. (1987) 'The theory and research behind neighbourhood watch: is it a sound fear and crime reduction strategy?', *Crime and Delinquency*, **33**, 103–34.

Rothenberg, T. (1995) 'And she told two friends: lesbians creating urban social space', in Bell, D. and Valentine, G. (eds) *Mapping Desire: Geographies of Sexualities*, Routledge, London.

Routledge, P. (1996) 'Critical geopolitics and terrains of resistance', *Political Geography*, **15**, 509–31.

Rowe, S. and Wolch, J. (1990) 'Social networks in time and space: homeless women in Skid Row, Los Angeles', *Annals of the Association of American Geographers*, **80**, 184–204.

Rowles, G. (1978) *Prisoners of Space? Exploring the Geographic Experience of Older People*, Westview Press, Boulder, CO.

Rowles, G. (1983) 'Geographical dimensions of social support in rural Appalachia', in Rowles, G. and Ohta, R. (eds) *Ageing and Milieu: Environmental Perspectives on Growing Old*, Academic Press, London.

Roy, D. (1973) 'Banana time: job satisfaction and informal interaction', in Salaman, G. and K. Thompson (eds) *People and Organisations*, Longman, London.

Rubinstein, J. (1973) *City Police*, Farrar, Straus and Giroux, New York.

Rucht, D. (1993) 'Think globally, act locally?', in Leifferink, J.D., Lowe, P.D. and Mol, A.P.J. (eds) *European Integration and Environmental Policy*, Belhaven Press, London.

Ruddick, S. (1996) *Young and Homeless in Hollywood*, Routledge, New York.

Ruddick, S. (1998) 'Modernism and resistance: how "homeless" youth sub-cultures make a difference', in Skelton, T. and Valentine, G. (eds) *Cool Places: Geographies of Youth Cultures*, Routledge, London.

Said, E. (1978) *Orientalism: Western Conceptions of the Orient*, Penguin, Harmondsworth.

Sanders, C.R. (1989) *Customising the Body: The Art and Culture of Tattooing*, Temple University Press, Philadelphia, PA.

Sangregorio, I. (1998) 'Having it all? A question of collaborative housing', in Ainley, R. (ed.) *New Frontiers of Space, Bodies and Gender*, Routledge, London.

Sardar, Z. (1995) 'alt.civilisations.faq: Cyberspace as the darker side of the West', *Futures*, **27**, 777–94.

Sarup, M. (1994) 'Home and identity', in Robertson, G., Mash, M., Tickner, L., Bird, J., Curtis, B. and Putnam, T. (eds) *Travellers' Tales: Narratives of Home and Displacement*, Routledge, London.

Sassen, S. (1991) *The Global City: New York, London, Tokyo*, Princeton University Press, Princeton, NJ.

Saunders, P. (1989) 'The meaning of "home" in contemporary English culture', *Housing Studies*, **4**, 177–92.

Saunders, P. and Williams, P. (1988) 'The constitution of the home: towards a research agenda', *Housing Studies*, **3**, 81–93.

Savage, M., Barlow, J., Dickens, P. and Fielding, T. (1993) *Property, Bureaucracy and Culture: Middle-Class Formation in Contemporary Britain*, Routledge, London.

Sayer, A. and Storper, M. (1997) 'Ethics unbound: for a normative turn in social theory', *Environment and Planning D: Society and Space*, **15**, 1–17.

Schlesinger, A. (1992) *The Disuniting of America*, Norton, New York.

Schulman, S. (1995) *Rat Bohemia*, Dutton/Penguin Books, New York.

Schuster, J.M. (1995) 'Two urban festivals: La Merce and First Night', *Planning Practice and Research*, **10**, 173–98.

Schwartz, H. (1986) *Never Satisfied: A Cultural History of Diets, Fantasies and Fat*, Collier Macmillan, London.

Schmitt, P. (1969) *Back to Nature: The Arcadian Myth in Urban America*, Oxford University Press, New York.

Scottish Federation of Housing Associations (1997) 'Obtaining a Fair Deal'. Briefing Paper No. 20. Available from the SFHA, 38 York Place, Edinburgh EH1 3HU, Scotland.

Seabrook, J. (1984) *The Idea of Neighbourhood: What Local Politics Should be About*, Pluto Press, London.

Seager, J. (1993) *Earth Follies: Feminism, Politics and the Environment*, Earthscan Publications, London.

Seager, J., Jones, F. and Rutt, G. (1992) 'Assessment and control of farm pollution', *Journal of the Institution of Water and Environmental Management*, **6**, 48–54.

Seamon, D. (1979) *A Geography of the Lifeworld: Movement, Rest and Encounter*, Croom Helm, London.

Selznick, P. (1992) *The Moral Commonwealth: Social Theory and the Promise of Community*, University of California Press, Berkeley, CA.

Sennett, R. (1993) [1976] *The Fall of Public Man*, Faber and Faber, London.

Sennett, R. (1996) *The Uses of Disorder: Personal Identity and City Life*, Faber and Faber, London.

Shakespeare, T. (1994) 'Cultural representation of disabled people: dustbins for disavowal', *Disability and Society*, **9**, 283–99.

Sharp, J. (1996) 'Gendering nationhood: a feminist engagement with national identity', in Duncan, N. (ed.) *Bodyspace: Destabilising Geographies of Gender and Sexuality*, Routledge, London.

Shields, C. (1994) *Stone Diaries*, Fourth Estate, New York.

Shilling, C. (1993) *The Body and Social Theory*, Sage, London.

Shoard, M. (1980) *The Theft of the Countryside*, Maurice Temple Smith, London.

Short, J. (1991) *Imagined Country: Society, Culture and Environment*, Routledge, London.

Short, E. and Ditton, J. (1996) *Does Closed Circuit Television Prevent Crime?* Scottish Office Central Research Unit, Edinburgh.

Sibley, D. (1988) 'Survey 13: Purification of Space', *Environment and Planning D: Society and Space*, **6**, 409–21.

Sibley, D. (1992) 'Outsiders in space and society', in Anderson, K. and Gale, F. (eds) *Inventing Places: Studies in Cultural Geography*, Longman, Harlow.

Sibley, D. (1995a) *Geographies of Exclusion: Society and Difference in the West*, Routledge, London.

Sibley, D. (1995b) 'Families and domestic routines: constructing the boundaries of childhood', in Pile, S. and Thrift, N. (eds) *Mapping the Subject: Cultural Geographies of Transformation*, Routledge, London.

Sibley, D. (1997) 'Endangering the sacred: nomads, youth cultures and the English countryside', in Cloke, P. and Little, J. (eds) *Contested Countryside: Otherness, Marginalisation and Rurality*, Routledge, London.

Sibley, D. and Lowe, G. (1992) 'Domestic space: modes of control and problem behaviour', *Geografiska Annala B*, **3**, 189–98.

Siegel, F. (1993) 'New York's public space'. Unpublished conference paper.

Silk, J. (1999) 'The dynamics of community, place and identity', *Environment and Planning A*, **31**, 5–17.

Silverstone, R. and Hirsch, E. (1992) *Consuming Technologies: Media and Information in Domestic Spaces*, Routledge, London.

Silverstone, R., Hirsch, E. and Morley, D. (1992) 'Information and communication technologies and the moral economy of the household', in Silverstone, R. and Hirsch, E. (1992) *Consuming Technologies: Media and Information in Domestic Spaces*, Routledge, London.

Sjoberg, G. (1960) *The Pre-Industrial City, Past and Present*, The Free Press, New York.

Skogan, W.G. and Maxfield, M.G. (1981) *Coping with Crime: Individual and Neighbourhood Reactions*, Sage, London.

Slouka, M. (1996) *War of the Worlds: the Assault on Reality*, Abacus, London.

Smith, A.D. (1995) *Nations and Nationalism in a Global Era*, Polity Press, Cambridge.

Smith, D.M. (1999) 'Geography, community and morality', *Environment and Planning A*, **31**, 19–35.

Smith, F. and Barker, J. (2000) ' "Out of school", in school: a social geography of out of school childcare', in Holloway, S. and Valentine, G. (eds) *Children's Geographies: Playing, Living, Learning*, Routledge, London.

Smith, G. and Jackson, P. (1999) 'Narrating the nation: the "imagined community" of Ukrainians in Bradford', *Journal of Historical Geography*, **25**, 367–87.

Smith, G.D. and Winchester, H.P.M. (1998) 'Negotiating space: alternative masculinities at the work/home boundary', *Australian Geographer*, **29**, 327–39.

Smith, L.T.J. (1989) 'Domestic violence: an overview of the literature', Home Office Planning and Research Unit, HMSO, London.

Smith, N. (1979a) 'Gentrification and capital: theory, practice and ideology in Society Hill', *Antipode*, **11**, 139–55.

Smith, N. (1979b) 'Toward a theory of gentrification: a back to the city movement of capital not people', *Journal of the American Planners' Association*, **45**, 538–48.

Smith, N. (1982) 'Gentrification and uneven development', *Economic Geography*, **58**, 139–55.

Smith, N. (1987) 'Of yuppies and housing: gentrification and social restructuring, and the urban dream', *Environment and Planning D: Society and Space*, **5**, 151–72.

Smith, N. (1990) *Uneven Development: Nature, Capital and the Production of Space*, Basil Blackwell, Oxford.

Smith, N. (1992a) 'Geography, difference and the politics of scale', in Doherty, J., Graham, E. and Malek, M. (eds) *Postmodernism and the Social Sciences*, Macmillan, London.

Smith, N. (1992b) 'New city, new frontier: the Lower East Side as Wild, Wild West', in Sorkin, M. (ed.) *Variations on a Theme Park: the New American City and the End of Public Space*, Hill and Wang, New York.

Smith, N. (1993) 'Homeless/global: scaling places', in Bird, J., Curtis, B., Putnam, T., Robertson, G. and Tickner, L. (eds) *Mapping the Futures: Local Cultures, Global Change*, Routledge, London.

Smith, N. (1996a) 'The production of nature', in Robertson, G., Mash, M., Tickner, L., Bird, J., Curtis, B. and Putnam, T. (eds) *FutureNatural: Nature, Science, Culture*, Routledge, London.

Smith, N. (1996b) *The New Urban Frontier: Gentrification and the Revanchist City*, Routledge, London.

Smith, S.J. (1984) 'Crime in the news', *British Journal of Criminology*, 24, 289–95.

Smith, S.J. (1986) *Crime, Space and Society*, Cambridge University Press, Cambridge.

Smith, S.J. (1987) 'Residential segregation: a geography of English racism?', in Jackson, P. (ed.) *Race and Racism: Essays in Social Geography*, Allen & Unwin, London.

Smith, S.J. (1989) 'Society, space and citizenship: a human geography for the "new times"', *Transactions of the Institute of British Geographers*, 14, 144–56.

Smith, S.J. (1999) 'Society–Space', in Cloke, P., Crang, P. and Goodwin, M. (eds) *Introducing Human Geography*, Arnold, London.

Smoult, T.C. (1994) 'Perspectives on the Scottish identity', *Scottish Affairs*, 6, 101–13.

Sobchack, V. (1995) 'Beating the Meat/Surviving the text, or how to get out of this century alive', in Featherstone, M. and Burrows, R. (eds) *Cyberspace, Cyberbodies, Cyberpunk: Cultures of Technological Embodiment*, Sage, London.

Soja, E. (1996) *Thirdspace*, Blackwell, Oxford.

Sommerville, P. (1992) 'Homelessness and the meaning of home: rooflessness or rootlessness', *International Journal of Urban and Regional Research*, 16, 529–39.

Soper, K. (1996) 'Nature/"nature"', in Robertson, G., Mash, M., Tickner, L., Bird, J., Curtis, B. and Putnam, T. (eds) *FutureNatural: Nature, Science, Culture*, Routledge, London.

Sorkin, M. (1992) 'See you in Disneyland', in Sorkin, M. (ed.) *Variations on a Theme Park: the New American City and the End of Public Space*, Hill and Wang, New York.

Sparks, J.R. and Bottoms, A.E. (1995) 'Legitimacy and order in prisons', *British Journal of Sociology*, 46, 45–62.

Spiegel, L. (1992) *Make Room for TV: Television and the Family in Postwar America*, University of Chicago Press, Chicago.

Squires, J. (1994) 'Private lives, secluded spaces: privacy as political possibility', *Environment and Planning D: Society and Space*, 12, 387–401.

Stables, J. and Smith, F. (1999) '"Caught in the Cinderella trap": narratives of disabled parents and carers', in Parr, H. and Butler, R. (eds) *Mind and Body Spaces: Geographies of Illness, Impairment and Disability*, Routledge, London.

Stacey, M. (1969) 'The myth of community studies', *British Journal of Sociology*, 20, 134–47.

Stacey, J. (1990) *Brave New Families: Stories of Domestic Upheaval in Late Twentieth Century America*, Basic Books, New York.

Stainton Rogers, R. and Stainton Rogers, W. (1992) *Stories of Childhood: Shifting Agendas of Childhood*, Harvester Wheatsheaf, Hemel Hempstead, Herts.

Stamp, J. (1980) 'Towards supportive neighbourhoods: women's role in changing the segregated city', in Wekerle, G.R., Peterson, R. and Morley, D. (eds) *New Space for Women*, Westview Press, Boulder, CO.

Stanko, E. (1985) *Intimate Intrusions: Women's Experiences of Male Violence*, Routledge and Kegan Paul, London.

Stedman Jones, G. (1971) *Outcast London: a Study in the Relationships Between Classes in Victorian Society*, Clarendon Press, Oxford.

Steele, V. (1996) *Fetish: Fashion, Sex and Power*, Oxford University Press, Oxford.

Stenning, P. and Shearing, C. (1980) 'The quiet revolution: the nature, development and general legal implications of private security in Canada', *Criminal Law Quarterly*, **22**, 220–48.

Stretton, H. (1976) *Capitalism, Socialism and the Environment*, Cambridge University Press, Cambridge.

Surette, R. (1985) 'Video street patrol: media technology and street crime', *Journal of Police Science and Administration*, **13**, 78–85.

Suttles, G. (1972) *The Social Construction of Communities*, University of Chicago Press, Chicago.

Sweetman, P. (1999) 'Anchoring the (postmodern) self? Body modification, fashion and identity', *Body and Society*, **5**, 51–76.

Sykes, G. (1958) *The Society of Captives*, Princeton University Press, Princeton, NJ.

Symanski, R. (1981) *The Immoral Landscape: Female Prostitution in Western Societies*, Butterworth, Toronto.

Synott, A. (1993) *The Body Social: Symbolism, Self and Society*, Routledge, London.

Takahashi, L. (1996) 'A decade of understanding homelessness in the USA: from characterisation to representation', *Progress in Human Geography*, **20**, 291–310.

Takahashi, L. (1998) Community responses to human service delivery in US cities, in Fincher, R. and Jacobs, J. (eds) *Cities of Difference*, Guilford Press, London.

Tanca, A. (1993) 'European citizenship and the rights of lesbians and gay men', in Waaldijk, K. and Clapham, A. (eds) *Homosexuality: a European Community Issue: Essays on Lesbian and Gay Rights in European Law and Policy*, Martinus Nijhoff, Dordrecht.

Tatchell, P. (1992) *Europe in the Pink: Lesbian and Gay Equality in the New Europe*, GMP Publishers, London.

Taylor, A. (1998) 'Lesbian space: more than one imagined territory', in Ainley, R. (ed.) *New Frontiers of Space, Bodies and Gender*, Routledge, London.

Teather, E.K. (ed.) (1999) *Embodied Geographies: Spaces, Bodies and Rites of Passage*, Routledge, London.

Thomashaw, M. (1995) *Ecological Identity: Becoming a Reflective Environmentalist*, MIT Press, Cambridge, MA.

Thorne, B. (1994) *Gender Play*, Rutgers University Press, New Brunswick, NJ.

Thrift, N. (1987) 'Manufacturing rural geography?', *Journal of Rural Studies*, **3**, 77–81.

Thrift, N. (1990) 'Transport and communications, 1730–1914', in Dodgson, R. and Butlin, R. (eds) *An Historical Geography of England and Wales*, Academic Press, London.

Thrift, N. (1995) 'A hyperactive world', in Johnston, R.J., Taylor, P. and M. Watts (eds) *Geographies of Global Change*, Blackwell, Oxford.

Thrift, N. (1996) *Spatial Formations*, Sage, London.

Thrift, N. and Leyshon, A. (1988) ' "The gambling propensity": banks developing country debt exposures and the new international financial system', *Geoforum*, **19**, 55–69.

Tivers, J. (1985) *Women Attached: The Daily Lives of Women with Young Children*, Croom Helm, London.

Tomaney, J. (1995) 'Reflections on the revolution in England', *Northern Economic Review*, **23**, 3–23.

Tonnies, F. (1967) [1887] *Community and Society*, Harper & Row, New York.

Travis, A. and Campbell, D. (1994) 'Security staff's 2,600 crimes', *The Guardian*, 9 December, p. 8.

Turkle, S. (1996) [1995] *Life on the Screen: Identity in the Age of the Internet*, Phoenix, London.

Turner, B. (1984) *The Body and Society*, Basil Blackwell, Oxford.

Turner, B. (1992) *Regulating Bodies: Essays in Medical Sociology*, Routledge, London.

Turner, B. (1996) *The Body and Society*, Sage, London.

Tyler, M. and Abbott, P. (1998) 'Chocs away: weight watching in the contemporary airline industry', *Sociology*, **32**, 433–50.

Unwin, T. (1992) *The Place of Geography*, Longman, Harlow.

URL 1: http://home.cybergrrl.com/dv/body.html Females Find It. UK Statistics on Domestic Violence. August 1999.

URL 2: http://www.dvsheltertour.rog/fact.html Victim Services. Domestic Violence the Facts. August 1999.

URL 3: http://www.fpsc.org.uk Family Policy Studies Centre. June 2000.

URL 4: http://nch.ari.net/rural.html National Coalition for the Homeless. Rural homelessness. August 1999.

URL 5: http://nch.ari.net/numbers.html National Coalition for Homeless. How Many People Experience Homelessness? August 1999.

URL 6: http://www.ojp.usdoj.gov/bjs/cvict_v.htm, US Department of Justice, Bureau of Statistics. Victim Characteristics. August 1999.

URL 7: http://www.fbi.gov/publish/hatecrime.htm, US Department of Justice, Federal Bureau of Investigation, Criminal Justice Information Services, Hate Crime Statistics. August 1999.

URL 8: http://www.onpatrol.com/cs.privsec.html, Spencer, S. Private Security. August 1999.

URL 9: http://www.aic.gov.au/media/981202.html Australian Institute of Criminology, Regulating Private Security in Australia. August 1999.

Urry, J. (1990) 'The consumption of tourism', *Sociology*, **24**, 23–35.

Valentine, G. (1989) 'The geography of women's fear', *Area*, **21**, 385–90.

Valentine, G. (1990) 'Women's fear and the design of public space', *Built Environment*, **16**, 288–303.

Valentine, G. (1992) 'Images of danger: women's sources of information about the spatial distribution of male violence', *Area*, **24**, 22–9.

Valentine, G. (1993) '(Hetero)sexing space: lesbian perceptions and experiences of everyday spaces', *Environment and Planning D: Society and Space*, **11**, 395–413.

Valentine, G. (1995) 'Out and about: a geography of lesbian communities', *International Journal of Urban and Regional Research*, **19**, 96–111.

Valentine, G. (1996a) '(Re)negotiating the heterosexual street', in Duncan, N. (ed.) *Bodyspace: Destabilizing Geographies of Gender and Sexuality*, Routledge, London.

Valentine, G. (1996b) 'Children should be seen and not heard? The role of children in public space', *Urban Geography*, **17**, 205–20.

Valentine, G. (1996c) 'Angels and devils: the moral landscape of childhood', *Environment and Planning D: Society and Space*, **14**, 581–99.

Valentine, G. (1996d) 'An equal place to work? Anti-lesbian discrimination and sexual citizenship in the European Union', in Garcia-Ramon, M.D. and Monk, J. (eds) *Women of the European Union: The Politics of Work and Daily Life*, Routledge, London.

Valentine, G. (1997a) ' "My son's a bit dizzy." "My wife's a bit soft": gender, children and cultures of parenting', *Gender, Place and Culture*, **4**, 37–62.

Valentine, G. (1997b) 'A safe place to grow up? Parenting, perceptions of children's safety and the rural idyll', *Journal of Rural Studies*, **13**, 2, 137–48.

Valentine, G. (1997c) 'Making space: separatism and difference', in Jones III, J.P., Nast, H.J. and Roberts, S. (eds) *Thresholds in Feminist Geography: Difference, Methodology and Representation*, Rowman and Littlefield, Oxford.

Valentine, G. (1998) 'Food and the civilised street', in Fyfe, N. (ed.) *Images of the Street*, Routledge, London.

Valentine, G. (1999a) 'A corporeal geography of consumption', *Environment and Planning D: Society and Space*, **17**, 329–51.

Valentine, G. (1999b) 'Eating in: home, consumption and identity', *The Sociological Review*, **47**, 493–524.

Valentine, G. (1999c) ' "Oh please, Mum. Oh please, Dad": Negotiating children's spatial boundaries', in McKie, L., Bowlby, S. and Gregory, S. (eds) *Gender, Power and the Household*, Macmillan, Basingstoke.

Valentine, G. (1999d) 'Imagined geographies: geographical knowledges of self and other in everyday life', in Massey, D., Allen, J. and Sarre, P. (eds) *Human Geography Today*, Polity Press, Cambridge.

Valentine, G. (1999e) 'Consuming pleasures: food, leisure and the negotiation of sexual relations', in Crouch, D. (ed.) *Leisure/Tourism Geographies: Practices and Geographical Knowledge*, Routledge, London.

Valentine, G. (2000) 'Exploring children and young people's narratives of identity', *Geoforum*, **31**, 257–67.

Valentine, G., Holloway, S.L. and Bingham, N. (2000) 'Transforming cyberspace: children's interventions in the new public sphere', in Holloway, S.L. and Valentine, G. (eds) *Children's Geographies: Living, Playing, Learning*, Routledge, London.

Valentine, G. and Holloway, S.L. (2000) 'Cyberkids?: Exploring children's identities and social networks in on-line and off-line worlds', Paper available from the authors.

Valentine, G., Holloway, S.L. and Bingham, N. (2000) 'The digital generation? Children, ICT and the everyday nature of social exclusion', Paper available from the authors.

Valentine, G. and Longstaff, B. (1998) 'Doing porridge: food and social relations in a male prison', *Journal of Material Culture*, **3**, 131–52.

Valentine, G. and McKendrick, J. (1997) 'Children's outdoor play: exploring parental concerns about children's safety and the changing nature of childhood', *Geoforum*, **28**, 219–35.

Vanderbeck, R. and Johnson, J.H. (2000) 'That's the only place where you can hang out: urban young people and the space of the mall', *Urban Geography*, **21**, 5–25.

Veness, A. (1993) 'Neither homed nor homeless: contested definitions and the personal worlds of the poor', *Political Geography*, **12**, 319–40.

Veness, A. (1994) 'Designer shelters as models and makers of home: new responses to homelessness in urban America', *Urban Geography*, **15**, 150–67.

Van Gelder, L. (1996) [1985] 'The strange case of the electronic lover', in Kling, R. (ed.) *Computerization and Controversy: Value Conflicts and Social Changes*, Academic Press, San Diego, CA.

Vidal, J. (1997) *McLibel: Burger Culture on Trial*, Macmillan, London.

Virilo, P. (1997) *Open Sky*, Verso, London.

Waaldijk, K. (1993) 'The legal situation in member states', in Waaldijk, K. and Clapham, A. (eds) *Homosexuality: a European Community Issue: Essays on Lesbian and Gay Rights in European Law and Policy*, Martinus Nijhoff, Dordrecht.

Wagner, P. (1984) 'Suburban landscapes for nuclear families: the case of Greenbelt Towns in the United States', *Built Environment*, **10**, 35–41.

Walker, L. (1995) 'More than just skin-deep: fem(me)ininity and the subversion of identity', *Gender, Place and Culture*, **2**, 71–6.

Walkovitz, J. (1992) *City of Dreadful Delight*, Virago, London.

Waltzer, M. (1986) 'Pleasures and costs of urbanity', *Dissent*, **33** (summer), 470–5.

Ward, C. (1990) [1988] *The Child in the Country*, Bedford Square Press, London.

Ward, N. (1995) 'Technological change and the regulation of pollution from agricultural pesticides', *Geoforum*, **26**, 19–33.

Ward, N. (1996) 'Surfers, sewage and the new politics of pollution', *Area*, **28**(3), 331–8.

Ward, N. (1996) 'Pesticides, pollution and sustainability', in Allanson, P. and Whitby, M. (eds) *The Rural Economy and the British Countryside*, Earthscan, London.

Ward, N., Clark, J., Lowe, P. and Seymour, S. (1998) 'Keeping matter in its place: pollution regulation and the reconfiguring of farmers and farming', *Environment and Planning A*, **30**, 1165–78.

Ware, V. (1992) *Beyond the Pale: White Women, Racism and History*, Verso, London.

Warrington, M. (1996) 'Running to stand still: housing the homeless in the 1990s', *Area*, **28**, 471–81.

Warner, M. (1994) *Managing Monsters: Six Myths of Our Time*, Vintage Press, London.

Warren, K. (ed.) (1994) *Ecological Feminism*, Routledge, London.

Waters, M. (1995) *Globalization*, Routledge, London.

Watkins, F. (1998) *Imaginings of 'Community': Contested Social Relations in an English Rural Village*, Ph.D. thesis, Department of Geography, University of Sheffield.

Watson, S. (1999) 'City politics', in Pile, S., Brook, C. and Mooney, G. (eds) *Unruly Cities? Order/Disorder*, Routledge, London.

Watt, P. (1998) 'Going out of town: youth, "race", and place in the South East of England', *Environment and Planning D: Society and Space*, **16**, 687–703.

Watt, P. and Stenson, K. (1998) ' "It's a bit dodgy around here": safety, danger, ethnicity and young people's use of public space', in Skelton, T. and Valentine, G. (eds) *Cool Places: Geographies of Youth Cultures*, Routledge, London.

Webber, M. (1963) 'Order in diversity: community without propinquity', in Wingo, L. (ed.) *Cities and Space*, Johns Hopkins University Press, Baltimore, MD.

Webster, C. (1996) 'Local heroes: violent racism, localism and spacism among Asian and white young people', *Youth and Policy*, **53**, 15–27.

Weiss, M. (1989) *The Clustering of America*, Harper and Row, New York.

Wellman, B. (1979) 'The community question: the intimate networks of East Yorkers', *American Journal of Sociology*, **84**, 1201–31.

Wellman, B. (1987) *The Community Question Re-evaluated*, University of Toronto, Toronto.

Wellman, B. and Leighton, B. (1979) 'Networks, neighbourhoods and communities: approaches to the study of the community question', *Urban Affairs Quarterly*, **13**, 363–90.

Werkele, G.R., Peterson, R. and Morley, D. (1980) *New Space for Women*, Westview Press, Boulder, CO.

Westwood, S. and Williams, J. (eds) (1997) *Imagining Cities: Scripts, Signs and Memory*, Routledge, London.

Weston, K. (1995) 'Get thee to a big city: sexual imaginary and the great gay migration', *GLQ: A Journal of Lesbian and Gay Studies*, **2**, 253–77.

Whatmore, S. and Boucher, S. (1993) 'Bargaining with nature: the discourses and practices of environmental planning gain', *Transactions of the Institute of British Geographers*, **18**, 166–78.

Whine, M. (1997) 'The Far Right on the Internet', in Loader, B. (ed.) *The Governance of Cyberspace*, Routledge, London.

Whitaker, B. (2000) 'Saudis claim victory in war for control of web', *The Guardian*, 11 May, p. 17.

Whittle, S. (ed.) (1994) *The Margins of the City: Gay Men's Urban Lives*, Ashgate, Aldershot.

Williams, D.R. and Kaltenborn, B. (1999) 'Leisure places and modernity: the use and mean-ing of recreational cottages in Norway and the USA', in Crouch, D. (ed.) *Leisure/Tourism Geographies: Practices and Geographical Knowledge*, Routledge, London.

Williams, R. (1973) *The Country and the City*, Chatto and Windus, London.

Willis, P. (1977) *Learning to Labour: How Working Class Kids Get Working Class Jobs*, Saxon House, Westmead.

Willson, M. (1997) 'Community in the abstract: a political and ethical dilemma?', in Holmes, D. (ed.) (1997) *Virtual Politics*, Sage, London.

Wilson, A. (1992) *The Culture of Nature: North American Landscape from Disney to the Exxon Valdez*, Basil Blackwell, Oxford.

Wilson, A.R. (2000) 'Getting your kicks on Route 66! Stories of gay and lesbian life in rural America, c.1950–1970s', in Phillips, R., Watt, D. and Shuttleton, D. (eds) *De-centring Sexualities: Politics and Representations Beyond the Metropolis*, Routledge, London.

Wilson, E. (1991) *The Sphinx in the City*, Virago, London.

Wilson, E. (1983) *What is to be Done about Violence against Women?* Penguin, Harmondsworth.

Wilson, W. (1987) *The Truly Disadvantaged*, University of Chicago Press, Chicago.

Winstanley, M. (1989) 'The new culture of the countryside', in Mingay, G. (ed.) *The Vanishing Countryman*, Routledge, London.

Wirth, L. (1938) 'Urbanism as a way of life', *American Journal of Sociology*, **44**, 3–24.

Wise, S. and Stanley, L. (1987) *Georgie Porgie: Sexual Harassment in Everyday Life*, Pandora Press, London.

Wolch, J. and Dear, M. (1993) *Malign Neglect: Homelessness in an American City*, Jossey-Bass, San Francisco, CA.

Women and Geography Study Group (1984) *Geography and Gender: an Introduction to Feminist Geography*, Hutchinson, London.

Women and Geography Study Group (1997) *Feminist Geographies: Explorations in Diversity and Difference*, Addison Wesley Longman, Harlow.

Wood, D. (1982) 'To catch the wind', *Outlook*, **46**, 3–31.

Wood, D. and Beck, R.J. (1994) *Home Rules*, Johns Hopkins University Press, Baltimore, MD.

Woods, G. (1987) *Articulate Flesh: Male Homo-eroticism and Modern Poetry*, Yale University Press, New Haven, CT.

Woods, G. (1995) 'Fantasy islands: popular topographies of marooned masculinity', in Bell, D. and Valentine, G. (eds) *Mapping Desire: Geographies of Sexualities*, Routledge, London.

Woods, M. (1998a) 'Mad cows and hounded deer: political presentations of animals in the British countryside', *Environment and Planning A*, **30**, 1219–34.

Woods, M. (1998b) 'Researching rural conflicts: hunting, local politics and action net-works', *Journal of Rural Studies*, **3**, 321–40.

Woodward, B. (1995) 'Is the modern 4WD vehicle "saviour" or "sinner"?', *Geo Australasia*, **17**, 26–36.

Woodward, R. (1998) 'It's a man's life! Soldiers, masculinity and the countryside', *Gender, Place and Culture*, 5, 3, 277–300.

Yearley, S. (1995) 'The transnational politics of the environment', in Anderson, J., Brook, C. and Cochrane, A. (eds) *A Global World?: Re-ordering Political Space*, The Open University, Milton Keynes.

Yeo, E. and Yeo, S. (1988) 'The uses of community from Owenism to the present', in Yeo, S. (ed.) *New Views of Co-operation*, Verso, London.

Yingling, T. (1997) *AIDS and the National Body*, Duke University Press, Durham, NC.

Young, I.M. (1990a) *Justice and the Politics of Difference*, Princeton University Press, Princeton, NJ.

Young, I.M. (1990b) *Throwing Like a Girl and Other Essays in Feminist Philosophy and Social Theory*, Indiana University Press, Bloomington and Indianapolis.

Young, I.M. (1990c) 'The ideal of community and the politics of difference', in Nicholson, L.J. (ed.) *Feminism/Postmodernism*, Routledge, London.

Young, M. and Willmott, P. (1962) [1957] *Family and Kinship in East London*, Penguin, Harmondsworth.

Yuval-Davis, N. (1997) *Gender and Nation*, Sage, London.

Zolo, D. (1997) *Cosmopolis: Prospects for World Government*, Polity Press, London.

Zukin, S. (1988) *Loft Living: Culture and Capital in Urban Change*, Radius, London [first pub. 1981]

Zukin, S. (1991) *Landscapes of Power: from Detroit to Disney World*, University of California Press, Berkeley, CA.

Zukin, S. (1995) *The Culture of Cities*, Blackwell, Oxford.

Zukin, S. (1998) 'Urban lifestyles: diversity and standardisation in spaces of consumption', *Urban Studies*, 35(5/6), 825–41.

Index